HANDBOOK OF COMPOSITE ORGANIC FARMING

Himadri Panda, obtained M.Sc. in Physical Chemistry with 'Drug' as a special subject (First Class First) from University of Indore, Indore (M.P.). Then completed research work in Rohilkhand University, Bareilly for Ph.D.. He is a well known writer and written about 50 Technical Books and 250 Technical/Review articles in various subjects of Industrial Chemistry which were published in National and International Journals. He is an active member of Indian Chemical Society, Institution of Chemists (India), Essential oil Association of India, India Pulp and Paper Technical Association, Oil & Colour Chemist's Association (U.K.), Chinese Forest Academy (China). He and his wife (Dr. Rakhshinda Panda) who is at present working as Research Officer in 'The Indian Turpentine & Rosin Co. Ltd', a U.P. Govt. Organisation, jointly awarded by Chinese Forest Authority for the best paper published in their Journal, and invited to China to deliver Key Note address in International Symposium on 'Forest Based Products'. He was also invited as Guest speaker in 'GreenTech Environment Excellence Award' and 'Indian System of Medicine' organised be Ministry of Health and Family Welfare, of India, New Delhi. At present he is the Technical Director of Devashish Consultants (P) Ltd and Technical Consutlant to Arundhati Industrial Research Laboratory. He was nominated as 'Technical Consultants' to Afro-Asian Countries for the year 1985-1990 by Government of India.

Dr. Dharamvir Hota is presently the emeritus professor in the department of agriculture science and economics, an appointment he has enjoyed since 1998. He was a senior researchers at the International Institute of Agricultural Science, where he worked on the use of genetically modified crops for structure determination of molecular modeling.

He has been teaching and doing research on agricultural systems in and around the country. He teaches soil science, organic agriculture and land use, and conducts research on the effects of organic and conventional farming systems on soil, crop productivity and quality, with publications appearing in various internationally reputed journals.

Himadri Panda, obtained M.Sc. in Physical Chemistry with 'Drug' as a special subject (First Class First) from University of Indore, Indore (M.P.). Then completed research work in Rohilkhand University, Bareilly for Ph.D.. He is a well known writer and written about 50 Technical Books and 250 Technical/Review articles in various subjects of Industrial Chemistry which were published in National and International Journals. He is an active member of Indian Chemical Society, Institution of Chemists (India), Essential oil Association of India, India Pulp and Paper Technical Association, Oil & Colour Chemist's Association (U.K.), Chinese Forest Academy (China). He and his wife (Dr. Rakhshinda Panda) who is at present working as Research Officer in 'The Indian Turpentine & Rosin Co. Ltd', a U.P. Govt. Organisation, jointly awarded by Chinese Forest Authority for the best paper published in their Journal, and invited to China to deliver Key Note address in International Symposium on 'Forest Based Products'. He was also invited as Guest speaker in 'GreenTech Environment Excellence Award' and 'Indian System of Medicine' organised be Ministry of Health and Family Welfare, of India, New Delhi. At present he is the Technical Director of Devashish Consultants (P) Ltd and Technical Consutlant to Arundhati Industrial Research Laboratory. He was nominated as 'Technical Consultants' to Afro-Asian Countries for the year 1985-1990 by Government of India.

Dr. Dharamvir Hota is presently the emeritus professor in the department of agriculture science and economics, an appointment he has enjoyed since 1998. He was a senior researchers at the International Institute of Agricultural Science, where he worked on the use of genetically modified crops for structure determination of molecular modeling.

He has been teaching and doing research on agricultural systems in and around the country. He teaches soil science, organic agriculture and land use, and conducts research on the effects of organic and conventional farming systems on soil, crop productivity and quality, with publications appearing in various internationally reputed journals.

HANDBOOK O
ORGANIC

HANDBOOK OF COMPOSITE ORGANIC FARMING

Edited by
Himadri Panda
&
Dharamvir Hota

GENE TECH

2026
Gene-Tech Books

ISBN 978-81-89729-81-3

Published by : **GENE-TECH BOOKS**
4762-63/23, Ansari Road, Darya Ganj,
NEW DELHI - 110 002
Phone: 011-43003222
e-mail: genetechbooks@yahoo.co.in

PRINTED IN INDIA

Preface

Our concern about environmental safety and sustainability of land productivity is increasing among the scientists, environmentalists and higher officials. With the increasing population, it is also becoming clear that food security for the millions will not be possible unless the available resources are efficiently utilized for increasing the productivity. The strategy adopted during the green revolution cannot be valid anymore under the present conditions. A new strategy of living with the nature and nurturing it of sustainable high productivity should be evolved. Though the use of chemical inputs cannot be altogether avoided, but their use in agriculture has to be minimized or rationalized. So organic farming shows us the way to effectively use the available natural resources for the benefit of the mankind. But unfortunately there are a lot of misunderstanding. A compilation of the available information with upgrading technology has been felt need of students, research workers, agriculturists, environmentalists and scientists in the field of agriculture. In this handbook almost all the aspects related to organic farming are covered with a balanced approach. The author demonstrates the newness of the organic concept for readers. This will help readers to discover easily the philosophical and technical differences between organic and conventional farming system. Though it is not exhaustive, which it is not meant to be, it is felt that book will give an overview on the subject. The author welcome suggestions from the readers for the improvement of the book in future editions.

Himadri Panda

Dharamvir Hota

Contents

Chapter 1

Integrated Plant Nutrition System (IPNS): An Approach Towards Organic Farming

The greatest challenge facing mankind in the 21st century is to produce the basic necessities of food, feed, fibre, fuel and raw materials from 0.14 ha or less land per caput. While the use of mineral fertilizers is the quickest and surest way of boosting crop production, their cost and other constraints frequently deter farmers from using them in recommended quantities and in balanced proportions. As a consequence of this and other constraints there seems to be no option but to fully exploit potential alternative sources of plant nutrients. Complementary use of available renewable sources of plant nutrients (organic/biological) along with mineral fertilizers is of great importance for the maintenance of soil productivity, *i.e.* soil structure, soil bioactivity, soil exchange capacity and water holding capacity. Results from various cropping systems and ecologies illustrate that positive interactions result from the integrated use of mineral fertilizers and organic/biological sources of plant nutrients within the framework of integrated plant nutrition system (IPNS).

In response to this need, FAO programmes on IPNS are receiving priority–an area in which FAO has been active for over a decade. This chapter provides a brief account of the concept and components of IPNS, present state of the art on the subject, initiatives and programmes of the FAO and some IPNS models which can be transferred across locations. As several of the results obtained in India have been discussed in other chapters in this volume, results included here are largely from other countries. Experimental results cited are more in the nature of examples and a detailed technical review of all available results is neither intended nor is it within the scope of this chapter.

IPNS Concept and Material

The basic concept underlying IPNS is the maintenance or adjustment of soil fertility and of plant nutrient supply to an optimum level for sustaining the desired crop productivity through optimization of the benefits from all possible Sources of plant nutrients in an integrated manner. The appropriate combination of mineral fertilizers, organic manures, crop residues, compost or N-fixing crops varies according to the system of land use and ecological, social and economic conditions. Contrary of the Low External Input (LEI) and Organic Farming approaches, the IPNS involve fertility and plant nutrition management for a targeted yield based not only on cropping and farming systems but also on distinct geographical areas or villages as a dynamic system. The IPNS approach can be modulated; a factor of targeted yields in any area according to land, water and climate potentials.

The cropping system rather than an individual crop, and the farming system rather than an individual field, are the focus of attention in this approach for developing IPNS practices for major agro-ecological zones and for various categories of farms. The best associations of various types of plant nutrients in different fields are identified for a balanced plant nutrition and high yield, at the same time sustaining soil fertility and controlling nutrient losses. It is envisaged that locally-available materials of plant or animal origin as by-products of agricultural activities be used or, where such materials are not abundantly available, *in situ* production of organics be attempted. In, areas and farms where non-cultivated areas are available (forests, pastures and fallows), biomass production is attempted for nutrient transfer to cultivated areas in a sustainable system. In areas and farms where land is scarce and limited to the fields, increasing biomass production in fallows and on the borders of the plots can be attempted. Introduction of green manuring, of living or dead, cover in the cropping system can also contribute to nutrient supplies.

Alternative sources of organic material which have considerable potential are: quick-growing leguminous shrubs grown as a part of the cropping system and incorporated into the soil at an appropriate stage as green manure; leguminous trees grown in hedgerows and their lopping used as mulch materials or incorporated into the soil of the cropped alleys between them; forage or food legumes properly inoculated with *Rhizobium* grown in the cropping sequence; and the use of Azolla of blue-green algae with wetland crops.

The following phenomena must be understood in the context of organic recycling and IPNS:

1. *In situ* recycling of crop residues rather than their non-agricultural use and wastage will bring back a certain amount of nutrients to the same field. However, they are not new sources and are deficient in the same nutrients (like P, S, K, Ca, Mg) as those of the soil;
2. *In situ* growing of leguminous trees and their biomass incorporation will bring atmospheric N into the system, arid occasionally, through their deep roots, nutrients leached beyond root zone and sub-soil mineralized nutrients are brought to the surface layer for the use of annual crops;
3. When biomass is brought from outside the plot or farm and/or cattle graze on uncultivated land, it is basically a transfer of nutrients from one place to another. So, except for N where biofixation can introduce nutrients into the soil-crop system, IPNS is a process which limits the losses of the system and not a process which supplies plant nutrients for the whole system.

State of the Art of IPNS

Attempts made in several countries to complement the use of mineral with organic sources of plant nutrients have generated useful, though limited information on the complementary and synergistic effects of these materials on the yield of crops. Some examples of results are described in the following sections.

Farmyard Manures (FYM)

In many countries such as India and China, cow dung is abundantly available for use as an organic source of plant nutrients. In Indian trials, the rice yield was increased by 450 kg/ha (47 per cent) by the application of 5.6 t/ha of FYM alone and by 600 kg/ha (63 per cent) when an equivalent amount of N (67 kg/ha of N) was applied as urea (Table 1.1). Applied together, the yield increase was 1120 kg/ha (118 per cent) over no N or no FYM (control plot). In China, under low soil fertility conditions, the FYM (15 t/ha) effect on wheat yield exceeded the effect of 70 kg/ha N. When these two sources of nutrients were applied together, the yield increase over no treatment control was 460 kg/ha (35 per cent), while the yield increase with 140 kg/ha N alone was only 300 kg/ha (23 per cent).

Table 1.1: Effect of Complementary Application of FYM and Mineral Fertilizers on the Grain Yield of Rice and Wheat

Country	*Crop*	*FYM (t/ha)*	*Fertilizer N (kg/ha)*	*Yield (kg/ha)*	*% Yield Increase Over Control*
India	Rice	0	0	950	–
		5.6	0	1400	47
		0	67	1550	63
		5.6	67	2070	118
China	Wheat	0	0	1330	–
		15	0	1570	18
		0	70	1530	15
		15	70	1790	35
		0	140	1630	23
		15	140	1990	50

In another set of long-term experiments conducted at different locations in India representing varying agro-ecological situations, positive effects of FYM were observed. Thus, integrated use of organic manures with mineral fertilizers improved the soil and crop productivity and increased the yield response.

Green Manures and Crop Residues

Green manuring with legumes has long been known to be beneficial for sustainable crop productivity. With intensification of agriculture and the advent of mineral fertilizers, farmers enthusiasm for this practice declined. The lack of interest arises from farmers' perception of inadequate economic return from the labour and from the time the green manure crop occupies the land. To overcome this objection, an innovative technique is being studied in some countries, namely to plant a food legume, harvest the pods just prior to full maturity and turn the biomass into the soil when still green.

One such attempt was made at the International Rice Research Institute (IRRI) was cowpea (*Vigna unguiculata*), of which the whole biomass was turned in at 75 per cent flowering stage and compared with the treatment where the grain was harvested before turning in the residues. Virtually no difference was found in the two years, either for upland or lowland situations, in the rice yield between the two processes. The gain from either method, averaged over the two years, was 800 kg/ha under lowland and 500 kg/ha under upland conditions, compared with either a pre-rice fallow or a pre-rice cowpea from which all the above ground parts had been removed (Table 1.2).

Table 1.2: Grain Yield of Rice as Affected by Pre-Rice Land Use

Sl.No.	Pre-rice Treatment	Lowland (kg/ha)		Upland (kg/ha)	
		1986	1987	1986	1987
1.	Fallow	4,800	4,500	1,200	1,300
2.	Cowpea incorporated into soil at flowering	5,700	5,400	1,600	2,000
3.	Cowpea grown, grain harvested and straw incorporated into soil	5,500	5,200	1,700	2,000
4.	Cowpea grown, both grain and straw removed	4,900	4,500	1,400	1,500

Under lowland conditions, incorporating the whole biomass of cowpea at flowering, and growing rice without any fertilizer produced yields which were equivalent to applying 36 kg/ha N in 1986 and 58 kg/ha N in 1987 to the rice crop grown after a previous fallow or a pre-rice cowpea from which all the above-ground parts had been removed. Under upland conditions, these advantages were equivalent to applying 70 kg/ha N in 1986 and 65 kg/ha N in 1987 after a pre-rice fallow. The root residues of cowpea were equivalent to 30 kg/ha N in 1986 and 15 kg/ha N in 1987 under upland situations. Farmers can thus choose to increase yields or save on fertilizer N or a combination, depending on the economic and agronomic conditions.

Similarly, results of joint research by the Centre for Soil Research, Indonesia and IRRI during 1986/87 showed that on lowland rice incorporation of *Sesbania rostrata* at 45 days, along with an application of 60 kg/ha N enhanced rice yield by 500 kg/ha over the application of 120 kg/ha N alone (Table 1.3).

Table 1.3: Effect of N Fertilization and *Sesbania rostrata* Green Manuring (SR) on the Yield of Lowland Rice (IR-36) in East Java, Indonesia

Fertilizers kg/ha			Grain Yield kg/ha
N	P_2O_5	K_2O	
0	0	0	2000
60	45	30	3700
0 + SR at 45 days	45	30	3100
60 + SR at 45 days	45	30	4600
120	45	30	4100

While beneficial effects of mulching are widely recognized, procuring mulch material in sufficient quantity is a serious practical problem in many areas. Management of crop residue as a source of mulch is, therefore, closely linked with cropping systems, tillage methods, agro-ecological conditions, farming systems and particularly, cattle population in the farm. A range of cultural practices is available, mainly in humid tropics, to provide adequate amounts of residue mulch for soil protection and fertility enhancement, *e.g.* cover crops, conservation tillage, after cropping, use of industrial by products etc. The sustainability of each practice depends on the local biophysical and socio-economic environment.

Legumes in Cropping Systems and Fertilizer Use

Forage legumes are well recognized as capable of improving soil fertility through symbiotic N fixation to the extent that the non-legumes subsequently grown benefit from the residual fertility.

Medicago or annual *Trifolium* species are estimated to contribute 50–70 kg/ha N for the subsequent crop.

There is evidence that grain legumes also benefit the subsequent crop through enrichment of soil fertility. In an experiment conducted over a three-year period at Milingano (Tanzania), P applied to mungbean (*Vigna radiata*) not only increased the grain yield of the legume by 70 kg/ha (25 per cent) but also of the subsequent maize by 1000 kg/ha (59 per cent). Phosphorus application to maize via the previous mungbean crop was as effective as direct P application to maize, and the additional yield of mungbean made the system economically more attractive.

Biological Sources of Nitrogen

Biological fixation of atmospheric N by heterotrophic organisms like cyanobacteria (blue-green algae) individually or in association with the aquatic fern *Azolla* has been in use in rice production systems in many countries. Extensive trials have been made with blue-green algae and *Azolla* in several countries. In an experiment conducted in India, when mineral fertilizers were used with blue-green algae on rice, the yield was increased by 90–220 kg/ha (3 to 5 per cent over a base yield of about 4000 kg rice/ha) by their combined application.

From work done with *Azolla* in the Philippines it was observed that top-dressing *Azolla* with a basal application of 30 kg/ha N increased rice yield by 800 kg/ha (31 per cent) over mineral N alone or 400 kg/ha (13 per cent) over *Azolla* alone (Table 1.4). The *Azolla* was equivalent to about 30 kg/ha N.

Table 1.4: Grain Yield of Lowland Rice as Affected by Mineral N and Incorporation of *Azolla* into the soil

Treatment	*Yield (kg/ha)*
T1 30 kg/ha N	2600
T2 *Azolla* top-dressed	3000
T3 30 kg/ha N + *Azolla* top-dressed	3400
% increase T2 over T1	15.0
% increase T3 over T1	31.0
% increase T3 over T2	13.0

The importance of management of *Azolla* in the process of its incorporation into the soil and its interaction with added fertilizer phosphorus has been shown. In an experiment conducted at IRRI, Philippines, it was noted that *Azolla* growth without P was poor. Its inoculation in the paddy field increased rice grain yield only in the plots which were not puddled. Mid-season puddling increased rice growth over shadowing the effectiveness of inoculation. *Azolla* growth was profuse in plots receiving P and the rice grain yield increased substantially with *Azolla* incorporation. An increase in rice yield of 1050 kg/ha (71 per cent) was obtained by a combination of *Azolla* inoculation, its mid-season incorporation (39 days after inoculation) and P addition over control. With only *Azolla* inoculation, but without its incorporation or P addition the yield advantage did not exceed 25 per cent. It thus becomes evident that proper management of *Azolla* and its incorporation in the soil are crucial factors determining its effectiveness (Table 1.5).

From this brief review presented and a wide analysis of documentation, the following conclusions can be drawn on the present state of understanding with regard to IPNS:

1. Organic/biological sources of plant nutrients complement mineral fertilizers in meeting nutrient requirements of crops. The magnitude of the contribution will vary according to sources and agro-ecological conditions.
2. In many situations, synergistic effects due to such combined application could be expected, thus increasing the fertilizer use efficiency (FUE).
3. Residual effects of added organic sources in the cropping system could also be expected together with an improvement in soil physical condition.
4. Under high input production systems, where productivity cannot be further increased with incremental use of mineral fertilizers alone, addition of organic sources could again increase yields through increased soil productivity and higher FUE.

Table 1.5: Grain Yield (kg/ha) of Rice as Affected by *Azolla* Inoculation, Method of its Incorporation and Phosphorus Addition

Azolla Treatment	*No P_2O_5*		*32 kg/ha P_2O_5*	
	No Puddling	*Mid-season Puddling*	*No Puddling*	*Mid-season Puddling*
No inoculation	1480	2360	1860	2250
Azolla inoculation	1850	2160	2020	2530
% difference due to *Azolla* inoculation	+25	–8	+9	+12

FAO Programmes in the Development of IPNS

FAO has taken a pioneering initiative in developing the IPNS approach within the following conceptual programmes:

Availability of Organic Sources and Potential of BNF

When planning programmes and activities in organic recycling, it is worthwhile, if not indispensable, to consider all the possible uses of a particular material and to examine if recycling it as a manure is really the best choice. Once this is clear, one can start looking into how the actually available organic materials can be recycled and used as manure/mulch/soil conditioner in an optimal way. Only then can sound ideas be developed on how to integrate the available nutrient carriers (organic, biological and mineral) in an environment-friendly way which is also socially and economically acceptable to the farmer.

To have a first hand appreciation of the potential availability of various sources, status reports for selected countries in the Asia and the Pacific Region and the Africa Region, covering various aspects of soil fertility and crop nutrition, were prepared. These documents show that the prospects of various organic and biological sources differ from country to country. It would, however, be necessary to quantify their actual availability for field application and, also, such assessments need to be extended to other countries. The following are the main nutrient sources, in order of priority (Table 1.6).

IPNS Models

After the identification of potential organic/biological resources, some simple models of nutrient scheduling underlying the concepts and application of IPNS for some selected cropping systems have been developed. Their details have been presented elsewhere. Attempts are being made to develop computer models for an IPNS calculation programme of plant nutrition recommendations, based on general agronomic assumptions. Farming systems should be observed and understood before proposing a new approach. Involvement of farmers in the whole innovation process is probably the most efficient way to build up sustainable and socially acceptable models adapted to different local situations.

Table 1.6: Main Nutrients Sources

No.	*Country*	*Main Organic and Biological Sources*
1.	Bangladesh	Animal wastes; BNF (*Rhizobium*); green manuring
2.	Indonesia	BNF (*Rhizobium*); recycling of legume crop residues and rice straw; animal wastes
3.	Nepal	Hill areas animal wastes; BNF (*Rhizobium*) *Tarai areas* BNF (*Rhizobium*); green manuring
4.	Pakistan	Animal wastes; BNF (*Rhizobium*); green manuring
5.	Sri Lanka	Recycling of rice straw and legume crop residues; BNF (*Rhizobium*)
6.	Thailand	BNF (*Rhizobium*); crop residues; agro-industrial wastes
7.	Burkina Faso	Animal wastes; crop residues; BNF (*Rhizobium*)
8.	Guinea Bissau	Crop residues, BNF (*Rhizobium* and *Azolla*)
9.	Madagascar	Animal wastes; crop residues, particularly rice straw; BNF (*Rhizobium* and *Azolla*)
10.	Rwanda	Animal wastes (in Butare and Gitarama regions); BNF (*Rhizobium*); crop residues
11.	Sudan	Animal wastes; crop residues; BNF (*Rhizobium*)
12.	Tanzania	BNF (*Rhizobium*); crop residues
13.	Zaire	Crop residues; forest leaves; BNF (*Rhizobium*)
14.	Zambia	Animal wastes (certain areas in southern, western and central provinces); crop residues; BNF (*Rhizobium*)

Field Trial Network

To generate further information on combined applications of organic/biological sources with mineral fertilizers in different agro-ecological situations and dominant cropping systems, some research institutions have been sub-contracted to initiate long-term experiments. To bypass the limitations on the number of treatments and trials that could be carried out by local research institutions in any country, and to benefit from the results of others having similar agro-ecological conditions, this field activity is being expanded into a collaborative network programme within the framework of Technical Cooperation between Developing Countries. These trials will complement the existing FAO Sponsored Fertilizer Programme in most of the countries. To start with, trials at fixed sites representing major agro-ecological conditions are being undertaken in India, Indonesia, Laos, Nepal, Pakistan, Tanzania and Thailand. In time and depending on financial resources, the network programme is proposed to be extended to other countries.

Results obtained from field experimentation can be validated against predictive models. Field experimentation can bring about a step-by-step improvement by identifying and dealing with each major constraint to plant nutrition and crop production. Moderate improvements in traditional farming may be the initial step in a long-term strategy aimed at transforming low-input subsistence farming into science-based profitable agriculture with high or moderate inputs. Thus, the above network activity is only catalytic and a beginning. Much more remains to be done to determine integrated plant nutrition and adapted practices for sustainable agriculture, according to the goals of target groups of farmers.

Promotion and Technology Transfer

Promotion of Symbiotic BNF Activities

Parallel to the field trials, the BNF technology is being developed through a number of technical cooperation projects. Major emphasis is on assisting national research institutions in the selection and production of the best *Rhizobium* strains, and in setting guidelines for maintaining high standards of inoculant quality. An expert consultation on *Rhizobium* inoculant quality was held in March 1991.

Technology Transfer

To facilitate the transfer of IPNS technology, a data bank on available results is being established.

Promising treatment combinations from the field trial network will be included in the national demonstration programme/block demonstrations within the framework of the FAO Fertilizer Programme field projects, currently operational in 21 countries.

Transfer of this technology for the benefit of small farmers.

To foster applications of technical results, FAO continues to publish publications and training materials and on appropriate methods of production, conservation and efficient use of organic/ biological sources along with efficient use of mineral fertilizers.

Future Research and Development Needs

Research Needs

It is apparent that there is a need for:

1. More information on integrated nutrient recommendations for cropping systems as a whole, taking into account balanced plant nutrition through the complementary synergistic effect of combined use of both mineral and organic/biological sources for sustained crop production and enhancing labour productivity.
2. Although general principles may be the same, technological packages for sustainable management of soil and plant nutrient resources are site-specific and depend on farming systems, farm size, availability of inputs, and socio-economic factors. Area specific and on-farm synthesis of packages is needed on the basis of components and sub-systems such as those described above. Agronomic productivity, economic profitability, and ecological compatibility of such packages needs to be assessed through appropriate research on well-defined, representative sites.
3. Recommendations for different agro-ecological situations, taking into account available nutrient resources, farming systems and the choice of improved agro techniques to alleviate soil constraints for plant nutrition.
4. The application of IPNS on a large scale requires some adaptive research, demonstration programmes backed by training in good management practices, and a policy improvement and infrastructure which stimulates the efficient use of both internal and off-farm inputs.

Development of National Strategy

To provide a framework for national strategies for the conservation and regeneration of soil productivity, the following activities are being initiated. These complement the ongoing activities on IPNS:

1. Selected case studies of national fertilizer and fertility situations and strategies with the following components:

   "National accounts" for plant nutrients, assessing supplies and losses and preparing balance sheets and projections by crop and homogenous agro-ecological zone, together with identification of critical areas and total future requirements at country level.

   (*a*) Identification and appraisal of all currently and potentially available sources of plant nutrients–organic and mineral–at country level. Review of technical and economic potentials. Assessment of options for use of these additional/alternative sources of nutrient supplies.

   (*b*) Formulation of alternative strategies to meet future demand for soil fertility improvement and maintenance, including policy options with regard to prices, incentives, subsidies and taxes.

   (*c*) Preparation of national action plans for the sustainable development and maintenance of soil productivity.

   A series of national programmes and projects may then follow, expanding upon experience gained through more than 25 years of activities of the FAO Fertilizer Programme.

2. Preparation of a Covenant on Good Fertilizer Use Practices.
3. Preparation of a block demonstration programme on sustainable agricultural practices, incorporating elements of sustainable development of soil productivity and crop production such as Integrated Nutrient Management, Integrated Pest Management, Improved Soil and Water Conservation and Integrated Crop Husbandry Management, and its implementation within the framework of the ongoing block demonstration activities of the FAO Fertilizer Programme and other related field programmes.

Conclusions

Soil productivity can be improved and maintained by integrated plant nutrition systems, which address two important problems: the low and decreasing crop yields caused by plant nutrient depletion and inefficient fertilizer use in the developing world, as well as the less extensive but similarly serious problem of pollution by excessive and inappropriate plant nutrient application in developed and locally in developing countries.

The reality is that there are limitations to both low input and high input systems. Solutions will most probably have to be tailored to individual agro-ecological and socio-economic situations and are likely to involve elements of both low-input and high input systems. In most cases, minimizing external inputs through optimal combinations of plant nutrient sources available within the farm will be one of the objectives. Increased efficiency, as well as the reduction of off-farm inputs, will contribute to environmental conservation and decrease or eliminate pollution of soils and groundwater.

Each country will need to use balance sheet approaches to match the needs of its present and future populations with the potentials of its land and water resources under different input levels, and build a long-term plan of action on the results. Depending on the situation in each country, this may imply changes in policy or legislation, prices, incentives or taxes, as well as institutions.

A strategy or action plan for the sustainable development of soil productivity through IPNS would involve the mobilization of advisory services as well as of land users' Associations. It may also require technical assistance, aid-in-kind and other supply options, and revolving funds and national production of mineral fertilizers.

Chapter 2
Preparation, Processing and Preservation of Organic Fertilizers

Preparation of Organic Manures

Raw Materials

Compost and farmyard manure are processed from agricultural organic wastes such as straw, leaves, paddy husk, groundnut husk, sugarcane trash, bagasse, cattle dung and urine and habitation wastes like city garbage, nightsoil, sewage and sullage and vegetable matter. The materials undergo intensive decomposition under thermophilic and mesophilic conditions in heaps or pits with adequate moisture and finally yield a brown to dark coloured humified material in four to six months which is more stable in form, valuable for replenishments of plant nutrients, maintenance of soil organic matter and in improving the physical and microbiological conditions of the soil.

Principles of Composting

Action of Microorganisms

The biodegradation process is carried out by different groups of heterotrophic microorganisms, bacteria, fungi actinomycetes and protozoa. The role of cellulolytic and lignolytic microorganisms in decomposition of crop wastes and residues is of prime importance. Microorganisms involved in the process derive their energy and carbon requirements from the decomposition of carbonaceous materials and for every 10 parts of carbon, 1 part of nitrogen is required for building up of their cell protoplasm. Fungi are more efficient in carbon assimilation than bacteria and actinomycetes. Thus carbon dioxide evolution is comparatively less when fungi are more active in biodegradation than bacterial.

When organic materials are broken down in presence of oxygen, the process is called as aerobic decomposition. Under aerobic conditions. living organisms which utilise oxygen, decompose organic matter and assimilate some of the carbon, nitrogen, phosphorus, sulphur and other nutrients for synthesis of their cell protoplasm. Heterotrophs derive energy from the decomposition of organic matter, resulting on production of carbon dioxide, humic substances and release of available plant nutrients. Carbon serves both as energy source and is also required for cell protoplasm, greater amount of carbon is assimilated than nitrogen. Generally about two-thirds of the carbon is respired/evolved as CO_2 and the remaining one-third is combined with nitrogen in the living cells.

Aerobic decomposition of organic materials is most common in nature and generally occurs in arable soils and in forest soil surfaces where animal droppings and organic residues are stabilised into humus, with involvement of different groups of microflora. In the aerobic process, there are no nuisance problems such as foul odour associated with it as is produced under anaerobic conditions due to intermediate compounds.

A great deal of exothermic energy is released during the oxidation of carbon to carbon dioxide. Organic materials in compost heaps or piles under proper insulation generate substantial amount of heat which increases the temperature up to 65–70°C. However, if the temperature exceeds 65–70°C, the microbial activity is decreased due to thermal kill of the microorganisms and the stabilization of organic matter is slowed down. Thermophilic organisms develop when the temperature exceeds above 45°C, and they thrive best in the temperature range of 45–65°C. The major reactions likely to occur under aerobic decomposition system are as follows:

Sugars $(CH_2O) \times + XO_2 \longrightarrow \times CO_2 + \times H_2O +$ Energy

Celluloses

Hemicelluloses

Lignins

Proteins (Organic N) $\longrightarrow NH_3 \longrightarrow NO_2^- \longrightarrow NO_3^-$

Organic sulphur $S + \times O \longrightarrow SO_4$

Organic phosphate $\longrightarrow H_3PO_4 \longrightarrow Ca\,(HPO_4)_2$
(Phytin, Lecithin)

Anaerobic microorganisms break down organic materials by a process of reduction in absence of oxygen. First a special group of acid producing bacteria, facultative heterotrophs degrade organic matter into fatty acids, aldehydes and alcohol, etc. Then a group of bacteria convert the intermediate products to methane, ammonia, carbon dioxide and hydrogen. Oxygen is also required for the anaerobic process but its source is chemical compounds and not free dissolved oxygen. Like aerobic process, the organisms use nitrogen, phosphorus and other nutrients in developing cell protoplasm. During this process the decomposition is not complete and there is less production of carbon dioxide and intermediates, like organic acid will occur in greater amounts. So is the case of nitrogen containing substances such as ammonia. Due to lesser microbial biomass production and carbon assimilation, there is greater production of methane. This type of fermentation takes place in gober-gas plants, waste stabilisation ponds, in marshy soils, in buried organic materials devoid of oxygen or with low oxidation-reduction potential. Intensive reduction of organic matter is also known as a putrefactive process accompanied by foul odours of hydrogen sulphide and of reduced sulphur containing organic compounds such as mercaptans.

As compared to aerobic process where the release of energy is much greater (484-674 K Cal/glucose molecule) only about 26 K Cal of energy per gram molecule of glucose are released. The energy of carbon is in the methane gas and the resultant energy from gober-gas plants in India is utilised for cooking purpose and can also be used in running engines. The biochemical reactions that occur in anaerobic decomposition of wastes are as follows:

$$(CH_2O) \times \longrightarrow \times CH_3COOH$$

$$CH_3COOH \longrightarrow CH_4 + CO_2$$

$$\text{Organic N} \longrightarrow NH_3$$

$$2H_2S + CO_2 + \text{Light} \longrightarrow (CH_2O) \times S_2 + H_2O$$

$$\text{Organic P} \longrightarrow \text{reduced P}$$

Factors in Process of Composting

The most important factors in decomposition of organic wastes are as follows:

1. Carbon-nitrogen ratio
2. Blending or proportioning of wastes
3. Moisture and aeration
4. Temperature
5. Reaction
6. Microorganisms involved
7. Use of inoculants
8. Calcium phosphate, and
9. Destruction of pathogenic organisms.

C : N Ratio

The carbon-nitrogen ratio of organic materials is the most important aspect of composting. The process of conversion of organic materials into manure is chiefly microbiological and is, therefore, influenced by the proportions of carbonaceous and nitrogenous materials that are present in organic wastes to start with. Microorganisms need carbon for growth and nitrogen for protein synthesis. It wag found that a C : N ratio of 30 of raw materials could be most desirable for efficient composting. The C : N ratio between 26 to 40 as reported by many workers, provide for rapid and efficient composting. If the organic material is poor in nitrogen or in other words, carbon nitrogen ratio is wide, biological activity diminishes and several successions/cycles of organisms may be required to degrade carbonaceous materials. Immobilised nitrogen is recycled on the death of some of the organisms. Thus, the limited nitrogen is recycled by reducing the carbon content of the organic wastes. At low C:N ratio, ammonia is formed which under favourable conditions can be further oxidised to nitrite and nitrate. Other nutrients, like phosphorus, potassium, sulphur and micronutrients are also essential for microbial growth.

Lower C : N ratios under unfavourable conditions cause increasing loss of ammonia and higher values may prolong the period of composting. High C : N ratios are generally caused by organic materials poor in nitrogen such as cereal residues of wheat, paddy, *jowar*, *bajra*, maize, sugarcane trash, stalks of cotton, jute and sawdust. Farm and urban wastes have been found to vary widely in C : N ratio ranging between 30 and 80. As the stock of available nitrogen gets exhausted, the activity of nitrogen fixing microorganisms predominates and thus there is a gain of nitrogen fixed from the atmosphere. With a C : N ratio of raw materials less than 30 : 1, the proportion of nitrogen is in excess of the requirements of microorganisms. Although the process of decomposition goes on uninterrupted the unassimilable nitrogen is lost in the form of ammonia gas.

Similarly, 20 to 40 per cent nitrogen is lost in the form of ammonia during preparation of organic manure from cattle-dung and night-soil. This implies that a thorough mixing of these fractions with the carbonaceous wastes cannot be attained in practice and there are always pockets of concentration of nitrogenous materials which cause an imbalance in the decomposition process and loss of nitrogen. It is obvious that simple nitrogenous compounds are more susceptible to loss which are easily decomposed by microorganisms and these compounds are of high utility to plants. Control of the loss and retention of this important plant nutrient in the final manure would be of practical value in obtaining a better quality manure.

The time of composting can be reduced by adding a nitrogen source of organic materials such as activated sewage sludge or by blending with organic residues richer in nitrogen such as legume residues and aquatic weeds (water hyacinth), slaughter-house wastes, green leaves, wastes of sugar, and wastes of antibiotic, yeast and paper industries.

Shredding and Blending

Composting process can be accelerated if raw materials are shredded into smaller pieces or ground since the material becomes more susceptible to bacterial attack due to exposure of greater surface area. The most desirable particle size for composting is less than 5 cm although larger sizes can be composted satisfactorily. However, the advantages of shredding may be uneconomical while composting on farms and in villages or by individual farmers.

The C/N ratio and moisture percentage are two important parameters which should be considered while mixing or blending different type of organic wastes for composting. There may not be any need of blending if the C/N ratio of organic materials is between 25 to 50, although 30 to 40 may be a good range. Substances poor in nitrogen such as saw dust, paper, straw, etc. can be mixed with comparatively nitrogen rich materials, *e.g.* legume residues, water hyacinth, slaughter-house wastes, fish scrap, night-soil, sewage sludge and biogas spent slurry to obtain a near optimum C/N ratio between 30 to 40. Similarly materials too dry and too moist can be blended together to obtain a desirable moisture level to avoid anaerobic conditions leading to bad odour.

Soil at 5 to 10 per cent is also sometimes added to compost to reduce the moisture content and to absorb ammonia in low C/N ratio materials. The dry soil may be added if sufficient dry materials are not available. Soils may be added to high C/N ratio organic matter to buffer acid conditions and to act as a diluent for retarding anaerobic process. It can also be used for improving the appearance of finished compost to give it more granular structure.

Moisture

If the amount of moisture in the compost pile is below 40 per cent (W/W), decomposition will be aerobic but slow. Aerobic decomposition can go on at any moisture content between 30 and 100 per cent if adequate turnings are provided. During aerobic composting high moisture content should be avoided so that anaerobic conditions are not created. The optimum moisture level of 50 to 60 per cent may be quite satisfactory for aerobic composting. However, it may be in range of 40 to 80 per cent depending on the nature of organic material to be composted. At IARI, when composting fibrous materials containing straw were composted aerobically. 80 to 85 per cent moisture was maintained. If anaerobic composting is to be carried out, the maximum moisture content is not important. The moisture content between 80 to 90 per cent may be maintained.

Temperature

With the multiplication of microorganisms in the composting mass, the heat of exothermic biological reaction is retained due to sufficiently large mass and when the temperature rises above 40°C, the mesophilic microorganisms are replaced by thermophilic. High temperatures are essential for destruction of pathogenic organisms and weed seeds. This generally occurs within 2 to 5 days of the start of the composting. The temperature in the middle of the pile goes up to 55° to 70°C and after this it gradually cools to ambient temperature. Decomposition is fastest in the thermophilic stage. The optimum temperature based on oxidation of organic matter into carbon dioxide and water has been found to be 60°C. Schulz showed that a maximum temperature of 71°C was optimal, but it was found that temperature should not exceed 70°C for long because decomposition will be slowed down by a thermal kill of microorganisms and only a few of the thermophilic organisms actively carry on decomposition above 70°C. However, while composting high C/N ratio materials such as wheat and paddy straw, *jowar* stalk and *jamun* leaf fall, etc. of C/N ratio 48 to 50 in cubic metre pits, it was observed by Gaur that temperature of the compost pile did not go beyond 52°C. This indicated that extent of rise of temperature in compost will depend on the type material being composted and probably also on the size of the heap.

Reaction

The initial pH in compost heaps is generally slightly acidic round pH 6 as is found in the cell sap of most of the plants. The production of organic acids during the early stages of composting causes further acidification (pH 4.5–5.0) but as the temperature rises, the pH increases to slightly alkaline (pH 7.5–8.5) reaction.

Microbiology

The microbial population changes during aerobic composting, Chang has described a typical pattern. The fungi and acid producing bacteria appear during the initial mesophilic stage. As the temperature increases above 40°C, these are replaced by thermophilic bacteria, actinomycetes and fungi. Spore-forming bacteria develop at temperature above 70°C. Finally, mesophilic bacteria and fungi reappear as the temperature falls down.

Many aerobic mesophilic bacteria initially present in the composting material multiply and show increased activity. As the temperature is raised, their numbers decrease due to change in environment. A minimum is reached at 55° to 65°C during composting of grass cuttings and straw but their number increases again as temperature drops below 45° to 50°C.

The role of bacteria may be involved in raising the temperature of the compost for the development of thermophilic micro-organisms which colonise. Mesophilic bacteria which flourish during limited time, consume the most readily degradable carbohydrates. During the process compounds rich in nitrogen such as proteins, amino acids, blood-meal and peptones are bio-degraded especially by spore forming *Bacillus* spp. Actinomycetes degrade starch actively and also bring about large losses of water soluble fractions. Thermophilic bacteria do not appear to be important in degradation of cellulose and lignin but attack protein, lipids and hemicellulose fractions. Lipids are degraded to a great extent whereas degradation or celluloses and hemicelluloses is comparatively slow and intermediate. Lignin are the most resistant to decomposition and persist for longer period and tend to accumulate.

Actinomycetes (*Thermomonospora curvata*) may be important in cellulose decomposition. *T. curvata* was the most frequently occurring actinomycete in municipal and mushroom composts, because therrmophilic actinomycetes can grow at higher temperatures than thermophilic fungi and they become dominant at the warmest stage.

Mesophilic fungi are present as the compost warms up moderately. These fungi are saprophytic sugar fungi which are quickly replaced by thermophilic fungi. The mesophilic fungi reappear in large numbers as the heap cools down below 40°C. Evidently they persist in outer layers during the thermophilic stage and reappear when the temperature drops. They can utilise cellulose and hemicellulose but not as efficiently as thermophilic fungi.

Thermophilic fungi occur when the temperature ranges between 40°C and 60 C. Like mesophilic fungi, they survive in the periphery of the heap when the temperature exceeds this range. The bacteria, fungi and actinomycetes isolated and reported from compost are presented in Table 2.1.

Table 2.1: Bacteria, Actinomycets and Fungi Isolated from Compost

BACTERIA	**FUNGI**
Mesophilic	Mesophilic
Cellumonas folia	*Fusarium culmorum*
Chondrococcus exiguous	*F. roseum*
Myxococcus virescens	*Stysanus stemonitis*
M. fulvus	*Coprinus cinereus*
Thiobacillus thiooxidons	*C. megacephalus*
T. denitrificans	*C. lagopus*
Aerobacter spp.	*Clitopilus pinsitus*
Proteus spp.	*Aspergillus niger*
Pseudomonas spp.	*A. terreus*
Thermophilic	*Geotrichum candidum*
Bacillus stearothermophilis	*Rhizopus nigricans*
ACTINOMYCETES	*Trichoderma viride*
Thermotolerant and Thermophilic	*T. (lignorum) harzianum*
Micromonospora vulgaris	*Oospora variabilis*
Nocardia brasiliensis	*Mucor spinescens*
Pseudonocardia thermophila	*M. abundans*
Streptomyces rectus	*M. variens*
S. thermofuscus	*Cephalosporium acremonium*
S. thermoviolaceus	*Chaetomium globosum*
S. thermophilus	*Glomerularia* sp.
S. thermovulgaris	*Pullularia* (')
S. violaceoruber	*Fusidium* spp.
Thermomonospora curvata	*Actinomucor corymbosus*
T. fusea	*Mucor jansseni*
T. glaucus	*Talaromyces (Penicillium) variabile*
Thermopolyspora polyspora	*Helminthosporium sativum*
	Aspergillus fumigatus

Contd...

Table 2.1–Contd...

Thermotolerant and Thermophilic	*Sporotrichum thermophile*
Humicola insolens	*S. chlorinum*
H. griseus var. *thermoideus*	*C. t.* 6 (*Mycelia sterilia*)
H. lanuginosa (*Thermomyces lanuginosus*)	*Stilbella thermophila*
Mucor pusillus	*Malbranchea pulchella* var. *sulfurea*
Chaetomium thermophile	(*Thermoidium sulfureum*)
Absidia ramosa	*Dactylomyces crustaceous*
Talaromyces (*Penicillium*) *duponti*	(*Thermoascus aurantiacus*)
T. thermophilus	*Byssochlamys* spp.
T. emersonii	*Torula thermophila*

Inoculants and Other Preparations to Compost

If C/N ratio is too wide, adding nitrogen can hasten composting by lowering the ratio. Microbial inoculations may also be beneficial in a compost heap poor in microorganisms for speeding up decomposition of materials rich in cellulose and lignin. It has been observed in sawdust compost that inoculation with spores of cellulolytic fungus, *Coprinus ephemerus* and addition of nitrogen, phosphorus and potash shortened composting of sawdust from usual 1–2 years to 3 months and produced a compost that did not immobilise nitrogen as compared to fresh sawdust.

Japanese patents for additives to accelerate decomposition of wood products have been reported recently. The addition of nitrohumic acid (humus or complex natural ligno-protein treated with nitric acid) to bark wastes and chicken manure (1 : 7 : 100) reduced the decomposition time from 180 days to 60 days. Decomposition of wood pulp was improved by urea and crude acetic acid distilled from wood. The recent work done in India on microbial inoculants for accelerating the compost-making has been discussed separately in this chapter.

Calcium Phosphates

Several investigators have reported that calcium phosphate increases the rate of decomposition and nitrogen conservation, but more than 2 per cent inhibited the decomposition rate. Chang found increased rate of decomposition occurring mainly in cellulose and hemicellulose fractions. The number of cellulolytic microorganisms increased with the addition of phosphate and conserved nitrogen by decreasing the number of denitrifying bacteria. Although N fixation may be increased by adding calcium phosphate, there is generally sufficient phosphate in the organic wastes for meeting the requirements of microorganisms which need about 5 to 20 per cent only as much P as N. However, our experiments have shown that application of rockphosphate is important for nitrogen conservation and enrichment of organic manures, while composting wide C/N ratio organic materials (paddy, *jowar* straw, *jamun* leaves). Gaur recorded that *Azotobacter* numbers in compost were greater in presence of added rock phosphate.

Gaur reported that addition of rockphosphate at 1.0 per cent (W/W) increased the decomposition of paddy straw by cellulolytic fungi (*Aspergillus* spp. supplemented with urea nitrogen) (Table 2.2). The carbon content was decreased by 17.2 per cent and nitrogen augmented by 31.4 per cent. As a result C/N ratio of 23.1 was obtained within 8 weeks as compared to 37.7 in the control. Increased amounts of rockphosphate (2 and 3 per cent) showed slight reduction in decomposition intensity. Mathur from Ranchi Agricultural College reported that quality of compost prepared from mixture of paddy straw, grass and water hyacinth was improved when rock phosphate was applied to it with and without pyrite. The chemical composition of the compost is given in Table 2.3. The results showed that low dose of rock phosphate (5 per cent) increased the nitrogen content by 30 per cent of the compost over the control. The enrichment of compost with rock phosphate besides increasing the phosphorus content, increased the concentration of calcium and micronutrients particularly of iron, manganese and zinc. Beneficial effect of addition of rock phosphate at an optimum rate during the initial stage of compost making has been reported at IARI and ICAR Co-ordinated Project on "Microbial Decomposition and Recycling of Organic Wastes."

Table 2.2: Effect of Rock Phosphate on the Decomposition of Paddy Straw by *Aspergillus* sp.

Treatments	*% Loss in Weight (2 Months)*	*Organic Matter %*	*Carbon %*	*Total N%*	*C/N Ratio*
Urea (0.5 per cent w/w) control	45.2	75.35	43.71	1.190	37.13
Urea + Rock phosphate (1.0 per cent w/w)	55.0	62.36	36,17 (–17.2)	1.564 (31.4)	23.12
Urea + Rock phosphate (2.0 per cent w/w)	52.9	62.64	36.33 (–16.9)	1.326 (11.24)	27.40
Urea + Rock phosphate (3.0 per cent w/w)	52.8	64.60	37.47 (–14.3)	1.275 (7.1)	29.39

Figures in parentheses show per cent loss or gain.

Table 2.3: Chemical Characteristics of Compost Prepared with Rock Phosphate

Components	T_1	T_2	T_3
Total Ca (%)	1.85	2.27	1.92
Total Mg (%)	0.37	0.51	0.38
Total Na (%)	0.06	0.19	0.09
Total K (%)	0.92	0.81	0.83
Fe (ppm)	5,080	12,720	28,200
Mn (ppm)	180	520	540
Zn (ppm)	60	95	80
Cu (ppm)	10	24	44
pH	4.40	5.55	5.80
Organic Carbon (%)	14.64	18.06	12.83
Total Nitrogen (%)	0.61	0.19	0.58
C/N ratio	24.00	22.80	22.10

T_1: 650 kg raw material (P_0)

T_2: 650 kg raw material + 32.5 kg Mussoorie rock phosphate (MRP)

T_3: 650 kg raw material + 32.5 kg (MRP) + 32.5 kg Amjhore Pyrite.

Destruction of Pathogens and Parasites

The destruction of pathogens and parasites of animals and humans is an important aspect and a problem when compost is prepared from materials containing sewage sludge and night soil. Aerobic composting at high temperatures is effective in killing pathogens. Gotaas recommended 60°C for "Thermal Kill" of common pathogens listed in the Table 2.4.

Table 2.4: Temperature and Time of Exposure Required for Destruction of Some Common Pathogens and Parasites

Organism	*Observations*
Salmonella typhosa	No growth beyond 46°C; death within 30 minutes at 55°–60°C and within 20 minutes at 60°C; destroyed in a short time in compost environment
Salmonella spp.	Death within 1 hour at 55°C and within 15–20 minutes at 60°C
Shigella spp.	Death within 1 hour at 55°C
Escherichia coli	Most die within 1 hour at 55°C and within 15–20 minutes at 60°C
Enlamoeba histolytica cysts	Death within a few minutes at 45°C and within a few seconds at 55°C
Taenia saginata	Death within a few minutes at 55°C
Trichinella spiralis larvae	Quickly killed at 55°C; instantly killed at 60°C
Brucella abortus or *Br. suis*	Death within 3 minutes at 62°–63°C and within 1 hour at 55°C
Micrococcus pyogenes var. *aureus*	Death within 10 minutes at 50°C
Streptococcus pyogenes	Death within 10 minutes at 54°C
Mycobaclerium tuberculosis var. *hominis*	Death within 15–20 minutes at 66°C or after momentary heating at 67°C
Corynebacterium diptheriae	Death within 45 minutes at 55°C
Necator americanus	Death within 50 minutes at 45°C
Ascaris alumbricoides eggs	Death in less than 1 hour at temperatures over 50°C

The extent and duration of high temperatures in compost heaps as well as antibiosis as a result of mixed population of microorganisms provide reasons to believe that pathogens and parasites are not able to survive the composting process. To achieve this at least 2 to 3 turnings of compost heap is suggested. Knoll found that 65°C for one day ensured elimination of *Salmonella* spp. Stranch reported the destruction of *S. enteriditis, Erysipelothrix, Rhusiopathiae* and the *Psittacosis* virus under similar conditions. *Bacillus anthracis* was destroyed only at temperature above 55°C for 3 weeks or longer at a moisture content of 40 per cent or more. He concluded that higher temperature may not be the only way to kill pathogens during composting but some are killed by competition with other microorganisms. Schraff (1940) and Scott (1953) found that resistant forms of parasites such as *Ascaris* eggs, cysts of *Entamoeba histolytica* and hookworm eggs were destroyed during aerobic composting. Scott found that at least 2 turns were necessary for the destruction of *Entamoeba histolytica* in 12 days and three turns were required to destroy the eggs of *A. lumbricoides* in 36 days. Anaerobic composting at the mesophilic temperature range does not effect satisfactory destruction of parasites in a relatively short time. Anaerobic composting when using sewage sludge or night soil should be preceded by aerobic composting at least for a week. The natural death of pathogens and parasites will occur in an anaerobic environment and the microbial antagonism will eventually eliminate them but this will generally take longer period of about six months.

Heaps versus Pits

When preparation of manure is carried out in over-ground heaps, the decomposition becomes mainly aerobic, with air freely accessible for microorganisms to carry out their activities. The rate of decomposition is accelerated if periodical turnings are given which also promotes physical disintegration of the raw materials. Owing to quick oxidation of organic matter, high temperatures (60–70°C) develop, thereby killing weed seeds and pathogenic organisms. Loss of organic matter in this process often amounts to above 50 to 60 per cent and decomposition may continue even after the C/N ratio is reduced to the optimum. Loss of nitrogen from 20 to 50 per cent invariably takes place in this method.

Moisture content necessary for microbial activity is effected in the heap method, with losses occurring because of exposure to the atmosphere and the high temperatures that develop in the manure heap. In dry climate, this would affect the rate of decomposition and necessitate addition of extra water periodically. During the rainy season, on the other hand, there may be too much of water in the heap to slow down decomposition and even cause loss of nutrients and organic matter by leaching, unless manure preparation is carried out under sheds. This is not always feasible where large quantities of manure are required to be prepared.

Preparation of manure in pits has several advantages over method of preparation in heaps over ground although it involves some initial expenditure on digging and permanent occupation of land for this purpose. The decomposition is mainly anaerobic, except in the surface few inches and proceeds slowly. Losses of organic matter as well as nitrogen are consequently reduced, amounting to about 25 and 20 per cent respectively, enabling a larger quantity as well as good quality of the manure to be obtained. However, the process of composting can be accelerated by providing a few turnings at periodic intervals which will create aerobic conditions as well. The materials are protected against loss of moisture by evaporation as well as excess of water in the rainy season and physical loss of nutrients by leaching. High temperatures do not develop in this case owing to the anaerobic nature and the slow rate of decomposition. But weed, seeds and pathogenic organisms are destroyed because of the toxic products of this type of decomposition. Little physical disiategration of material such as straw and refuse, however, takes place in this system but this is not necessary from scientific standards of assessment of quality of manure of which a final C/N ratio of less than 20 : 1 indicates its matuıity and fitness for application to land. The manure may remain stored in pits until required without much loss due to very restricted decomposition after it is ready.

Utilisation of Urine Fraction

The liquid excreta of animals contains nitrogen in the form of urea, phosphorus, potash and several other nutrients in water soluble form which are of high utility to plants. Stress has always been laid on proper collection of this fraction and its utilization by adding urine-soaked litters or urine-earth to the manure-pit. However, in the process of its microbial fermentation with carbonaceous materials, the original simple compounds of nitrogen are converted into complex microbial proteins or into ligno-proteins by chemical reaction. There is also considerable loss of nitrogen in the form of ammonia due to the ease of decomposition of urea by numerous microorganisms. An overall reduction in the manurial value of this fraction thus takes place when it is incorporated in manure heaps. The low C/N ratio of urine and the simple nature of its nitrogenous compounds suggests that its separate collection and direct utilisation as manure would be a more effective alternative if practical methods of collection and conservation of this liquid manure are evolved. Work on this aspect carried out at the Indian Agricultural Research Institute and National Dairy Research Institute showed that urine-soaked earth can be treated with superphosphate or gypsum to control the decomposition of nitrogen and then quickly dried to stop bacterial activity or urine-soaked ground paddy husk or saw-dust can be composted along with dung with a good quality manure. The plant nutrient constituents thus remain conserved in their original form and the manure obtained is highly effective. Alternative materials like dry leaves, dry dung, etc. can be used in conjuction with soil in cattle-sheds to soak urine more effectively and processed as above. In *pucca* cattle-sheds where urine and cattle-shed washings are collected in cess pits, the liquid, manure may be similarly treated with superphosphate or a mineral acid cake hydrochloric acid and then soaked and dried in absorbing materials of the type mentioned.

Results obtained at the Indian Agricultural Research Institute have shown that the method can be effectively extended and applied to the collection and utilisation of large quantities of human urine which has not been exploited so far. A simple design of urinals evolved by the Institute (IARI) enables urine to be collected in absorbing materials contained in a drum or a pit dug for the purpose. The decomposition of urine is controlled by acidulating the material with superphosphate or hydrochloric acid. The material can be repeatedly soaked and dried to obtain concentrated manures. Dehydrated human urine manures have been found to be highly effective in stimulating growth and yield of crops. The proposition of separate collection and utilisation of cattle and human urine in the above manner offers a new yet unexplored field for efficient utilisation of an inexhaustible and rich source of manure available in the country.

Table 2.5: Effect of Human Urine Dehydrated in Different Materials on Wheat (Manures Applied at 112 kg/ha)

Treatment	*Yield in g from 30 Plants*		
	Grain	*Straw*	*Total*
No manure	11.5	23.8	35.3
Urine in soil	26.6	33.9	60.5
Urine in saw-dust	27:4	37.5	64.9
Urine in charcoal-dust	28.0	37.3	65.3
Ammonium sulphate	22.8	51.6	54.4
C.D.	5.7	4.8	7.6

Preparation of Farmyard Manure

Organic waste materials that generally go in production of farmyard manure including dung, urine, soaked earth or litter, remnants of cattle food, ash and other household and kitchen wastes. The existing method of preparation by dumping the small daily collection in a heap or shallow pit leaves much to be desired and yields a low-quality manure.

Sectional Filling

One of the effective methods of preparation of farmyard manure that has been found to yield superior quality manure is the process called 'Sectional Filling of Trenches', which includes all the advantages of carrying out of decomposition in pits and suits well to the conditions existing in this country of having to deal with a comparatively small daily collections. The method consists of digging suitable-sized trenches depending on the number of cattle as shown in Table 2.6.

Table 2.6: Methods of Digging Trenches

No. of Cattle	Length (m)	Breadth (m)	Depth (m)
2 to 5	6.5	1.0	1.0
6 to 10	8.0	1.2	1.0
11 to 20	10.0	1.4	1.0
Above 20	10.0	1.6	1.0

Filling of mixed wastes from the cattle-shed is recommended from one end of the trench in a section, one metre in length. A movable partition made from cotton stalks or such other material between two bamboo poles is used for making the section to be filled. The daily collection is dumped in this and covered with a thin layer of soil. This is continued until the section is filled to a level of about 50 cm above ground level. The top is then covered with earth and the partition is moved one metre further along the length to form the next section for filling as before. Leaves and other farm wastes like paddy and groundnut-husks, straw, etc. could be added to increase the quantity of manure if sufficient litter from the cattle-shed is not available. The manure is ready in about three months and contains about 2 per cent nitrogen on dry weight basis. The quantity of manure obtained per pair of animals amounts to 10 to 12 tonnes per year.

Composts

The principles of production of farmyard manure from cattle-shed wastes have been extended for the preparation of 'composts' in which a variety of carbonaceous and nitrogenous waste materials may be utilised. Thus, farm-wastes like straw, chaff, leaves, paddy and groundnut-husks, sugarcane-trash, bagasse and other agricultural as well as industrial and habitation wastes of all sorts like street-refuse, night-soil, slaughter-house wastes, sewage and sullage water, sludge, etc. can be pressed into service to yield manures.

In India, pioneering work in this field was carried out by Howard at Indore and Fowler at Bangalore, who standardized conditions for degradation of leaves, straw and town-refuse, utilizing dung and nightsoil as starters for promoting bacterial activity.

Indore Method of Composting

In this method the waste materials are spread out the moistened with dung or night-soil slurry. These are then built up into heaps, four to six metres in length, one metre in breadth and one metre in height. Under these conditions, the rate of decomposition is rapid and high temperatures develop quickly. The process is accelerated if turnings are given, whereby aeration and mixing of the materials and moistening at this stage, if necessary, cause more or less total disintegration, yielding a brown homogenous manure in about three months. Under this aerobic process of decomposition losses of organic matter and nitrogen are heavy, amounting to 40 to 50 per cent of the initial. The manure resembles the traditional farmyard manure in appearance and properties and it has been shown that decomposable organic materials of any type can thus be processed to supplement the supplies of manure. This method of manure preparation, however, involves considerable labour in building up the heap to proper shape and periodical turnings that are given and becomes expensive and impracticable where large quantities of materials have to be processed. However, the turnings of the mass are not necessary and decomposition can go on to the desired extent if adequate moisture level is maintained. The average composition of manure prepared by the Indore method has been found to be 0.8 per cent nitrogen, 0.3 per cent phosphoric acid and 1.5 per cent potash.

Fowler worked out the process of 'activated compost' in which fresh raw materials are incorporated in an already fermenting heap so that the established and large microbial population brings about quicker decomposition The process is useful particularly where offensive materials like night-soil are to be quickly and effectively disposed of.

Compost from Habitation Wastes–Bangalore Process

Pioneering work in preparation of manure in pits has been carried out in India by Acharya particularly on the utilisation of town-refuse and night-soil. What is called the hot fermentation method of manure production or the Bangalore process has by now been adopted in many towns in India, solving effectively one of the most difficult problems of sanitary disposal of these offensive wastes and yielding a high quality organic compost manure in the bargain. The operative part of the process can be adapted to the prevailing system of collecting refuse and night-soil. The method carried out in practice is as follows:

The compost production depot is located just on the outskirts for convenient transport of street-refuse and night-soil to the pits. In large towns, there may be two or three such depots to serve different localities. The compost depot should accommodate about 200 trenches of dimension as under, which can take the dumpings for one year (Table 2.7).

Table 2.7: Trenches Dimensions

Population (in '000)	*Length (m)*	*Breadth (m)*	*Depth (m)*
Below 10	4.5	1.5	1.0
10 to 20	6.0	2.0	1.0
20 to 50	9.0	2.0	1.0
Above 50	10.0	2.5	1.0

The trenches are dug in rows, with a space of 1 to 5 m between trenches. Roads of suitable width are provided between rows for the carts to approach and unload the materials inside the trenches. At first the refuse is dumped into the trench and spread out with rakes to make a layer of 15 cm. Night-soil is then discharged over this and, if necessary, spread out with long wooden spades in a layer of about 5 cm. This is then covered with a 15 cm layer of refuse. Night-soil and refuse thus follow in alternate layers until the pit is filled to 15 cm above ground level, with a final layer of refuse on the top. This may be dome-shaped and covered with a thin layer of soii. In some towns, night-soil comes mixed up with refuse in which case the material is simply dumped to fill the trenches. Sullage water, if collected in carts as in some towns, may also be emptied over the layers of refuse. In fact, the system provides a method of disposal of every kind of waste, that the town may have, including slaughter-house, carcasses of animals, sewage, etc.

The decomposition of dumped materials in pits takes place largely in the absence of sufficient air except in the surface foot or so. This anaerobic decomposition is comparatively slow but markedly less wasteful. The difference in recovery of organic matter and nitrogen by this method as compared with the aerobic method in over-ground heaps is given below for manure prepared from town-refuse and night-soil.

Table 2.8: Percentage Recovery

	Heap	*Trench*
Organic Matter	53	78
Nitrogen	54	74

In this method, no high temperatures develop in the lower layers and since the material does not receive turnings, disintegration into a homogeneous mass of manure does not take place. Even so, the C/N ratio is reduced to less than 20 : 1 in about six mouths and the manure is then ready for use. Deeper in the trenches, the material still possesses an unpleasant smell and it is desirable to dig it out and pile it up outside for some days before disposal.

In the absence of facilities for digging trenches, town-refuse and night-soil compost can be prepared in over ground heaps of one metre height, one metre width and any convenient length, with refuse and night-soil, placed in alternate layers, as in trenches and a final refuse layer on the top. Night-soil carts empty the material along the length of heap. No turnings need be given to the heaps and the material decomposes more quickly than in pits and can be used after three to four months.

The basic aspect of these processing treatments in manure production is that after the initial active decomposition and loss of excessive carbon has taken place, the proportion of nitrogenous night-soil and organic carbonaceous material is just proper and it can be used as manure. Often, the night-soil and refuse layers can be identified when the manure is dug out. The material gets mixed up only while it is being taken out, but because the mixture is in proper proportion, it is safe to apply as manure. Further, decomposition is carried out in the soil itself and because of the presence of adequate nitrogen this causes no adverse effect and instead soon releases nitrogen for crops to assimilate.

In villages individual latrines are limited to only a few rich families, hence night-soil collection and its compost making is not popular. It can, therefore, be a great step in improving the existing insanitary conditions to provide suitable community or individual laterines, preferably Wardha-type trench latrines to villagers to utilise human-wastes. Before putting Wardha-type laterines in use, a 7.5 cm layer of refuse (grass, leaves, straw, weeds, trash, etc.) is spread at the bottom of the trench. A stock of refuse or dry earth is left to cover the faeces. After the material rises to a height below 15 cm of the top, the trench is sealed for decomposition and another trench put in use. Manure of good quality is ready in four to six months.

Preparation of Synthetic Compost

The basic principles underlying the microbiological process of conversion of organic wastes into manure have been extended for preparation of synthetic composts, using inorganic fertilizers as sources of nitrogen for decomposition of carbonaceous materials. It has been found that organic nitrogen in the form of dung, required by microorganisms, can be completely substituted by inorganic nitrogen compounds like ammonium sulphate and urea, which are utilised equally effectively for decomposition and well fermented manure, resembling farmyard manure, can thus be obtained. This opens up a vast field for utilisation of large quantities of various organic waste materials where supplies of dung are either short of requirements, or not available at all, as on mechanized farms. The basic principles of C/N ratio in manure preparation can be applied to add nitrogenous fertilizers in sufficient quantity to reduce the ratio to about 30–40 : 1 and then allow the material to be decomposed and disintegrated until the C/N ratio is reduced to 20 : 1. The adco process of preparation of synthetic compost worked out by Hutchinson and Richards is based on this principle.

This process of production of compost from organic wastes of wide C/N ratio is worked with fertilizers like ammonium sulphate or urea. They provide complete or partial nitrogen for microbial decomposition in place of natural materials like dung or night-soil which may not be available at all or may be in short supply. Nitrogen is added to make up a total of 1.2 per cent on dry weight basis of the material. Thus straw containing 0.4 per cent nitrogen requires an additional 0.8 per cent nitrogen, *i.e.* about 4 kg of ammonium sulphate per 100 kg of straw. For neutralizing acidity, which may develop in this system, the process includes addition of lime at the rate of 5 kg to the above material. The material to be composted is spread out in layer in a heap or pit and sufficiently moistened. This is sprinkled with the fertilizer solution and then with lime. Superphosphate may be added to fortify the phosphatic contents of the manure. The treatment is continued layer-wise until the heap or pit is filled to size and allowed to ferment. The manure becomes ready for application in about four to six months and resembles farmyard manure in its action on soil and plant growth.

Mechanical Compost Plants

Although the major quantities of compost (310–350 million tonnes) at present used in agriculture are prepared in villages either by traditional or improved methods (Indore or Bangalore methods), these methods cannot be suitably used for processing of large quantities of organic refuse of bigger cities.

The mechanised/semi-mechanised plants will serve two pertinent objectives-sanitary disposal of city refuse and production of organic manure. It is estimated that by 1979 nearly 50 million tonnes of municipal wastes will be generated annually in India. Moreover, Indian city wastes contain about 50 to 80 per cent of compostable organic materials which can be processed conveniently in compost plants. During Sixth Five Year Plan, Ministry of Agriculture and Irrigation timely launched a programme to set up 35 composting plants in selected cities with a population of 3 lakhs and above. Five compost plants have already been set up one each at Ahmedabad, Baroda, Bangalore, Kolkata and Jaipur and the remaining 30 plants are expected to be set up during Sixth Five Year Plan (1978–83). The intake capacity of these plants already set up is about 125 to 200 tonnes garbage per day and can produce 60 to 70 tonnes of finished compost per day if the plant runs in shifts.

The costs of production is rather high due to mechanisation of the processes and at some places the plants were running at under-utilised capacity due to non-availability of required quantities of garbage. The selling price of finished compost has been kept between Rs 50 to 80 per tonne depending upon the composting system adopted and plant capacity. The cost may be reduced by installation of plants with less mechanisation.

The compost prepared from city wastes is bulky and low in plant nutrients. It contains 0.5 to 0.6 per cent N, 0.6 per cent P_2O_5 and 0.5 per cent K_2O with a C/N ratio of 20 : 1. There is a need to enrich this manure in respect of nitrogen and phosphorus and reduce its bulkiness involving low-cost technology. The work carried by Gaur showed that by application of 10 kg low-grade rock phosphate containing 20 per cent P_2O_5 per tonne of organic wastes (crop wastes and leaf fall, etc.) and inoculation with microbial fertilizers improved the nitrogen, phosphorus and humus content appreciably. This can be adopted in improving the compost prepared by mechanical compost plants.

Use of Composted Versus Uncomposted Materials

Direct application of carbonaceous materials of wide C/N ratio to soils adversely affects plant growth. This is owing to the utilisation of all soil nitrogen by microorganisms, engaged in the process of decomposition. Mixed organic wastes containing adequate nitrogenous materials like dung and night-soil can, however, be directly applied to land and allowed to decompose *in situ* in soil without causing deficiency of nitrogen and the consequent depressing effect on plant growth. If the C/N ratio of the materials is restored to 30 : 1 or less, the decomposition in soil is faster owing to dilution and large variety of microbial population of the soil. Positive advantages that accrue from the decomposition carried out in the soil are: (*a*) no loss of nitrogen in the form of ammonia takes plate in the presence of the large quantity of soil; (*b*) carbon dioxide evolved in large quantities during decomposition assists in enhancing the available plant nutrients of the soil; and (*c*) production of compounds possessing soil aggregating properties during decomposition by microorganisms, imparts good structure to the soil. The succeeding crop thus derives greater benefit from this manner of application than from the same raw materials after composting. It is, however, obvious that there are some limitations to the use of raw waste materials in practice. Materials like long-straw and street-refuse may be difficult to incorporate without prior disintegration. Offensive materials like night-soil cannot be handled in raw condition. Composting destroys pathogenic organisms and weed seeds that may be present in such materials. Under intensive cropping, the gap between one crop and the other may be too short for direct application of organic materials to soil and they may have to be composted before application. It has, however, often established that whenever possible, direct incorporation of materials like straw, leaves, etc. sufficiently in advance of sowing, with supplementary application of fertilizer nitrogen in adequate quantity, results in marked improvement in soil fertility with significant effect on crop yield.

Night-Soil Poudrettes

The disposal of night-soil is often a serious problem in towns where facilities for its composting are not available or refuse in adequate quantity is not collected. Trenching the material as such takes a long time for it to dry and causes nuisance of foul smell and profuse fly-breeding. Dehydration of night-soil by mixing it with any absorbing material like soil, ash, dry broken leaves, paddy-husk, etc. is a highly effective method of disposal of this offensive waste which yields poudrette manures of high manurial value. One of the most effective materials for this purpose is sawdust, which has high dehydrating capacity as well as the property of absorbing foul smell. The material can be used whenever available to treat night-soil and obtain an organic manure of high quality. Owing to its low C/N ratio, utilisation of night-soil as such without fermentation yields higher efficiency of response than when it is composted. The question of survival of pathogenic organisms in preparation and use of poudrette manures of this type is, however, a matter warranting consideration and further study.

Super-Digested Compost

Recent trends in the preparation of farmyard manure and composts include the use of superphosphate which is sprinkled over the raw materials as they are being filled in pits, at the rate of about five per cent. It has been found that the treatment makes organic manures of this type more balanced and markedly reduces losses of nitrogen that. take place during decomposition. The availability of phosphorus, thus intimately mixed with organic matter, is higher than when this fertilizer is used as such and this is of particular advantage to acidic, calcareous and heavy soils in which fixation of phosphorus greatly reduces the efficiency of this plant nutrient. Results of work by Walunfkarhar and Acharya illustrated the effect of super-digested compost on crop growth are given in the Table 2.9. Yield of *marua* was significantly better with application of super-digested compost than with separate application of the two.

Table 2.9: Average Yield of *Marua* in g per Pot

Sl.No.	Treatment	Grain	Straw	Total
1.	No Manure	7.4	23.8	31.2
2.	Compost (2½ tonnes dry matter/hectare)	10.3	33.0	43.3
3.	Superphosphate (P_2O_5 equal to No. 2)	8.9	30.8	39.7
4.	Super-digested compost	12.3	41.6	54.1
5.	Compost plus superphosphate (as in No. 4)	10.3	32.6	42.9
	S.E.	0.35	1.73	3.18

Subbiah by the use of tracer technique, showed that the movement of applied phosphorus in soil generally increased when organic matter in the form of compost was also applied with the fertilizer and this was further enhanced when applied as super-digested compost. The increase in movement due to super-digested compost over mere combination of superphosphate and compost was higher in laterite, red and calcareous soils.

Control of Losses of Nitrogen During Manure Preparation

It has been mentioned that the process of decomposition in the preparation of organic manures results in significant losses of nitrogen in practice, which reduces the quality of final manure that is obtained. The losses ranging up to 20 per cent take place although they are heavy in heaps and are considerably reduced in pits. It is, therefore, always desirable to prepare manure in pits to obtain a better quality and quantity of manure.

Loss of nitrogen during fermentation takes place mainly in the form of ammonia gas and methods have been worked out to retain this in the manure by the use of acidic materials to fix nitrogen chemically or by absorbants which physically hold nitrogen. Thus addition of superphosphate at the rate of about five per cent or a heap mineral acid like hydrochloric acid at one per cent have been found to conserve nitrogen in the manure. Materials like charcoal-dust, sawdust, gypsum and even soil added at the rate of five to ten per cent reduce nitrogen losses considerably. The inclusion of this simple treatment in the process of manure preparation would result in obtaining a superior quality of manure in which a substantial amount of nitrogen is held in forms that are highly available to plants.

Enriched Organic-based Manures

Fermented organic manures are low grade materials and, under Indian conditions, have to be applied in large bulk to obtain significant effects on crop growth and yield. The total nitrogen content in them does not exceed 1.5 per cent at best and with the small quantity of dung used by farmers, it is of the order of 0.5 per cent. The percentage availability of nitrogen for plants in such manures ranges from 5 to 20 per cent with the result that low responses are obtained when applications are made which are inadequate to make good the primary deficiency of nitrogen in our soils.

At the Indian Agricultural Research Institute, a new approach to this problem has been worked out which enables such low-quality manures and other organic materials to be effectively utilized. It has been found that the rate and extent of liberation of nitrogen from applied organic manures is adequate for normal plant growth when their C/N ratio is less than 10 : 1 or the nitrogen content is more than 2.5 per cent. In practice, the C/N ratio of organic materials can be reduced by adding the requisite amount of nitrogen in the form of any nitrogenous fertilizer to it and thereby increasing the value of the denominator instead of decreasing carbon in the numerator by fermentation. Thus, a low-quality manure containing 0.5 per cent nitrogen and a C/N ratio of 20 : 1 can be treated with ammonium sulphate or urea solution to add 2 per cent more nitrogen and raise the total nitrogen content to 2.5 per cent. This would immediately reduce the C/N ratio to 8 : 1 and the enriched manure can then be utilised effectively. This method can utilise large quantities of other organic waste materials of wide C/N ratio like dry dung, paddy and groundnut-husks, sawdust, tobacco-waste, wheat-chaff and any material in condition to be easily incorporated in soil. This method enables the process of composting to be eliminated thus saving time and labour.

Enriched organic manures of the above type are processed and get ready in a short time and possess many advantages. Losses resulting from their composting are avoided and the total quantities of organic matter and nitrogen taken initially are conserved and utilised. The decomposition taking place in the soil in the presence of adequate nitrogen incorporated will cause no adverse effects and sufficient nitrogen, in excess of microbial requirements, is liberated for optimum growth of crops.

Variations in the preparation of these enriched manures include incorporation of more nitrogen to obtain concentrated manures containing 5 to 7 per cent nitrogen and supplementing this with superphosphate and potassium sulphate to get mixed organic-mineral fertilizers of any desired grade to suit specific requirements of soil or crops.

Preparation of Enriched Manures

Enriched organic-based manures are prepared by taking requisite quantities of nitrogenous and other fertilizers and dissolving them in sufficient water. The solution is then treated with 100 ml of commercial hydrochloric acid per quintal of the final manure for controlling decomposition of the added nitrogen and its retention in the manure. This is then absorbed in the calculated quantity of the organic material, mixed thoroughly and dried in the sun. The dry enriched manure can then be bagged and stored for use.

Microbial Method for Hastening Compost Preparation

Compost is bulky in nature and contains on an average 0.5 to 1.0 per cent N, 0.6 pet cent P_2O_5 and 0.5 per cent K_2O. There is a need to reduce its bulkiness and period of composting to make it more acceptable to the farmer. Moreover, the quality of manure should be improved to supply more nitrogen and phosphorus and humus content per unit weight. Researches conducted by Gaur in the Division of Microbiology, IARI have shown that it is feasible to hasten the period of composting by proper inoculation with cellulolytic microorganisms as well as improve nitrogen and phosphorus contents of organic fertilizers with low-cost technology. Gaur reported that mesophilic cellulolytic fungi when used as inoculant in composting of plant residues (straw, leaf fall) can hasten the composting process and improve the quality of compost. The carbon content of compost mass was lowered by 25 per cent of the initial raw material (48 per cent) organic carbon during the first month of composting in pits. The loss was more due to fungal inoculation, *Penicillium* spp. showing the maximum decomposition of 46 per cent followed by *Aspergillus* spp. with 37.5 per cent. As a result, C/N ratio within one month was decreased ranging from 19.6 to 23.6 due to inoculation with cellulolytic fungi whereas in uninoculated control, it was 31.4 (Table 2.10). Subsequently also the decreases in C/N ratio were greater due to inoculation.

The data (Table 2.11) on recovery percentage of compost on oven dry basis indicated that bulkiness of organic manures decreased by 2 to 8 per cent due to inoculation. The efficient culture reduced bulkiness by 8 per cent. The total nitrogen content showed increase on the other hand, the increase being more in inoculated series. Maximum of 17 per cent increase was recorded with *Aspergillus* spp. (RF_1) followed by *Aspergillus* spp. (IF_1) and *Penicillium* spp. (RF_2) with 15.6 and 14.2 respectively as compared to uninoculated control.

Table 2.10: Effect of Cellulolytic Culture in Hastening the Process of Composting

Treatments	*C : N Ratio of Compost Sampling Interval in Months*		
	I	*II*	*III*
Control	31.4	22.4	20.0
Aspergillus sp. (IF_1)	23.6	–	16.8
Trichoderma viride (IF_2)	23.1	19.5	18.9
Aspergillus sp. (RF_1)	20.6	16.9	14.3
Penicillium sp. (RF_2)	19.6	16.8	14.8

Initial C/N ratio 45.7; Weight of raw material 32 kg/pit.

Table 2.11: Final Yield of Compost and Nitrogen Status in Compost Material

Treatment	*Final Yield of Compost (kg)*	*% Recovery*	*Total N %*	*Total N in*		*% Increase Over Control*
				Finished Compost Material (g)	*Per Tonne of Finished Compost (kg)*	
Raw material	–	–	1.05	(336)	(10.5)	–
Uninoculated	19.6	61.25	1.41	276	14.1	34.29
Aspergillus sp. (IF_1)	18.9	59.06	1.63	308	16.3	55.24 (15.6)
Trichoderma viride (IF_2)	18.1	56.56	1.33	241	13.3	26.67 (–5.6)
Aspergillus sp. (RF_1)	17.0	53.12	1.65	281	16.5	57.14 (17.0)
Penicillium sp. (RF_2)	16.9	52.81	1.61	272	16.1	53.33 (14.2)

Weight of organic material composted 32 kg/pit.

Figures in parentheses show increase or decrease over uninoculated control.

The studies indicated that due to inoculation of compost heaps by cellulolytic cultures the period of composting of dry and wide C/N ratio materials can be reduced by 4 to 6 weeks. These findings have been confirmed by the studies carried out at other centres.

Lost Cost Technology for Enrichment of Compost

Recent experiments conducted at IARI have shown that inoculation with *Azotobacter chroococcum* and phosphate solubilising microorganisms (*Aspergilus awamori* or *Bacillus polymyxa*) improved the manurial value of compost. Inoculation of compost with nitrogen fixer and P solubiliser was done when the temperature had stabilised around 30–35°C after one month of composting. Low-grade rock phosphate (10 kg/tonne of raw material) was applied at the initial steps of filling up of compost pits.

The results indicated that 31.5 per cent carbon was decomposed in control during the first month whereas the loss in different treatments ranged between 33.3 to 41.5 per cent. This was further increased during second month showing maximum loss of 52.4 per cent of organic carbon in rock phosphate, *Azotobacter* and P solubilizer inoculated series (Table 2.12). In this treatment C/N ratio was reduced to 14.1 and 12.0 during 1st month and 2nd month after inoculation respectively.

Table 2.12: Changes in Carbon (%), Total Nitrogen (%), C/N Ratio During Enrichment of Compost by *Azotobacter* and P Solubilizing Cultures (Sampling interval in months after inoculation)

Treatment	1st Month				2nd Month			
	Carbon	Nitrogen	C/N Ratio	Humus %	Carbon	Nitrogen	C/N Ratio	Humus %
Control	31.4	1.36	23.0	7.8	28.5	1.38	20.6	4.85
Control + Rock Phosphate @ 10 kg P_2O_5/tonne	30.7	1.61	19.0	9.15	27.5	1.77	15.5	6.90
Azotobacter + P Solubilizer	30.5	1.51	20.1	9.30	26.6	1.73	15.4	6.00
Rock Phosphate + *Azotobacter* + P Solubilizer	26.9	1.89	14.1	9.0	21.9	1.82	12.0	6.95

Initial Sample: Organic carbon 46.0%, Total nitrogen 0.86%, C/N ratio 53.3.

There was appreciable gain in nitrogen content per tonne of finished compost in control as well as inoculated series ranging from 60.5 to 111.6 per cent over the initial raw material (Table 2.13). Addition of powdered rock phosphate alone resulted in an increase of 28.3 per cent over control. Inoculation with *A. chroococcum* and *A. awamori* amended with rock phosphate showed an increase of 32.0 per cent in nitrogen content over the control resulting in gain of 22 kg N/5 tonnes of finished compost. It was also noted that humus content increased in all the treatments including control (Table 2.11). The best effect in humus content 9.3 per cent was observed in case of *A. chroococcum* + *A. awamori* after 1 month of inoculation.

Table 2.13: Final Yield of Compost and Nitrogen Status in Compost Material

Treatment	Weight of Raw Material (kg/pit)	Final Yield of Compost (kg)	% Recovery	Total N %	Total N in Finished Compost Material (g)	Total N in Per Tonne of Finished Compost (kg)	% Increase Over Control
Raw material	32	–	–	0.86	275	8.6	–
Control	32	11.85	37.03	1.38	164	13.8	60.46
Rock phosphate @ 10 kg P_2O_5/tonne	32	12.30	38.44	1.73	213	17.3	101.16
Azotobacter + P solubilizer	32	11.45	35.78	1.77	203	17.7	105.81
Rock phosphate + *Azotobacter* + P solubilizer	32	13.00	40.63	1.82	236	18.2	11.03

In another study the number of *Azotobacter chroococcum* in compost was increased due to its inoculation and use of rock phosphate. The mean numbers of nitrogen fixers of nine weekly intervals per gram of dry compost in different treatments were as follows: Control 1,380, rock phosphate 3,590; *Azotobacter* 6,070; rock phosphate + *Azotobacter* 21,010. This also resulted in an increase in the nitrogen content of conlpost by 25 to 44 per cer.t over control. These findings clearly show that quality of compost can be improved by inoculation with *Azolobacter* and P solubilisers suitably amended with rock phosphate.

Fly-Breeding in Manure Preparation and its Control

One of the disturbing aspects of manure preparation from organic-wastes particularly town-wastes, is the profuse fly breeding that takes place in this medium. In over-ground heaps, owing to the heat that develops inside, fly maggots migrate and concentrate on the surface few cm all over. If the material is given turnings, as in the Indore method, the maggots get substantially destroyed, otherwise flies develop profusely. In pits, the top layer of refuse or soil becomes the seat of migration of all maggots where they finally develop into flies.

The emergence of flies from such manure pits has been observed from the 6th day after filling and goes on till the 12th day or so, after which it ceases. Based on these observations, a simple and highly effective method of destroying flies, while they are resting in the maggot or pupae stage in the surface layer, has been worked out. The entire surface area of the pit is covered on the 6th day after filling with rags, refuse, straw, cotton or other stalks, sawdust, paddy or groundnut-husks or any such combustible waste and set on fire, which may continue to smoulder for about an hour. The maggots and pupae lodged on the few centimetres of upper surface are completely destroyed by the heat. The operation may be repeated on the 12th day to obtain absolute control of fly breeding. The treatment leaves a heat layer of ash on the top which is no more attractive to the flies.

Oil-Cakes

Oil-cake is a valuable by-product of the oilseed extraction industry and contains significant quantities of proteins, of value as animal feed if edible and as nitrogenous manures otherwise. There is always a residue of oil present in them depending on the process of extraction as shown below:

	Oil per cent
Country *ghani* Extracted	10–12
Hydraulic Press	8–10
Steam Expeller	5–8
Solvent Extract	1–2

The presence of oil slows down decomposition and the process of mineralization of their nitrogen in soil. Since this also gets decomposed in due course, the overall formation of mineral nitrogen is not materially affected in quantity. Their C/N ratio is narrow varying from 3 to 10 and about 60 to 75 per cent of the nitrogen is easily mineralizable and available for plants.

Effective Utilization of Mahua-Cake

Mahua-cake is the poorest of oil-cakes with a nitrogen content of 2.5 per cent and has been found to be much less effective than others even when applied in larger quantities. It has been found that nitrogen of this manure mineralised to a small extent in the soil owing to the presence of toxic saponins. The efficiency of the manure can be increased by moitening it and allowing it to ferment for some time to decompose the toxic substances. An effective treatment worked out at the IARI involves treatment of the material with ammonia solution which not only increases the nitrogen content of the manure but enables the original nitrogen to be effectively mineralized and utilized by plants. The treatment could be extended to enrich other oil-cakes and utilize them more effectively.

Utilization of Slaughter-House Wastes

Blood

Blood is the most valuable of slaughter-house wastes which yields a rich proteinaceous meal on dehydration. It is useful as a feed or as a nitrogenous manure. The existing crude methods of its collection and processing in India result in wastage of a large part.

In slaughter-houses, blood is obtained partly in the form of spongy clots and partly as blood serum in liquid condition. These are dried by heating in large open pans over a slow fire, which is a time-consuming and highly offensive process. The Indian Agricultural Research Institute has worked out a simple method for quick treatment of blood to obtain blood-meal which can be used as feed or as a manure.

Blood-meal from serum is prepared by adding one per cent of lime and then heating it till the temperature rises to 80°C. At this stage, blood begins to coagulate by its reaction with lime, and further slow heating is continued with constant stirring until the flocculated blood separates out and the colour changes from red to ash. The remaining water is decanted off and the coagulated blood is passed through a quarter-inch mesh expanded metal sieve and dried in the sun. There is no loss of nitrogen in the process and the manure contains 10 to 12 per cent nitrogen.

Blood clots are processed separately by cooking them for a few minutes boiling in 0.5 per cent hydrochloric acid solution. The clots harden and rise up to the surface. They are then taken out, rubbed over an expanded metal seive and dried. The manure contains 14 to 15 per cent nitrogen.

Blood serum can also be absorbed as such in materials like dry dung powder, sawdust, dry leaf powder and such other materials and then dried in the sun. Such products may contain 4 to 5 per cent nitrogen. The use of these absorbing materials on the floor of slaughter-houses would enable blood to be directly absorbed and offers one of the simplest methods of effective collection and utilisation of this waste and keeping slaughter-houses clean.

Waste-meat

This waste in slaughter-house consisting of intestines, offal and other flesh cuttings is hardly utilized in our country. Their disposal and use in compost pits has been suggested where they can serve as a source of nitrogen required for the process of decomposition of other wastes. The proposition of utilization of meat wastes as such and obtaining concentrated meat-meal is more useful to consider. The method of cooking these in steam digesters under pressure can be adopted for the purpose but this requires expensive equipment. A simple method worked out at the IARI for treatment of such materials consists of cutting them into pieces and cooking in 5 per cent hydrochloric acid solution until they become soft. The material is then washed and rubbed over an expanded metal sieve and dried in the sun. The product may contain 8 to 10 per cent nitrogen and can be used as poultry feed or manure. The cost of processing works out to Rs 2.50 per tonne.

The method can also be adopted to utilize the entire flesh of carcasses of all animals that die in towns and villages and can be organized on a countrywide scale for the exploitation of a valuable source of manure.

Bones

Bones from slaughter-houses, carcasses of all animals and from meat processing industry constitute a valuable waste which could be exploited more fully in the country. The existing methods of disintegration of bones involve crushing them in hammer mills by which raw bone-meal is obtained as a by-product or by digesting them in steam digesters, under pressure which yields steamed bone-meal in fine powdery condition. The methods require expensive equipment and are not always suited for adoption and treating small collection in villages and remote places.

Chemically Processed Bone-Meal

The IARI has evolved a purely chemical process of disintegrating bones and obtaining a sterilized bone-meal in fine condition which can be used as a feed or as manure. The process can be worked both as a cottage industry to treat small collections, suiting conditions in this country or on a large scale, if required.

Fresh or dry bones are taken in a suitably-sized container and soaked in sufficient water to cover them. Caustic soda is added to this to make a 5 per cent solution in the quantity of water taken. The bones are stirred once or twice daily until they disintegrate completely to a paste in 14 days. The supernatent is then decanted and utilized again and the disintegrated bones are washed twice with water. The material is finally taken up with water and treated with 10 per cent hydrochloric acid solution until it is neutral to litmus paper. The supernatent is then decanted and the bone paste filtered through muslin cloth. This is then dried in the sun and it yields a sterilized bone-meal in fine condition. The product contains about one per cent nitrogen and 27 per cent phosphoric acid. The cost of production, excluding the value of bones works out to be Rs 60 per tonne. The process is patented.

Fish-Meal

Fish manure processed by drying non-edible fish or wastes from fish industry is a well-recognized, balanced organic manure containing both nitrogen and phosphorus in significant quantities. Fish-meal is also produced in India by cooking in steam digesters and then drying. The manure, however, has an extremely offensive smell which stands in the way of its wider use. At the IARI, it has been found that the objectionable smell can be completely smothered by the use of two litres of pipe-oil emulsion per tonne. The long coast-line of the country offers a great scope for fuller exploitation of this highly valuable manure for increasing agricultural production.

Leather-Wastes

Tanned and untanned leather wastes are obtained from tannery and leather processing industries where their disposal becomes a serious problem. Leather-wastes can be processed by heating in steam digesters and then grinding to powder. The product obtained contains 7 to 8 per cent nitrogen and has been used as a nitrogenous manure. The IARI has a patented process for preparing leather-meal in which the material is soaked in water and then cooked in 5 per cent hydrochloric acid. The leather disintegrates into a plastic mass which is washed in water and dried and ground to powder. The acid treatment also helps in breaking down complex nitrogenous compounds and yields a manure containing 7 to 8 per cent nitrogen of which 70 to 75 per cent is mineralizable. The method can be adopted on a cottage industry scale, or large-scale according to the needs.

Hair and Wool-Wastes

Large quantities of animal-hair wastes become available in tanneries, which have found no use so far. Human hair can be collected in a large quantity, from hair-cutting saloons. These have also not been exploited so far. Large quantities of wool-waste also become available from wool refining, cloth and carpet industries which are not properly utilized. These materials contain about 15 per cent nitrogen in the form of complex keratin and do not decompose in soil when used as such. No estimate is available of the quantities of these wastes that can be collected but it is considered that their utilization in agriculture as manure after suitable processing would provide a substantial additional source of a rich organic manure. A patented process worked out at the IARI involves treatment of these materials with a solution containing 8 per cent caustic soda on the weight of the material. This brings about their disintegration to a paste which is then treated with 10 per cent hydrochloric acid to neutralize the alkali and re-precipitate the proteins. The liquid is decanted and the spongy solid is rubbed over an expanded metal sieve and dried. The product contains 12 to 15 per cent nitrogen and has been found to be a good nitrogenous manure.

Chapter 3
Development of Biofertilizers

At present India produces about 117 million metric tonnes of food grains for its growing population of more than 800 million people. During the recent plan the target to produce is additional 30 million food grains and 5 million tonnes of pulses and oil seeds. In order to meet the food requirement of ever increasing population, the nitrogen fertilizer requirement for crop production by 2000 AD is estimated to be about 11.4×10^6 tonnes as against 3.9×10^6 tonnes. At present the waste gap can not be filled up merely through the production of synthetic nitrogenous fertilizers due to scarcity of high cost of raw materials such as fossil fuels. Biological nitrogen fixation is the key to sustain agricultural productivity application of biofertilizers in the field is the viable alternative. Biofertilizer as a living fertilizer composed of microbial inoculants or groups of micro-organisms which are able to fix atmospheric nitrogen, the micro-organisms are known as biological nitrogen fixers. They are grouped into free living bacteria (*Azotobacter* and *Azospirillium*) and the blue green algae and symbionts such as *Rhizobium*, and *Frankia* and *Azolla*.

In fertilizer manufacturing factories, nitrogen is fixed industrially, by means of the Haber-Bosch process requiring H_2 gas, very high temperature and anormous energy. Industrially fixed nitrogen has been used precipituously. It was produced 4.0 lakh metric tonnes in 1905. This was increased to 3.5 million tonnes of nitrogen fertilizer was produced industrially. Meanwhile the contribution of biologically fixed nitrogen has not changed. The estimate of chemically fixed nitrogen for 1976 is 42 million metric tonnes. It is projected that by 2000, if we continue to rely entirely on increased use of chemical fertilizers to achieve the food production level more demand for fertilizer nitrogen to results and around 100–200 million metric tonnes would be needed to produce at an average cost of rupees.

Materials of Biological Origin

Indian soils are usually poor in organic matter as well as in nitrogen. The materials of biological origin which are commonly used to maintain and improve soil fertility may be grouped into two main categories:

1. Biofertilizers and
2. Green manures

Biofertilizers

The term biofertilizers or which can be more appropriately called 'microbial inoculants' can be generally defined as a preparation containing live or latent cells of efficient strains of nitrogen fixing, phosphate solubilising or cellulytic microorganisms used for application of seed, soil or composting areas with the objective of increasing the numbers of such microorganisms and accelerate certain microbial process to augument the extent of the availability of nutrients in a form which can be easily assimilated by plant. In large sense, the term may be used to include all organic resources (manure) for plant growth which are rendered in an available form for plant absorption through microorganisms or plant associations or interactions.

Classification

Biofertilizers can be broadly classified into three categories (Table 3.1)

Table 3.1: Classification

Biofertilizers (BF)

Nitrogen fixing Biofertilizer (NBF)		Phosphate Mobilising Biofertilizer (PMBF)		Cellulolytic or Organic matter decomposer (OMD)	
NBF for Legumes *Rhizobium*	NBF for cereal, *Azotobacter* *Azospirillium* *Pseudomonous*	Phosphate Solubiliser *Bacillus*	Phosphate Absorber VA-Mycorrhiza (VAM)	Cellulolytic organism Cellulomonas	Lignolytic organism *Arthrobacter*
Azolla, BGA *Glomus*	*Aspergillus*	like	spore	*Trichoderma* *Agaricus*	

Potential of Biofertilizers in Crop Production in Indian Agriculture

Soil harbours a range of microorganisms which bring about a number of biochemical reactions in soil. The soil organisms are classified into two broad groups *i.e.*, soil flora and soil fauna; these are again subdivided, depending upon their size such as micro and macro flora. Soil microflora includes bacteria, fungi, actinomycetes, algae, etc. Of these groups, bacteria are the most abundant followed by actinomycetes, and fungi; algae are found under specific situations. The biomass and population of these microorganisms in soil are found under specific situations. The biomass and population of these microorganisms in soil have been given in the Table 3.2.

The presence of these microorganisms make the soil a living and active system. These microorganisms play a significant role in the life cycle of plants and animals through a number of processes such as decomposition, solubilization, nutrient fixation and supply of plants (Table 3.3).

Table 3.2: Important Soil Microorganisms, their Number and Biomass

Microorganisms	*Average Number (in lakhs) per g of Soil*	*Average Biomass (in Kg per ha)*
Bacteria	1000	500
Fungi	10	1000
Actinomycetes	100	750
Algae	0.01	150

Table 3.3: Important Biofertilizers Agents

Group	*Scientific Name*	*Activity*
Bacteria	*Bradyhizobium* sp.	Symbiotic nitrogen fixation
	Rhizobium sp.	Symbiotic nitrogen fixation
	Azotobacter sp.	Asymbiotic nitrogen fixation
	Azospirillium sp.	Asymbiotic fixation
	Pseudomonas striata	Phosphate solubilization
	Bacillus subtilis	Phosphate solubilization
	Xanthomonas sp.	Phosphate solubilization
	Nitrosomonas sp.	Phosphate solubilization
Fungi	*Aspergillus niger*	Phosphate solubilization
	A. awamori	Phosphate solubilization
	Fusaruim sp.	Phosphate solubilization
	Penicillium sp.	Phosphate solubilization
	Vesicular arbuscular mycorrhizae (VAM)	Phosphate solubilization
Algae	*Anabaena, Nostoc, Calothrix*	Biological nitrogen fixation
Fern	*Azolla* sp.	Nitrogen fixation
	A. caroliniana	Nitrogen fixation
	A. nilotica	Nitrogen fixation

Biofertilizer means using living organisms as fertilizer, either to fix atmospheric nitrogen or to solubilize mineral nutrients like phosphorus. The microbial inoculants have attained special significance in modem agriculture. Keeping in view the increasing cost of chemical fertilizers and poor purchasing capacity of Indian farmers several microorganisms which can be used in agriculture are being listed in Table 3.4.

Chemically Fixed Nitrogen Versus Biologically Fixed Nitrogen

With the fast depletion of fossil fuel and exorbitant cost of chemical fertilizers, efforts should be made for augmenting biological nitrogen fixation by microorganisms. On an average, one acre of grain legumes (like soybean, bean or pea) provide sufficient protein for 1000–2000 days for one person, whereas an average acre of plant materials converted to animal protein like beef and poultry provides it only for 75–250 days. India being one of the less affluent nations, where the per capita income is low and consequently, the intake of vegetable protein is more in comparison to animal protein. Accurate estimates of annual turnover of nitrogen in the biosphere vary from 100 to 200 million tonnes. The amount of nitrogen fixed by various legumes irrespective of soil and other conditions has been given in the Table 3.5.

Table 3.4: Important Microorganisms Constituting Biofertilizers

Microorganism	*Nutrient Fixed (Kg ha/year)*
Actinorrhizae (*Frankia* sp.)	150 kg nitrogen ha
Algae	25 kg N_2/ha
Azolla	900 kg N_2/ha
Azospirillum	10–20 kg N_2/ha
Rhizobium	50 to 300 kg N_2/ha
Azotobacter	0.026 to 20 kg N_2/ha
Mycorrhizae	Solubilize food phosphorus (60 per cent)
Phosphate solubilizing bacteria and fungi	Solubilize about 50–60 per cent of the fixed phosphorus in the soil

Table 3.5: Estimates of Nitrogen Fixed by Some Legumes

Crop	*Nitrogen Fixed (kg per hectare)*
Alfalfa	100–300
Berseem	120–150
Chickpea	26–63
Clusterbean	37–196
Clover	100–150
Cowpea	53–85
Blackgram	38–50
Greengram	50–55
Groundnut	112–152
Lentil	35–100
Peas	59–80
Pigeonpea	60–200
Soybean	49–130
Sesbania	69–90

Above every hectare of land, there are 78,000 tonnes of inert nitrogen gas (N_2). Nitrogen is the most limiting nutrient for increasing crop productivity. Only a few procaryotic organisms are able to fix nitrogen directly through a biological process. Annual biological nitrogen fixation (BNF) is estimated to be around 175 million tones of which close to 79 per cent is account for by terrestrial fixation.

Synergistic Interaction Between Biofertilizing Agents

Biofertilizing agents such as *Rhizobia*, mycorrhizae, *Azotobacter, Azospirillum*, etc. are member of soil-biota or soil biosphere; they live together since the introduction of life in the universe. Hence they have developed a sort of natural relationship with each other and they maintain it unless disturbed by human beings by growing crops, application of fertilizers or pesticides. Crop growth in the soil releases several organic substances which may selectively eliminate certain microorganisms or reduce their activities or even augment the activities of the soil biota. Since various biofertilizing agents are selective, such as *Rhizobia* fix N_2 from atmosphere in the root zone of the legumes or phosphate solubilizing bacteria (PSB) or fungi (PSF) solubilize phosphate in the soil; this results in the simultaneous application of *Rhizobia* and PSF or mycorrhizae. The question is always raised whether the simultaneous application may not jeopardize the activity of the individuals. Experiments were conducted on the simultaneous application or blending application of biofertilizing agents and it is generally concluded that the application has, by and large synergistic action (Table 3.6).

Table 3.6: Interaction of Various Biofertilizing Agents

Crop	*Inoculation of Biofertilizing Agents*	*Positive (+) or Negative (–)*
Wheat	*Azotobacter* (AZOTB) + *Rhizobium* (Rh)	+
Peanuts	*Azospirillum* (AZOS) + *Glomus* sp + *Rhizobium*	+
Onion	*Beijerinckia mobilis* + *Aspergillus niger* + *Glomus fasciculatum*	+
Tomato	*Azotobacter vinelandi* + *G. fasciculatum* + *Rhizobium* sp.	+
Chickpea	*G. caledonicum* + *Rhizobium*	+
Soybean	*Glomus* + *Gigaspora* + *Rhizobium* sp.	+
Grasses	*Azotobacter brasilense* + *Glomus* sp.	+

Biofertilizers are apparently environment and farmers friendly renewable source of non-bulky, low cost organic agro-input. While *Rhizobium, Blue Green Algae* (BOA) and *Azolla* are crop specific, bio-inoculants like *Azotobacter, Azospirillum, Phosphorus Solubilizing Bacteria* (PSB), *Vesicular Arbuscular Mycorrhiza* (VAM)etc. could be regarded as broad spectrum biofertilizers. Their use has so far not received desired attention. This is mainly because of inadequate awareness both among the extention workers and farmers regarding their utility, their short shelf life, lack of ready availability in time and in desirable quality, and inconsistency in results with their use. However, proper education of the extension workers, dealers and farmers about their significance and economic feasibility of their application both by seed treatment or to soil, could go a long way in prompting their adoption. These could effectively supplement the nutrient requirement of crops through chemical fertilizers, to meet the soaring demand for food, fibre and fuel.

A biofertilizer is an organic product containing a specific microorganism (microbial inoculant) in concentrated form (10^7 to 10^9 per gram) which is derived either from the nodules of plant roots or from the soil of root zone (rhizosphere). They possess unique ability to fix atmospheric nitrogen either by living symbiotically with the roots of leguminous plants or non-symbiotically (free living) or to transform native soil nutrients such as phosphorus, zinc, copper, iron, sulphur etc. from the non-usable (fixed) to usable form through biological processes. Biofertilizers are apparently environmental friendly, low cost, non bulky agricultural input which could play a significant role in plant nutrition as a supplementary and complimentary factor to mineral nutrition.

Biofertilizing Agents and Plant Disease Control

Biofertilizing agents control the plant pathogenic fungi directly as well as indirectly. Directly they parasitize the pathogens: application of rhizobium culture on the legume seeds control seed borne fungi such as *Colletotrichum, Ascochytu, Helminthosporium.* etc. The rhizobia produce a toxic substance when they multiply on the seed and rhizosphere. Phosphate solubilizing fungi such as *Aspergillus niger* and other *Penicilla* produce antibiotic substances and thus kill the pathogenic fungi. Indirect killing of the plant pathogens is achieved by producing healthy seedlings and phytoalexins. Application of mycorrhizae produce better root systems which overcome the attack of root rotting and soil borne pathogens. Numerous reports are available that applications of biofertilizers in the soil stimulate and augment the activity of saprophytic microorganisms.

Biofertilizers and Insect Control: Legume inoculation indirectly protect plants from the attack of insects.

Brief Account of Beneficial Microorganisms

Rhizobium

Symbiotic N_2 fixation by *Rhizobium* in legumes contribute substantially to total biological nitrogen fixation. Rhizobium inoculation is well known agronomic practice to ensure an adequate nitrogen of legumes in lieu of N-fertilizer. Different species of rhizobium are classified into two groups *viz.*, (*i*) slow growing rhizobia (under the genus *Bradyrhizobium*) and (*ii*) the fast growing groups (under the genus *Rhizobium*).

Why inoculation? Although, native rhizobia are present in the soil but all of them are not capable of forming nodules. Some of the strains are highly effective in this respect while others are partially or completely effective. It is also reported that native flora gradually lose their efficiency. Hence, the artificial inoculation with tested effective strains, should be taken up as a comparatively means, cheap insurance for obtaining optimum yield.

Carrier-based inoculants are usually coated for the introduction of bacterial strains into soil. However, alternative inoculation methods are necessary where seed treatment with fungicides and insecticides is needed or where seeds like groundnut and soyabean can be damaged when the inoculants is used with an adhesive. Direct contact with the acidic fertilizer can also be harmful for rhizobium.

Apart from application with seeds, the normal carrier-based inocula can also be separately applied. While all methods of application were successful under favourable conditions, "Liquid" and "Granular" methods were superior to seed inoculation under adverse conditions.

Azotobacter and *Azospirillum*

Many genera and species of N_2 fixing bacteria have been isolated from the rhizosphere of various cereals. Thus mainly belong to *Azotobacter* and *Azospirillum* genera.

These are free living bacteria and fix atmospheric nitrogen in cereal crops without any symbiosis. They fix 15–20 kg/ha nitrogen per year. *Azotobacter* sp. also has ability to produce antifungal compounds against many plant pathogens. They also increase germination and vigour in young plants leading to an improve stand in crops. There is evidence in the literature that *Azotobacter* biologically control the nematode diseases of plants.

Phosphate Solubilizing Microorganisms

Phosphorus is also one of the major element required for plant growth and higher yields. This element is necessary for the nodulation by *Rhizobium* and even to nitrogen fixers, *Azolla* and BGA. The phospho-microorganism mainly bacteria and fungi make available insoluble phosphorus to the plants. The root fungus association or mycorrhiza has high potential in accumulating phosphorus in the plants. Mixture of charcole and soil is satisfactory material for these microorganisms in order to prepare commercial inoculants. It is reported that microphos cultures increase yield upto 200–500 Kg/hectare and thus 30 to 50 kg superphosphate can be saved.

Next to nitrogen, phosphorus is the vital nutrient for plants and microorganisms. Superphosphate is one of the common forms of phosphate fertilizers. These fertilizers are prone to be fixed in the soil.

Rock phosphate is one of the basic raw materials for phosphatic fertilizers. Direct application of rock phosphate is limited to acidic soil, while in other types of soil the applied phosphate becomes insoluble within a short time. Monocalcium phosphate are converted to dicalcium phosphate which is slowly available to plants. Under such conditions large amount of phosphorus is fixed in the soil which is unavailable to the plants.

Several soil bacteria and a few species of fungi possess the ability to bring insoluble phosphate in soil into soluble forms by secreting organic acids. These acids lower the soil pH and bring about the dissolution of bound forms of phosphate.

Vesicular Arbuscular Mycorrhizae (VAM)

The symbiotic association between plant roots and fungal mycelia is termed as mycorrhiza (Fungal roots). These fungi are found associated with majority of agriculture crops. They are ubiquitous in geographic distribution occurring with plants growing in arctic, temperate and tropical regions alike. These fungi are obligate symbionts and have not been cultured on nutrient media.

VAM fungi infect and spread inside the root. They possess special structures known as vesicles and arbuscules. The arbuscules help in the transfer of nutrients from the fungus to the root system and the vesicles, which are saclike structures store P as phospholipids. VAM have been associated with increased plant growth, and with enhanced accumulation of plant nutrients, mainly P, Zn, Cu and S mainly through greater soil exploration by mycorrhizal hyphae. It has also been suggested that VAM stimulate plant growth by physiological effects other than the enhancement of nutrient uptake or by reducing the severity of diseases caused by the soil pathogens.

Azolla

Azolla is a floating fresh water fern inside which grows the nitrogen fixing BGA Anabaena. It contains 3.4 per cent nitrogen and also produces organic matter in soil. This biofertilizer is used for rice cultivation in different countries such as Vietnam, China, Thailand, Philippines. It is also used in fish culture ponds these days. This can be easily grown in cooler regions. There is need to develop tolerant strains to high temperature salinity, and pests and disease resistance for its wider adoption. Field trial indicated that rice yields are increased by 0.5–2 t/ha due to *Azolla* application. In China, about 18 per cent increase has been observed. In fish culture ponds, application of 25 kg/ha per day *Azolla* biofertilization has been worked out. Recent studies have revealed potentialities of *Azolla* as a nitrogenous fertilizer in carp culture ponds. Application of different doses of *Azolla* in fish culture ponds shown that a minimum of 25 kg N/ha/yr could be provided through application of 10–12 tonnes of *Azolla*/ha/year. There are six species of *Azolla–A. caroliniana, A. nilotica, A. mexicana, A. filiculoides, A. microphylla* and *A. pinnata*. It grows in ditches and stagnant water. This fern usually forms a green mat over water. The plant has a floating, branched stem, deeply bilobed leaves and true roots which penetrate the body of water. The leaves are arranged alternately on the stem. Each leaf has a dorsal and ventral lobe. The dorsal fleshy lobe is exposed to air and contains chlorophyll. It has an algal symbiont (*Anabaena azolla*) within the central cavity. *Azolla* is readily decomposed to NH_4 which is available to the rice plants. It has also been observed that N contents of rice receiving *A. mexicana* are 10 and 35 kg/ha respectively higher than that in controls.

Blue Green Algae

Blue-green algae constitute an important group of microorganism capable of nitrogen fixation. Most of the species possessing nitrogen fixation ability to the order Nostocales and Stigonematales. Over 100 species of blue-green algae are known to fix atmospheric nitrogen. These have been found to be very effective on the rice and banana plantation. In field condition, overall increase in the gram yield of rice is amounted to about 586 kg/ha. In case of crops other than rice, algalization increased nearly 34 per cent yield.

India is one of the countries where agro-chemical conditions appear to be favourable where blue-green algae technology has been put forward. In some parts of the country, production of BGA inoculants has been commercialised. Producing inoculum in artificially controlled conditions is well defined, but relatively expensive. On the other hand open air soil culture is more simple, less expensive and easily adaptable by tile farmers. Field scale production of algae biofertilizer is also possible, 20–25 kg dry algae can be obtained on 40 m field. Adopting this method, 15 t/ha of vet BGA can be obtained by the farmers. Farmers can also produce algae for countryard of the house.

Plant Growth Promoting Rhizobacteria (PGPR)

A group of rhizosphere bacteria that exert a beneficial effect on plant growth is referred as PGPR. They belong to several genera, *e.g. Actinoplanes, Agrobacterium, Alcaligenes, Amorphosporangium, Arthrobacterm Azotobacter, Bacillus, Cellulomonas, Enterobacter, Erwinia, Flavobacterium, Pseudomonas, Rhizobium* and *Bradyrhizobium, Streptomyces* and *Xanthomonas. Bacillus* spp. are appealing candidates for biocontrol because their endospores are tolerant to heat and desiccation. Currently *Pseudomonas* strains also suppress major plant pathogens. Take all, an important root disease of wheat caused by *Gaeumannomyces graminis* var. *tritici* is controlled by *fluorescent Pseudomonas.*

PGPR are thought to improve plant by colonizing the root system and pre empting the establishment of suppressing deleterious rhizosphere microorganisms on the roots. The plant growth promoting microorganisms improved potato growth and yield in short-but not long-rotation soils, primarily by suppressing cyanide-producing deleterious rhizosphere microorganisms. Large populations of bacteria established on planting material and roots become a partial sink for nutrients in the rhizosphere thus reducing the amount of C and N available to stimulate spores of fungal pathogens or for subsequent colonization of the root.

In field trials with wheat, potato, sugarbeet and zinnia conducted showed significant yield increases varying from–7 to 136 per cent with an average increase of 7–35 per cent in different crops over the control. Seed treatment with *B. subtilis* increased yield of carrot by 48 per cent, oats by 33 per cent and groundnut upto 37 per cent and it has been marketed as a treatment for groundnut in USA. A multitude of factors could account for inconsistent results, given the complex interactions among host, inoculated organisms, other rhizosphere organisms and the environment which are discussed earlier. Research with PGPR in India is at a low level.

Status of Biofertilizer in India

As mentioned earlier the total potential of biofertilizers is very high and existing production units do not have the capacity to meet the requirement at the right time. Most of the production units do not have adequate instruments, production techniques, skilled staff, etc. for production of quality based inoculants. The result is the production of poor quality products which are directly responsible for the failure so there is a need for upliftment of biofertilizer industry in India.

Biofertilizers differ from fertilizer in the sense that the former do not directly supply any nutrient to crop plant and are the cultures of some specific bacteria and fungi. Bacterial cultures like *Rhizobium, Azospirillum, Azotobacter* have the ability to fix atmospheric nitrogen which in turn increases nitrogen supply to crop. Bacterial cultures of *Pseudomonas* and *Bacillus* species and fungal cultures of *Pencillium* and *Aspergillus* species help to convert insoluble phosphates into soluble (plant soluble) phosphate and thus improve phosphate availability to the plants. Similarly, fungi like vesicular arbuscular mycorrhizae (VAM) increased nutrient uptake particularly of P, Zn and other micronutrients due to increased contact of roots from larger soil volume. It is important to mention that so far biofertilizers available are able to take care of nitrogen only. Therefore, other plant nutrients have to be applied from other sources.

At present, in India there is a gap of about 10 million tonnes of plant nutrients between removal by crops and replenishment through fertilizers. Supply of nutrients from organic manure has not so far been able to fill up this gap and also in years to come nutrient supply from the source is unlikely to improve due to competitive demand for alternate uses like fuel and fodder. Of late, biofertilizers are being promoted as an important component in supplementing plant nutrient need of the country. It is a good beginning but we still have a long way to go.

India is one of the important countries in biofertilizer production and consumption in the world. The present production capacity of different biofertilizers production units in the country is about 4500 tonnes per annum. The maximum production capacity is in Agro Industries Corporation followed by State Agriculture Departments, National Biofertilizer Development Centre, State Agriculture Universities and private sector. Among the different states, the maximum production capacity is in Tamil Nadu followed by M.P., U.P., Gujarat, Maharashtra. The actual production of biofertilizer during 1994–95 was about 2000–2500 tonnes.

The installed capacity of BF production in different fertilizer industries in about 400 t/annum. Main BF producing companies are GSFC, MLF, SPIC etc. Some more fertilizer companies (IFFCO, KRIBHCO, NFL, RCF etc.) are likely to start production soon and many more are planning to join the biofertilizer business.

Based on the area under different crops and dose of biofertilizer to be applied, the National Biofertilizer Development Centre (NBDC), Ghaziabad and Biotech Consortium India Ltd. (BCIL) have estimated the total requirement of biofertilizers (*Rhizobium, Azotobacter, Azospirillum* and *Blue Green Algae*) to be about 5.07 and 3.44 lakh tonnes, respectively. The recent estimated potential demand of different kinds of biofertilizers by Government of Tami Nadu are *Rhizobium* 35 thousand tonnes; *Azospirillum* 482 thousand tonnes; *Azotobacter* 162.61 thousand tonnes: *Blue-Green Algae* 267.72 thousand tonnes, *Azolla* 20.38 thousand tonnes and phosphate solubiliser 275.51 thousand tonnes. The total of all these amounts to be 12.44 lakh tonnes which is significantly higher than the estimates of NBDC and BCIL, mainly because they did not indicate phosphate solubiliser and their estimates for *Azospirillum* were also low.

Production technology of biofertilizer is relatively simple and its installation cost is very low compared to chemical fertilizer plants. Organised promotional and marketing strategy for successful biofertilizer business is very important. Most of the biofertilizer units lack in this respect. Those who have very good organised marketing network, have done excellently well. For example, fertilizer company like GSFC has more than 200 farm information centres-cum-depots situated in remote areas. These are manned by experienced agricultural graduates to provide the technical know-how, conduct the demonstration and give before and after sales services to the farmers. In order to provide BF upto village level, GSFC has established its own distributor's network. MLF and SPIC have also well organised themselves in this respect.

Table 3.7: Estimated Installed Capacity for Biofertilizers in Different States

State	*Installed Capacity (t)*
Tamil Nadu	1100
Madhya Pradesh	700
Gujarat	450
Maharashtra	250
Andhra Pradesh	175
Karnataka	40
Haryana	75
Orissa	75
West Bengal	72
Rajasthan	90
Punjab	75
Total	3102

Response of biofertilizers have been obtained in cereals, millets, pulses, legume oilseed, sugarcane and cotton grown under different agro-climatic conditions. In general, increase in yield due to biofertilizer application was of the order of 15–20 per cent. It is found that biofertilizer like *Rhizobium* can supply 20–25 kg N/ha. Of late, complex biofertilizers have been reported to be more effective in increasing nutrient supply and crop yield.

To promote biofertilizer use, fertilizer industry is engaged in a number of activities like (*i*) field experiments on Research & Development (R&D) farm (*ii*) field experiments on farmers' fields (*iii*) farmers' meetings and field days (*iv*) seasonal campaign (*v*) farm youth training programme (*vi*) audio visual display to educate farmers (*vii*) preparation of distribution of literature on biofertilizer in local languages (*viii*) distribution of paper stickers to institutional agencies, dealers, etc. (*ix*) publicity through local newspapers and (*x*) publicity through exhibitions, fairs, festivals etc. Though there had been a lot of improvement in biofertilizer development and promotion in supplementing plant nutrient supply, the bottlenecks which still need to be sorted out are short shelf life, lack of suitable carrier material, suitable of high material, susceptibility to high temperature, problem in transportation and timely distribution in remote areas etc. Besides standard packaging and quality control measures etc. will help to develop healthy market in the time to come. There is urgent need to work out aggressive promotional and marketing strategies for gaining momentum to BF use in the country.

Thrust in Research and Development

The rising cost of chemically fixed nitrogen fertilizer and the massive inputs of non renewable resource will be devastating. As an alternative there is urgency of major research effort in biological nitrogen fixation. Nitrogen in the form of ammonia is used by plants and microbes as a building block for the synthesis of amino acids and of other nitrogenous compounds. The conversion of atmospheric

Nitrogen into ammonia by the nitrogen fixing microbes, mostly certain bacteria and blue green algae is called biological nitrogen fixation. Although leguminous plants have been used extensively in agriculture for centuries, it is possible to exploit these nitrogen fixing plants meaningfully. The use of new leguminous crops as well as the breeding of more vigorous and effective traditional leguminous crops can be helpful in food production. As the bacteria are the agents of fixation in bacterial plant association, plant breeding has been neglected to enhance the effectiveness of this association.

Rhizobial technology and soil microbiology are the neglected areas. It would be a right type technology that all farmers, could use Legume-Rhizobial combinations constitute, a built-in nitrogen source for food crop production. Approximately 75 per cent of the nitrogen utilised by legume comes from the soil, with only 25 per cent from BNF. It is necessary to increase the rate of nitrogen fixation achieved, which would have a multiplier effect on the enhancement of crop productivity. There have been many research frontiers in BNF. The two genetic systems involved are micro-organisms and the higher plants and their greater potential for improving the both. In nature, certain nitrogen fixing agents do not perform well. As a result their nitrogen fixing efficiency remains less though their abundance is much. It is necessary to make them highly competitive through genetic engineering. It will help them to establish an association with the plants in competitions with the organisms occurring naturally in the ecosystem. Wherever possible, the agents should have a broad spectrum in their specificity so that they can associate with a variety of plants. Thus the competitive strains developed could be introduced into the soil for better interaction of plant bacterial association.

Although nitrogen fixers are present in the soil, enrichment of soil with effective strains is much beneficial for the crop yields. Chemical fertilizers have temporary effect while biofertilizers have permanent effect without any production problem. Use of composite biofertilizers can increase soil fertility. The cost is too much low for the biofertilizers and its proper use does not only include a correct application of the inoculant to the seed-soil but also good crop management and provision of additional capital nutrients for-the plants and inoculum. Scientists have proved that the biofertilizers are cost effective, cheap and renewable source to supplement the chemical fertilizers. The rhizobia inoculant has been found to be effective for pulses and legume oilseeds like soyabean, groundnut and the blue green algae for low land paddy in our country. Much of the world today is concentrating on rhizobial technologies and the culture of other organisms having superior nitrogen fixing capabilities. As companion the legumes are being used as the soil improving crops and intercrop or in rotation with cereal grains and other high energy crops as is being practiced in China, South America, Africa, Philippines and in India. They are becoming apart of cropping systems research in total farming systems. Active linkages of scientists engaged in basic fundamental research in biological nitrogen fixation with those in mission oriented, applied and problem solving research would speed up the process. Considering the prospects of biofertilizers in the country, the biofertilizer development centres are being established both in government and private sector. It is possible to establish joint venture in biofertilizer agro based industry.

Chapter 4
Nitrogen Fixing Microorganisms: Symbiotic

Nitrogen is one of the most important essential plant nutrient to complete the life cycle of plants, which is required in a huge amount by the plants. It is abundantly available in the atmosphere. Approximately 78 per cent of total atmospheric gases by volume and 75.5 per cent by weight is nitrogen. The large percentage of nitrogen is in the form of gas or molecular nitrogen (N_2) which is entirely unavailable to crop plants in such form. For making it available to plants it is essential that this molecular nitrogen may be covered into combined nitrogen (NH_4^+). This nitrogen conversion phenomenon is termed as biological nitrogen fixation. The organisms responsible for such type of nitrogen fixation are known as "nitrogen fixers" or "Diazotrophs", *e.g. Rhizobium, Azotobacter, Azospirillum, Clostridium.* Blue Green Algae etc.

Nitrogen fixing organisms (NFOs) have very little in common with each other apart from being prokaryotic and diazotrophic. They may be symbiotic, nonsymbiotic or form casual associations with other organisms; they may fix N under aerobic, anaerobic or microaerophilic conditions. Some organisms are capable of fixing N independently of other species, however, some can grow and fix N at a higher rate in the vicinity of other organisms which can provide them with their growth requirements. They are to be found in greater numbers in association with, for example, the rhizosphere of plants. These associations may vary from rather loose, non-specific attractions between plants and bacteria to those in which the plants seems to be able to encourage the growth of more restricted range of microorganism in its vicinity. The associations are often specific. Some associations are such where both partners undergo considerable morphological and biochemical differentation in forming the N_2-fixing relationship.

Nitrogen is the most limiting nutrient in food production. The biological nitrogen cycle is responsible for a turnover of 10^8–10^9 t N per hectare on the earth in which biologically fixed N is one of the inputs. Nonsymbiotic and associative N fixation is considered to occur at magnitudes that may be of agronomic significance. The apparent potential for biological nitrogen fixation (BNF) associated with cereals exceeds its present utilization.

Since Winogradsky, a German microbiologist in 1893 established that *Clostridium pastaurianum* which fix atmospheric N and Beljerinck, a Denmark microbiologist described the first *Azotobacter*, the list of N-fixers is increased gradually. Many different genera and strains of N_2-fixing bacteria can be isolated from the soil and the roots. A list of organisms which fix nitrogen in soil has been given. Since in agriculture few organisms are studied for their N-fixing qualities from the list, hence, detail discussions was confined to very few of them (Table 4.1).

Table 4.1: Important Nitrogen Fixing Organisms

Nature/Organism(s)	*Remarks*
Anaerobic	
Clostridum pasteurianum *C. butyricum*	Isolated from soil, fresh water and strictly anaerobic; contribute about 33 kg nitrogen/yr/ha in rice field;
Methanosarcina barkeri	Strictly anaerobic but the contribution to N-fixation is unknown
Methanococcus thermolithotrophicus	
Anaerobic (Partial or Microaerobic)	
Enterobacter cloaceae	Isolated from soil and water; partial aerobic
Klebsiella pneumoniae	Isolated from water and soil and roots of grasses
Azospirillium lipoferum	
Aerobic	
Azotobacter vinelandii *A. paspali*	Isolated from soil, leaf and roots of plants; fixes atmospheric N- as free living bacteria;
Beijerinckia derxii	All strictly aerobic but some species can tolerate low O_2
Photosynthetic Bacteria	
Rhodospirillum rubrum	Aerobic and aquatic
Symbiotic Diazotrophs	
Rhizobium spp.	Legumes
Bradyrhizobium japonicum	
Frankia sp.	Non-legumes

Biological Nitrogen Fixation

Biological nitrogen fixation involves the reduction of nitrogen to ammonia. This becomes incorporated into the organic nitrogen of the N-fixing organisms. Some free-living diazotrophs excrete NH_4 or amino acids into soil but the greatest contribution to the ammonia content of the soil comes from the death and subsequent purification of plant and animal material. Ammonia in solution exists as the NH_4^+ ion which remain in the upper layer of soil and is readily available for utilization by plants and microorganism. This utilization may take the form of association, but in competition with association is the process of nitrification in which some bacteria oxidise NH_4^+ to nitrite and nitrate. The oxidation of NH_4^+ to nitrite is mediated by the bacterium *Nitrosomonas* and nitrite to nitrate by *Nitrobacter*. Both bacteria are obligate aerobes. Nitrite in the soil is less tightly bound than ammonia and there, it is subjected to run off. Nevertheless, nitrite is more easily absorbed than ammonia by, plants from the soil, however, nitrate which is absorbed by the roots of plants, must first be reduced to ammonia before it can be incorporated into the organic compounds that the plant requires. This assimilarity reduction of nitrate (which can occur both in the roots and the leaves) occurs in two main stages and involves two enzymes, nitrate reductase and nitrite reductase (Figure 4.1).

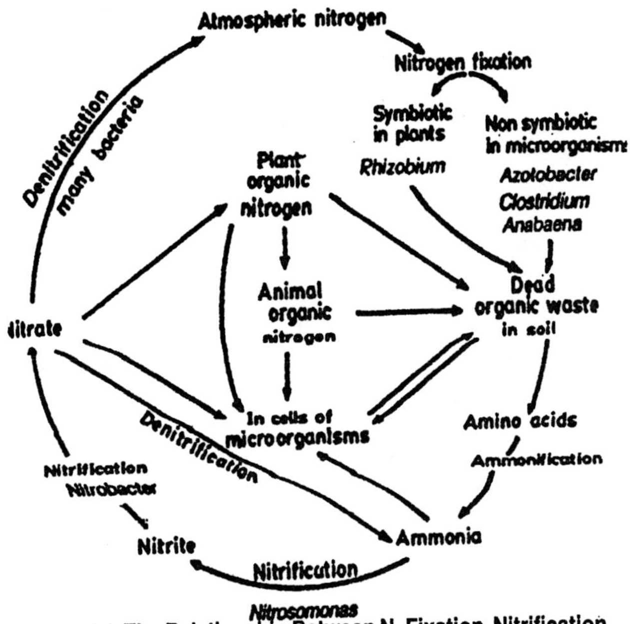

Figure 4.1: The Relationship Between N_2 Fixation, Nitrification Assimilatory Nitrite Reduction and Denitrification

The accurate estimates of annual turnover of nitrogen in the biosphere vary from 100 to 200 million tonnes. The ratio between chemically fixed N and biologically fixed N ranges approximately from I: 4 to I: 2.5 and within biological fixation, the legume fixation is equivalent to or at least ½ that of industrial fixation. The amounts of kg N fixed/ha/yr for individual legumes has been estimated is: 125–335 for alfa-alfa, 85–190 for red clover, 80–150 for peas, 65–115 for soyabean, 65–130 for cowpea and 95–155 for vetch.

Types of Biological Nitrogen Fixation

The following are the three types of Biological-Nitrogen-Fixation:

Symbiotic Nitrogen Fixation

Symbiosis is a process which expresses the mutual beneficial partnership between two organisms. The fixation of molecular nitrogen by this process is termed as symbiotic nitrogen fixation *e.g.* Legume-Rhizobium-symbiosis and *Anabaena-Azolla*-symbiosis.

Non-Symbiotic Nitrogen Fixation

This type of N-fixation is performed by the organism who live freely in the soil without any association with any other organisms *e.g., Azotobacter.*

Semi-Symbiotic or Loose-symbiotic Nitrogen Fixation

This is neither true symbiotic nitrogen fixation not non-symbiotic nitrogen fixation but rather it is in between two. This fixation is generally found in C_4 plants (High-photo-synthetic capacity like maize: jowar, bajra, sugarcane, rice and the responsible organism is *Azospirillum*. This bacterium preferentially grows in the rhizosphere of these crop plants and fixes said atmospheric nitrogen which made available to plants.

Symbiotic nitrogen fixation is performances in the legume root nodules by the *Rhizobium*, originally known as *Bacillus radicicola* which was isolated by Beijerinck in 1888. Nodules are the outgrowth on the root system of leguminous crops as the interacting effect of *Rhizobium* and host legume. Effective nodules are vigour in size, pinkish in colour and attached mainly on tap root system, which fixes higher amount of nitrogen than ineffective nodules.

Factors Affecting Nitrogen Fixation

Inorganic Nitrogen

Low concentration of nitrogen as NH_4^+ or NO_3-to the extent of 10–15 kg N ha^{-1} increases the nodulation, however, higher concentration of N reduces nodulation and N-fixation.

Soil pH

Acidic condition is deterious for legumes and also for colonization of *Rhizobia*. Soil acidity also results in the deficiency of Ca, Mg, K and Mo which indirectly effect nodulation. Toxicity of Fe and AI in acidic condition is responsible for lower N-fixation. Detrimental effect of acidity can be improved by liming the acidic soils and pelleting the inoculated seeds. The optimum range for better nodulation and N-fixation is neutrality to slight alkalinity. In high salt content soil, growth of legumes and N-fixation become reduced.

Temperature

The most congineal temperature for bacteriod tissue has been reported from 20–30°. The temperature above and below this range depressed the nodulation and N_2-fixation.

Light

Light through photosynthesis affects the nodulation and N-fixation. Sixteen hours light in the temperature range of 15–20°C has been reported better for N-fixation under controlled condition.

Growth Substances

Different types of growth substances affect nodulation and N-fixation differentially. Lower levels of IAA conducive for early nodulation while higher concentrations of NAA inhibit nodulation.

Use of Biocides

It is generally observed that all biocides (fungicides, weedicides, insecticides etc.) at the recommended dose do not adversely affect nodulation. However, persistence of these chemicals in soils at the high concentration may reduce the nodulation. DDT at lower concentration of 1–10 ppm has been reported to stimulate nodulation.

Genus–Rhizobium

The genus Rhizobium was established in 1889 by Frank based on its ability to form nodule on the root of legume. The present classification of *Rhizobium* depends mainly on the work of Fred and Coworkers.

The genus *Rhizobium* alongwith the genera *Agrobacterium* and *Chromabacterium* comprise the family Rhizobiaceae and divided into three main groups: *Azorhizobium* produces stem nodulation in plant *Sesbania rostrata*; the second group is Bradyrhizobium with two well defined species *B. japonicum* and *Brudirhizobium* sp. produce nodules in the roots of soyabean, lupin, cowpea, green gram, red gram, chickpea and groundnut. The members of this groups are slow growing strains producing alkali on Yeast Manitol Agar (YMA). The third group is *Rhizobium* establishing great relationship in pea, lentil, bean, lathyrus, berseem and lotus. They are classified as fast growing acid producing on the Yeast Manitol Agar (YMA), the important species are *R. phaseoli, R. trifolii, R. leguminosarum* and *R. meliloti*. It is opined that the DNA-base ratios of fast growing species. The lack of homology between the DNA of *B. japonicum* and DNA of representative strain of *R. leguminosarum* is sufficient to justify the separation of *Rhizobia* into two genera.

The *Rhizobia* are gram negative, unicellular cell size less than 2 micron wide and short medium rod, pleomorphic, motile with peritrichous flagella and accumulate poly β-hydroxy butyrate granules. The physiological studies indicated that the bacterium uses monosaccharides and disaccharides as carbon sources which are supplied by legumes through photosynthates. They are aerobic and the growth media is YMA.

Rhizobia

Significance role of *Rhizobium* in fixing atmospheric N and hence enhancing soil fertility, has been well established since time immemorial, however, commercialisation of their use is a relatively new venture. In view of their apparent low cost and risk free application leading to yield advantage both for the current crop as well as for the subsequent crop grown on residual fertility, their large scale application in intensive farming calls for systematic efforts. Their use is especially important in view of the fact that in the foreseable future it may not be at all possible for the country to be self-sufficient in the production of nitrogenous and phosphatic; fertilizers. *Rhizobium*–a symbiotic bacterium, colonises on the roots of specific legumes to form root nodules which fixes atmospheric nitrogen. In a way, the nodules act as mini factories for ammonia production. *Rhizobium* has been widely studied and extensively used as a biofertilizer because of its proven ability to fix and contribute N for the benefit of a leguminous host plant. In view of crop specificity of a *Rhizobium*, selection of an appropriate strain as an inoculant for a given host genotype is of utmost importance for desirable results. Dawdle and Bohlool suggested a simple strategy to develop an improved practical inoculant. The various steps of this strategy are (*i*) use of a dominant, competitive indigenous strain specific to a particular geographic region which is adapted due to natural selection, (*ii*) isolation of selected indigenous strain, its addition to the seed in high number and its subsequent use in the same region under favourable environmental conditions, and (*iii*) mutants with increased effectivity may be isolated from the indigenous strains which remains as competitive as the wild type.

The *Rhizobium* legume association could fix upto 100–300 kg N/ha in one crop season and in certain situation leave substantial N for the following crop. The N advantage through this symbiosis is greater in soils poor in available N. The response of pulses inoculation with *Rhizobium* varies widely but, by and large, has been found to be positive. Field experiments conducted at several locations and on different pulses, revealed yield increases ranging between 9 to 16 per cent. The response of *Rhizobium* is also believed not only to be crop specific but cultivar specific as well. In field experiments carried out on groundnut at ICRISAT, Hyderabad, *Rhizobium* strain NC-92 was found to give best results only with the groundnut cultivar Robot. These findings stress the need to isolate and multiply cultivar specific elite *Rhizobium* strains suitable for different crop varieties, for obtaining enhanced productivity. Also, response to *Rhizobial* application can vary widely depending upon the soil *Rhizobial* status as well as *Rhizobial* strain. In soils with higher than 1000 *Rhizobia*/gm, the response is minimum, while in soils with less than 100 *Rhizobia*/g, the response could be extremely high.

Rhizobium/Legume Symbiosis

Legumes are belong to family leguminosae with subfamilies *Caesalpinioodeae, Mimosoideae, Papilionoideae* and *Sewartziozeae.* The extents of nodulation in the family is not known exactly, but is least in *Caesalpinioodeae,* where only about 30 per cent of the species are nodulated and greatest in the *Papilionoideae.* With certain genera, for example, Cassia, not all species may be nodulated, possibly, reflecting heterogeneity in the genus.

Nodulation does not occur in the field until the first leaves are unfolding and N-fixation does not normally commence until the plant can safely divert a proportion of its photosynthate to this end. This requires synchronisation of plant and *Rhizobial* development at a number of stages, begin in the soil.

Seeds of legumes do not usually carry their own indigenous population of potentially symbiotic bacteria, is well established. This indicates that infection by *Rhizobium* after germination of the host is dependent upon the availability of suitable *Rhizobia.* Earlier it was thought that N-fixation can only occurs in symbiotic association with legumes and N-fixation by free living *Rhizobia* was therefore, not possible. It is reported *Bradyrhizobium japonicum* can fix nitrogen undermicroaerobic conditions. The various stages of *Rhizobium* legume symbiosis is discussed.

Methods for Study of Legume Root Nodulation

Isolation

Wash the root system of the test legume plant in water, a well-formated; healthy pinkish nodule on the tap root is carefully cut out with a portion of the root attached to the nodule. The nodule is surface sterilized for 5 minutes in 0.1 per cent mercuric chloride in water and repeatedly washed with sterile water to remove the chemical. The nodule is then washed in 70 per cent ethyl alcohol for 3 minutes followed by more washings with sterile water. The nodule is now crushed with a sterile glass rod in a small aliquot of sterile water. Serial dilutions of the suspension are then made so as to obtain sparse and distinct colonies when an aliquot of the appropriate dilutions is plated on YEMA medium. After six to seven days of incubation, distinct colonies of bacteria are picked up and transferred to agar slants for further identification. Several strains of *Rhizobium* of varying effectiveness, often associated with a related genus *Agrobacterium,* are isolated from the same plant. Obviously, further screening of such isolated becomes an essential step.

Differentiation of Rhizobium from its Common Associate Agrobacterium

Congo Red Medium

In Congo red incorporated agar medium (2.5 ml of a 1 per cent solution per litre of yeast extract manitol agar), rhizobia stand out as white, translucent, glistening, elevated and comparatively smaller colonies with entire margins in contrast to stained colonies of Agrobacteria concentration. Hofer's alkaline medium is other medium for differentiating these two bacteria. *Agrobacteria* tolerate higher pH while *rhizobia* is unable to do so. Therefore, growing bacterial isolates on an alkaline medium would serve as a useful criterion for distinguishing the two allied genera. *Lactose agar* is another medium for differentiation. *Agrobacteria* utilize lactose to form a reduced product, ketolactose, through the enzyme ketolactose. When Benedict's reagent is poured over agar medium containing lactose on which nodule bacteria are growing, the formation of yellow colouration due to Cu_2O indicates the presence of agrobacteria.

Tests for Nodulation

Rhizobium isolates with a range of effectiveness, occurrence of non-nodulating and also pathogenic strains among the isolates can be obtained. Therefore, testing strains for their ability to nodulate and for their effectiveness under sterile conditions is of prime-importance.

Small seeded legumes are grown on Agar-agar, vermiculite or sand. Any nitrogen-free plant nutrient is used to make agar slants. Seeds are surface sterilized 0.1 per cent silver nitrate and sown on the substrate in tubes. On germination, they are inoculated with a suspension of the bacterial isolate under test prepared in a small aliquot of the nutrient solution. The moisture content of the tubes should be checked and the nutrient solution replenished when necessary. Experiments which may run for months, the plants are examined for nodulation. A criterion for distinguishing the effectiveness is to record the relative dry weight of nodulated plants.

For larger seeded legumes isolated are tested for their efficiency by growing the inoculated legume in sterilized sand with N-free plant nutrient solution. The data on the appearance of plants in terms of colour and vigour, the number of nodules formed on the root system, dry weight and nitrogen content of the plants are collected. Temperature and light conditions be kept at optimum levels favourable for maximum performance of legume *Rhizobium* symbiosis. A standard assembly consisting of a bottomless screw capped bottle and a glass jar designed to permit automatic and continuous water flow to the growing root system can be used for the routine testing of isolates of *Rhizobium* for efficiency in symbiosis. Alternatively, sterilized sand or soil in earthenware or porcelain pots can also be used to grow large.

Infection Test

Seedlings may have to be, examined for the percentage of infection for virulence and improving nodulation. This is carried out by growing plants on agar slants containing N free medium and inoculating them with desired strains. Seedlings are fixed in 4 per cent formalin and examined for infection threads in root hairs and emergence of nodule primordial on roots. Depending on the extent of infection, isolates may be rated for virulence. A simple device for continuous observation of the progress of infection has been designed by Subba Rao.

Nodulation

This growing tip of a 3-day old seedling (grown aseptically) is cut and the blunt end implanted in a small vial containing organic constituents (glycine, B-vitamins and sucrose) in solidified agar. The tip of the excised root is suspended in nitrate-free mineral salt medium in a petri dish or in a test tube into which a suspension of *Rhizobium* is added. In this way, the organic constituents are provided to the growing root through the cut end and the growing end of the root receives mineral constituents along with *Rhizobium* culture. Nodules develop after 2–3 weeks when lateral roots are fairly well developed.

Tissue and Cell Cultures

Tissue cultures in plant morphogenetic studies are well known. Explants meristem can be grown into an callus mass on a suitable medium. New plantlets are transplanted to soil, they grow into mature plants bearing flowers and fruits. Tissue culture techniques have extensive use in development of resistant varieties. In recent past, root calli have been used to study the interactions of plant tissues with *Rhizobium* or *Azotobacter.* A mass of undifferentiated callus or a Suspension of isolated cells is inoculated with nitrogen fixing bacteria such as *Rhizobium* or *Azotobacter* and incubated under controlled conditions. Samples of calli or cells could also be cut into thin sections and observed under light or electron microscope.

Studies have demonstrated that (*i*) Infection thread-like structures and bacteria or bacteriod containing cells could be seen in callus tissues or isolated cells inoculated with *Rhizobium;* (*ii*) Definite nirogenase activity could be detected in inoculated calli or cells indicating the establishment of symbiosis in such plant cell-bacteria associations; (*iii*) When cells of *Azotobacter* and carrot were grown on N-free media, not only the bacterial cells entered isolated cells of carrot but such cells showed nitrogenous activity; and (*iv*) such studies are extended further, the day may not be far off when 'nitrogen fixing', plants are raised as clones through plartlets regenerated from bacteria incorporated cells of tissues.

Acetylene Reduction Assays

The enzyme nitrogenase, in addition to reducing nitrogen to ammonia can reduce certain other compounds like acetylene. This assay. is based on the nitrogenase-catalyzed reduction of acetylene to ethylene. It is simple, sensitive and inexpensive. The advantage of the technique includes facility for large number of *in situ* assays, economy, simplicity, effective gas exchange prior to incubation, easy removal of the gaseous end products and sensitive analysis with a flame ionization detection system after separation by gas chromatography. The technique is 10^3 times more sensitive than ^{15}N tracer technique. Samples are incubated in serum vials with purified acetylene at temperature, incubation period and illumination. After incubation, the gas from the container is removed for ethylene analysis by gas chromatographic (GC) fractionation followed by detection by flame ionization. After correction for the small amount of acetylene present as an impurity in ethylene the amount of nitrogen fixed is determined by the extent of reduction of acetylene to ethylene. Moles of nitrogen fixed is obtained by division of moles of acetylene reduced to ethylene by a factor of 6 electrons for reduction of nitrogen and 2 for acetylene.

Use of ^{15}N to Measure Biological Nitrogen Fixation

While ARA offers a rapid though indirect way of measuring nitrogenous activity and thus the ability of a biological system to fix atmospheric nitrogen, any critical evaluation of results must pass the ^{15}N test. The technique formulated involves the use of ^{15}N as a tracer. The amount of nitrogen fixed by a system is determined as the percentage increase of the tracer in samples incubated with the labelled gas. An enrichment of 0.015 or greater atom per cent ^{15}N in the sample (expressed as 0.015 atom per cent excess) is usually taken as an extensively in the analysis of nitrogen fixation in whole cells and in cell-free extracts.

Multiplication of Rhizobia: Root Hair Curling

On germination of legume seeds they release chemical substances which attracts nature or introduced *rhizobia;* the chemicals include amino acid, homoserine, which stimulate *Rhizobium leguminosarum* growth. The study was conducted with pea seeds. The tap root of pea does favour *rhizobial* muitiplication. However, as soon as lateral roots begin to emerge; release various chemicals which favours *rhizobium.* The first nodules, is common with those of most similar legumes, are formed on the tap root. Once a desired population of *rhizobia* has been built up, these can begin to influence and possibly infect the host plant. The first symptom is deformation of root hairs. Three types of deformation has been described: (*i*) branching (*ii*) moderate curling (*iii*) marketed curling. Not all hairs on a root respond and sometimes growth right through a *rhizobial* colony can occur without turning a hair. There are several theories proposed on root hair curling. The suggestion that auxin (Indole acetic acid) is an essential component. The curling may be due dialysable heat stable products of *Rhizobia* may be responsible. A curl could delimite an area when some vital substance accumulate. It is also possible that entry holes made in the crook of a curl may be more easily plugged to prevent entry of pathogens. Root nodules contain several bacteria other than *Rhizobia* but not pathogenic species. Before entry can be effected, *Rhizobium* of the right type must be correctly aligned on the hair. At this stage, *Rhizobia* are frequently seen to be embedded in mucilage, a material of legume origin, which may cover root to soil and also provide food for bacterial growth. The *Rhizobial* and host interactions was compared to antigen and antibodies in animals and it is predicted that some specific substance is produced which is identified as lectin and it is believed that this can contribute to legume *Rhizobial* specificity.

Lectins of host are responsible for binding to root hair surface and rhizobium. The binding sites on the parts are structurally and antigenically similar polysaccharides.

Formation of Infection Threads

The aligning of desired *Rhizobia* on root hairs is the first steps of complex interactions; the next step is the entry of *Rhizobium* into host cells. When desired bacterium firmly attached to the root hair a refractile spot becomes visible in the root hair wall, this indicates structural changes. After this infection threads become visible. This consists of one or more roots of bacteria aligned within a sheath of host cell wall origin. it grows down the root hairs. The infection thread grows from the root hair cell between and through cells of the root cortex often into considerable branching.

Nodule Formation

Nodules developments in the roots start once the cortical cells have been infected by *Rhizobia*. The development is of two types:

1. Those in which infection of new cells occurs by means of infection threads which are long and highly branched continuing infection of newly formed cells is depended upon the spread of infection threads. The nodules formed in this way are elongated, branched and designated as interminate; these are found in peas and clover and typical of temperate region.
2. In this type infection thread may not appear (peanut) and branched and ramify extensively (soybean). In this, nodules are more or less spherical in shape and are limited in size. Such nodules are determinate.

After the infection thread passes through the outer layers of cortical cells, the cells of the inner root cortex show increased RNA contents and begin to divide. Cell divisions are generally opposite protoxylene points in the root stele. The cell division occurs due to the presence of auxin and cytokinins together with some as yet unidentified factor. The advancing infection thread travels inner cortical cells, endodermis and possible the pericycle, the cells continue to divide in various planes and become infected. *Rhizobia* are released from the confines of the thread wall and develop into N-fixing bacteroids. The host cells lose their meristematic activity, the infection thread branches and infects the growing cells cut off by the nodule meristem. A continuous sequence of newly infected cells is formed. This process continues until long after the nodule has emerged from its subtending root. The outer layers of the nodule do not become infected; instead they form the cortex, within which form vascular traces, which connect to the state of the subtending root. There is a nodule endodermis continuous with the root endodermis and additionally, an endodermis around each vascular trace.

The nodules have all the bacteroid containing cells formed from cells individually infected by a branch of the original infection thread. According to Dr. Newcomb of UK, when an infection thread penetrates a cell it contains bacteria embedded in a matrix, mucopolysaccharide of bacterial origin. This in turn is surrounded by plasmalemma continuous with that of the infected cell. At the top of infection thread, the matrix appears to break through the host wall, releasing bacteria. In turn the bacteria break free from their own matrix and become individually surrounded by membrane envelops in the host cell cytoplasma, where they enlarge and differentiate into the N-fixing, bacteroids. The bacteria when differentiating into the N-fixing bacteroids show increasing invagination of the surface membrane and loss of mesosomes. The cell wall altered in structure and much permeable, largely due to alteration, and make it difficult to separate bacteroids. Changes in bacterial cell wall structure may be under host cell control. The outer wall may disappear and leave protoplasts. Such cell may lose viability. A relation between N- fixing activity and the presence of haemoglobin has been found. Haemoglobin is rare in the plant Kingdom; the two components appear to be coded by a different symbionts; haem by the *Rhizobium* and apoprotein by the host. The exact location of haemoglobin is still a matter of dispute but evidence is accumulating that it may be in the host cytoplasm.

According to Dr. J.I. Sprent (UK) the life of nodule is limited. The infected cells degenerate, turn greenish brown in the process. This has been attributed to the breakdown of haemoglobin into bile pigments. The studies also indicated that membrane envelops break down and cell contain degenerate as ageing proceeds. The cells may become invaded by free living *Rhizobia* which may later be killed by the action of phytolysosomes. Besides N-fixation, an efficient gaseous exchange system is clearly available. The pathways for aeration have been described in nodules of several crops. Proceeding from the soil, via lenticals on the surface to intercellular spaces in the centre of the nodule. Lenticals seem to be a common feature of the surfaces of determinate nodules, although the development is totally depend on water supply. Aerobic condition is necessary for N-fixation but in anaerobic condition N-fixation has been observed in soyabean.

Ineffective nodules are white, since they do not contain haemoglobin. Actually they are pasity on the host. Riboflavin is responsible for ineffective nodule. The development and effectiveness of nodule is under genetic control between host and *Rhizobium*.

In legume roots *Rhizobia* are incorporated in cells to form the N-fixing root nodules. This is extremely well balanced symbiotic system in which the host supplies the microsymbiont with all its carbon compounds and receives NH_3 in exchange leaving the *Rhizobia* free to concentrate on fixing N. An efficient O_2 supply to the active bacteroids is required. This is achieved with the help of leghaemoglobin cytochrome P450 and by branch respiratory pathway.

Cross Inoculation Group

Cross inoculation group refers to a group of related legumes which can be nodulated by a particular species of *Rhizobium* isolated from any member of that group. Legumes has a definite preference for some specific bacterium and in turn the bacterium has preference for certain legumes. Later it became clearly apparent that there exists groups of legumes each of which were nodulated by a different type of *Rhizobium*. Each of these groups of legumes and their specific *Rhizobia* constituted a cross-inoculation group and it generally held (*i*) that with each cross inoculation group would nodulate all other members of that group, and (*ii*) that *Rhizobia* isolated from one plant is cross inoculation group would not nodulate plants from other groups. Such observations develop the concept that plants mutually susceptible to nodulation by a specific *Rhizobia* constituted a cross inoculation group. This response of nodulation is fixed character and could be used as a parameter of classification of the bacterium

(Table 4.2). This has been agreed upon that most of the legumes in India are infected by cowpea group except that of chickpea which behaved differently, thus indicate chickpea rhizobias is from different group. Bergey's manual of Systematic Bacteriology is 1984 proposed a different classification (Table 4.3). Because the cross inoculation grouping system is not perfect, since the bacterium has often been found to cross-infect between groups. Such type of behaviour has been found what is known as promiscuity in groundnut, cowpea and pigeonpea. It has been observed that *Rhizobia* isolated from soyabean can nodulate cowpea but not the vice versa. Nodulation of pigeonpea by *Rhizobium japonicum* has been reported by Singh.

Table 4.2: Cross Inoculation Groups of *Rhizobium*

Rhizobium Species	*Cross Inoculation Group*	*Host*
R.trifolii	Clover group	*Trifolium*
R. meliloti	Alfaalfa group	*Melilotus, Medicago*
R. phaseoli	Bean group	*Phaseolus*
R. lupini	Lupine group	*Lupinus, Ornithopus*
R. leguminosarum	Pea group	*Pisum, Vicia, Lens*
*R. japonicum**	Soyabean group	*Glycine*
Rhizobium sp.	Cowpea group	*Vigna, Arachis*

*: Slow growing rhizobia.

Table 4.3: New System of Classification of Root Nodule Bacteria

Bacterium	*Biovars*	*Host*
Rhizobium leguminosarum	(a) *Viciae*	*Vicia*
	(b) *Trifolii*	*Trifolium*
	(c) *Phaseoli*	*Phaseolus*
Rhizobium meliloti	–	*Medicago*
Rhizobium loti	–	*Lotus*
Rhizobium friedii	–	*Glycine*
Bradyrhizobium		
B. japonicum	-	*Glycine*
Bradyrhizobium sp.	–	*Cicer, Cajanus, Vigna*

Legume plants of one or more genera or species which develop root nodules in association with the same varieties or species of *Rhizobium:* example: under per group *R. leguminosaram* can nodulate pea, lentil and lathyrus. However, Nobbe demonstrated with cultures that *R. lupini* did not nodulated with the pea but would do so with the lupin organism, and that beans and pea did not nodulated with isolated from Lupinus or Robinia. Atkinson found that the organisms from *Vicia sativa* and Dolichos bean would not cross infect. Further Hiltner demonstrated that nodule bacteria from *Trifolium pratense* would not nodulate *Medicago sativa.* These observation lead to the theory that separate *Rhizobium* species only nodulate specific related leguminous plants. Thus Fried recognised seven groups, *viz.,*

1. *Rhizobium japonicum*: Also known an *Bradyrhizobium japonicum;* the bacterium produces nodules in soyabean, *Lupinus* sp. and cowpea miscellany. Bacteroids in nodules are longer than normal cells, slender and with only occasionally branched and swollen forms. They are slow growing bacteria (Soybean grain).
2. This is fast growing acid forming species of the pea group which mainly forms nodule in pea, broad beans, lentils and vetch. Bacteroids in nodules are commonly irregular with X, Y, star and club shaped form (Pea grain).
3. *Rhizobium lupini*: This species cause nodule formation on *Lupinous* sp. and *Ornithopus* sp. Bacteroids are vacuolated rods and seldom branched (Lupin grain).
4. *Rhizobium meliloti*: The bacterium causes nodule formation in sweet clover, alfalfa and fenugreek; they are rod shaped, branched and grows fast on yeast extract medium (Alfaalfa gr).
5. *Rhizobium phaseoli*: The bacterium produces nodules on temperate species of Phaseolus. Bacteroids are rod shaped and often vaculated with few branched forms (Bean gr).
6. *Rhizobium trifolium: Trifolium* spp. nodules are produced by this bacterium. Bacteroids are pear swollen, vacuolated, rarely X and Y shaped (Clover gr).
7. *Rhizobium* sp.: Nodulate cowpea group. Bacteroids are irregular shaped (Cowpea gr).

Fungicide Enhancement of Nitrogen Fixation: Current practices used for legume production include inoculation of seeds with *rhizobia* and treatment of the seed with fungicides to reduce seed rot and seedlings damping-off resulting from infection by soil-borne and soil dwellers pathogens. Since the fungicides are biocides and are not specific for the pathogens, they also inhibit the soil biota *viz.* fungi, bacteria and other groups of microorganisms. Thus, seed protection and seed inoculation are frequently incompatible. Such in, compatibility are particularly serious when the fungicides are applied to seed of legumes, because the consequent suppression of species of *Rhizobium* may result in nodulation failures of reduction in quantities of N_2 fixed by legume-*Rhizobium* symbiosis. Besides this chemofobia (fear of chemicals) which is prevailing among cultivators that pesticides have toxic effects on legume nodulation is justifiable. Reasons for decline in *rhizobial* activity in relation to microorganism.

Colonization of the rhizosphere by *rhizobia* is probably important in determining extent of nodulation of legumes. This necessitate that *rhizobia* should proliferate as a first step for successful nodulation and this has been stressed by several microbiologists. A correlation has been established between failure to nodulate and low *rhizobial* count in the rhizosphere; there are several other reasons which affect *rhizobial* population and colonization of plant.

There are several ways to overcome these factors like using booster dose, increasing the number of *rhizobial* propagules and use of fungicides, singly or simultaneously.

Experimental evidence indicated that the population of *rhizobial* increased rapidly around germinating seeds and developing roots of soybeans, but the numbers of these bacteria fell abruptly. The decline was delayed by treatment of the seeds with benomyl so that, as the root system grew and opportunities for *rhizobial* infection of root hair arose, the population of *rhizobia* was maintained at higher levels than on plants not receiving the fungicide. The beneficial effect of benomyl was known and evident on first day. Such a stimulation was not restricted to fungicide treated seeds because a similar beneficial influence was noted by soil amendment with fungicides/antibiotics. This enhanced colonization associated with the use of fungicides and a fungicide resistant *rhizobia* strain may be account for greater yield and large number of nodules on the several plants receiving the fungicide.

Effect of benomyl and the antibiotics on colonization by *rhizobia* probably results from their effects on microorganisms antagonistic to the *Rhizobium*. Microbial antagonisms towards *rhizobia* and their symbiotic performance has been demonstrated by several studies: *Trichoderma viride* reduced the *rhizobial* population in nonsterile soil. Other organisms responsible are *Bdellovibrio* spp., bacteriophages, lytic microorganisms and myxobacteria. These are potentially harmful to the *rhizobia*. The enhanced development of *Rhizobium japonicum* in the *rhizosphere* after the addition of streptomycin and erythormycin to soil suggests that prokaryotes are important in limiting the growth of *Rhizobium* spp. the lesser number of protozoa after addition of these prokaryotic inhibitors probably result from the lesser number of bacterial prey than a toxicity to the predators.

Development of resistance of *Rhizobium* towards fungicides used for the seed treatment and soil application, at a concentration for which they are used to control is desirable approach to enhance *rhizobial* colonization in the *rhizosphere*. Several studies indicated that the tolerance or resistance towards benomyl, thiram and mancozeb in *Rhizobium meliloti* (alfalafa strain), *R. phaseoli* (Bean strain), *Rhizobium* (Cowpea). *R. trifoli* (red clover) and *R. japanicum* (Soyabean) has been developed automatically if these bacteria are continuously grown in the medium containing these fungicides. It is concluded that tolerant strains can detoxify these protectants, it is possible that the bacterial inoculants and the fungicides can be applied to legume seeds together without the letter destroying the former.

In a study it was observed that application of thiram to seeds inoculated with a thiram resistant strain of *Rhizobium* resulted in significant increase in dry weight and nitrogen contents of the plants compared with inoculation or thiram treatment alone. The young plants from such treatment were found to bear more nodules than untreated ones. Nodulated beans plant begins to fix nitrogen about 4 weeks after planting, so that an increase in number of nodules containing an effective strain of *R. phaseoli* could result in greater amounts of nitrogen being available to the plants treated with thiram during early part of growth. Inoculation of thiram treated seeds with a thiram-resistant strain of *R. phaseoli* resulted in higher numbers of *Rhizobium* in spermosphere and *Rhizosphere* of kidney beans as compared with seedlings that arose from untreated seeds.

The tolerance or resistance was developed in *Rhizobium* strains because the detoxified the protectants it is possible that the bacterial inoculants and fungicides can be applied to legume seeds together without the letter destroying the former. Hence the fungistatic agents might be employed to control fungal pathogens without jeopardizing the potential capacity of the *rhizobia* for nitrogen fixation following root hairs penetration.

The likeliest explanation for the decline of *rhizobial* population is grazing by protozoa. The data show that these organisms replicate rapidly as the seeds germinate, and they reach numbers sufficiently large to account for decline in counts of total bacteria as well as *rhizobia*. Benomyl application on seed drastically reduces the decline of *rhizobia* around soybean roots is because of inhibition of protozoa by benomyl. Similar to this thiram also inhibit protozoa and increased the *rhizobial* counts around bean roots. Protozoa have an anologus influence when added to sterile soil together with the mixture of other microorganisms.

Broad spectrum fungicides applied to soil or seed not only have an effect on root nodule bacteria, but also they may influence the interaction among the soil microorganisms and *rhizobia*. An inhibition by the fungicides of soil or *rhizosphere* inhabitants that complete with lyse, produce inhibitors against, parasitize or prey on *rhizobia* may prove beneficial because suppression might overcome the natural biological control of the *rhizobia*. Hence, a fungicide resistant strain of *Rhizobium* might proliferate more extensively in the legume *rhizosphere* and nodulate more frequently should its potential antagonists be inhibited by the added fungicides.

It is concluded from the several studies that grazing by protozoa and competition with bacteria limit the growth of *rhizobia* in soil apart from plants. Regardless of the precise contribution of proliferation on root surfaces and in soil under the direct influence of plants, the results of this study show that colonization can be enhanced by suppressing the predators or competitors by use of antimicrobial agents. Benlate and Bavistin systematic fungicides–Benlate or Bavistin at 0.15 per cent and non-systemic fungicides-thiram, captan and mancozeb at 0.3 per cent are being recommended for the control of seed borne diseases. Besides these antibiotics are also used for the seed treatment for the control of bacterial diseases. It is recommended that in seed treatment of legume's judicious use of these fungicide should be made.

Stem Modules

Certain aquatic and water tolerant legumes are known to bear nitrogen fixing nodules on stems. These stem nodules are caused by *Rhizobium*. Only some species of *Aeschnomene* and *Sesbania* are known to bear stem nodules and have been investigated in this regard. Recently, renewed interest on these stem nodulating plants has been evidenced in view of the potential benefit that could be achieved if similar stem nodulation of legumes, N-fixing stem nodules have been recorded in warry lecellate barks of various trees inhabited by entero-bacteria. Stem nodules offer a challenging and convenient system to investigate the inter-relationship between photosynthesis and BNF. The cortex of stem nodules is green due to chlorophyll and the bacteriod tissue is pink possibly due to leghaemoglobin tissue for the process of nitrogen fixation can easily be traced and measured with accuracy by the use of labelled carbon. The stem nodules can easily be peeled off at the base for laboratory experiments. The genetic basis for the stem nodulation habit may have to be understood in great detail so as to induce stem nodules in cultivated crops to secure better BNF of atmosphere N.

The bacterium, *Aeschynomene indica*, a gram-negative, coccoid prokararyote wan elaborate internal membrane system reminiscent of certain photosynthetic bacteria has been discovered. Ultra structurally, this organism bears a striking resemblance to the purple photosynthetic bacteria of the genus *Rhodopseudomonas* such as *R. capsuoata* and *R. spharoides*.

Genus–Frankia

The genus *Frankia* belongs to family Frankiaceae order Actinomycetales. The associations between non-legumes and *Frankia* have been given the name actinorhizal associations. Because this association is far less specific than *Rhizobium*, current practice is to refer to the generic name only (*i.e. Frankia*) without attempting to define the species. Dr. Callatiam and Cownkers from UK cultivated *Frankia* for the first time in 1978 in pure culture. The actinomycete could reduce acetylene was a further advance. when grown in complex liquid media containing fixed N compound. *Frankia* proliferates as a filamentons mat consisting of branched, septate hyphae. However, in response to N deprivation, short, branched filaments develop into swollen terminal vesicles which are capable of reducing acetylene. These spherical vesicles, about 3–5 mm in diameter are developed by *Frankia* which have so far been examined in N-limiting pure culture. However, within the host, a variety of vesicle shape is detected, varying from the spherical vesicles observed in pure culture to complete absence of vesicles. Studies on the morphology of nodules in various host indicate different shape and separations. The uniform appearance of vesicles in pure culture, therefore, contrast with the variation in vesicle morphology which occurs in the host and suggests that the host may have a part to play in determining vesicle shape. The vesicles possess an outer layer called the vesicle envelop, which is multilaminate. The envelop functions as a barrier to the diffusion of O_2 into the vesicle and thus protects nitrogenase from inactivation by O_2. The concentrations of oxygen that determines vesicle occurrences and shape.

A number of non-leguminous shrubs and trees (*e.g. Alnus, Myrics, Hippophae, Casurina* and *Eleagnus*) are nodulated by the nitrogen fixing Actinomycetes, *Frankia*, with Frankiaceae (Actinomycetales), the only actinomycete reported to fix nitrogen to date. The so called actinorhizal plants include 173 species of seventeen genera belonging to eight families and the number is growing all the time so the range of hosts is much more wider than for *Rhizobium*. Although there is some host specificity *Frankia* is much more promiscuous in the range of hosts infected than in Rhizobium. For this reason no species have yet been created.

Frankia nitrogen fixing activity has been shown in a free living stage *in vitro* but its actual contribution to soil nitrogen economy comes mainly through its capacity to establishing nitrogen fixing symbiosis with over 200 plant species all of them of a woody nature but for two species of *Datisca*, non-leguminous higher plants belonging to some 23 genera. The nitrogen fixed in the root nodules of these symbiosis is in part translocated eventually to the rest of the plant. Then the nitrogen goes back to the soils in the annual decomposition of the plant material and through the root exudates, both important for the growth of associated plant communities. *Frankia sumbiosis* also provides other nutrients to the plants, which contribute to soil fertility. Deciduous trees conserve nitrogen by mobilizing leaf nitrogen in autumn with subsequent storage of the nutrient in twigs and branches, from where it is made available for growth the following spring. Such a phenomenon is not observed in alders plants probably owing to the fact that they can fix atmospheric nitrogen and obtain lesser benefits from such conservation mechanisms. Experimental evidence indicates that much nitrogen is present in the leaves at leaf fall.

Actinornizae are pioneer species of poor soils, and at present their commercial potential is restricted to forestry and land reclamation. Some of the trees such as alder are used in agriculture as wind breaks. Rates of nitrogen fixed can be comparable to legumes (upto 150 kg nitrogen per hectare per year for alder) but they are frequently less.

Nodulation seems to follow a pattern similar to that observed in legumes but is much less clearly understood. Nitrogen fixation occurs in terminal swellings of the actinomycete hyphae, called vesicles. Each infected cell contains one vesicle cluster containing several hundred spherical vesicles. Nodule size can range from few millimeters to several centimeters in diameter. They are not pink in colour, but a leghaemoglobin-like molecule has recently been detected in nodules from most but not, all actinorhizae investigated. It is recently proved possible to grow *Frankia ex planta* in pure culture. Although it is often described as a microaerophile, nitrogen fixation in pure culture appears to be relatively insensitive to oxygen. All the isolations have been made from nodules. To date no-one has succeeded in isolating *Frankia* from soil stage in its life cycle. All isolates produce stores in large sporangia (upto 60 mm in diameter) when grown in pure culture, but only some of them produce spores inside the nodule.

Biofertilizer Role

A biofertilizer role of actinorhizal plants has been demonstrated by the works of Dr. G. Bond (Glasgo University) and Dr. A.I. Virtanen of Helsinki University, using isotopic N_2 determination. They reported that a sizable amount of N_2 was released from the biomass of actinorhizal plants from the amount of nutrients incorporated to the soil through leaf fall. Leaf nitrogen from actinorhizal plants ranges from 1.8 per cent to 3 per cent. The N_2 content at leaf fall lowers C/N ratio in litter leading to its faster decomposition in soil and making the nutrients, available to other crops growth.

Alder leaf litter takes one to four years to decompose completely whereas it takes some seven to ten years for a complete decomposition of pine needles. In Myrica gale decomposes 40 per cent biomass and releases 10 per cent N_2 after five years. The slow rate of decomposition is due to higher legnin content of this plants. The *Frankia* nodulated plants possess a nitrogen reductase system which absorbs mineral nitrogen and stores in the leaves, roots and nodules.

Genus–*Azolla*

Introduction

Azolla is a small aquatic fern generally found is omnipresent in nature in shallow ponds with stagnant water in winter season. Morphology of *azolla* consists of a floating branched rhizome with small alternate bitobed leaves and roots hanging down in water. Although spores are present but reproduction is mostly through vegetation. The important characteristics of *azolla* is that a blue green algae (BGA), each leaf has dorsal and ventral lobe. The dorsal lobe is green and has a blue green algal symbiont (*Anabaena azollae*) within a central cavity. The heterocyst of the symbiont *Anabaena* is the site of N_2 fixation.

The congenial temperature for the growth of *azolla* lies between 15°C to 30°C. Favourable pH is 6 to 7. It grows less at alkaline pH and stops growing in acidic soil. It can be grown in shallow ponds or low lying lands during October to February in plain. A layer of *azolla* growth over one hectare of land produces about 10 tonnes of green matter which may give about 30 kg N/ha for growing rice. *Azolla* is very rich in nitrogen containing about 4 to 6 per cent N on dry weight basis. It also has sufficient quantity of phosphorus and potassium. Average chemical composition of *azolla* on dry weight basis is given in Table 4.4.

Table 4.4: Average Chemical Composition of *Azolla*

Nutrient	*On Dry Weight Basis Per cent Content*
Nitrogen	4.5
Calcium	0.7
Iron	0.16
Manganese	0.15
Magnesium	0.55

The iinportance of *azolla* as an organic input in rice cultivation was first demonstrated in North Vietnam in 1957 and subsequently its potentiality has been recognized in U.S.A., Indonesia, Japan, Philippines and China respectively. *Azolla* grown in the rice fields and ploughed under, so as to facilitate its quick decomposition however in some areas it is grown as a dual crop alongwith the rice. *Azolla* plants are triangular or polygonal in shape and the leaves consists of green or purple chlorophyllous dorsal lobe which contains algal symbionts (BGA). *Azolla pinnata* is commonly found in India, out of six sp. distributed throughout the world and it has been seen that this sp. from Vietnam grows well in summer as it keeps lustre green vigour throughout the year. Under most favourable conditions, it has been observed that as a result of symbiosis, it has been observed that as a result of symbiosis between blue green algae and *Azolla* upto 900 kg N/ha/year is fixed. Phosphorus application happens to be one of the most critical factor to exploit its full potential. Different strains have differential nitrogen fixing ability depending upon agro-climatic conditions.

Morphology and Taxonomy

The genus *Azolla* belongs to the order Salviniales has been assigned to a monotypic family Azollaceae and subdivided into two genera on the basis of their reproductive morphology *viz.*, *Euazolla* and *Rhizosperma*. There is no morphological differences in the species and they are widely distributed in tropical and temperate fresh water ecosystem throughout the world. *Azolla* is indigenous to some rice fields and can grown in others when inoculated. The various species *viz.*, *A. caroliniaba* (North America), *A. filiculoides* (South America), *A. microphylla* (tropical and subtropical America) are belongs to subgenus *Euazolla; A. pinnats* is distributed in tropical and subtropical Asia and coastal Africa and *A. nilotica* (Africa) are belongs to subgenus *Rhizosperma*. The individual *Azolla* plants are 1–5 cm but *A. nilotica* which becomes as long as 15 cm. The plant consists of a branched floating stem that bears alternatively arranged overlapping leaves and true roots. Each leaf has two lobes, the dorsal and ventral. The later lobe contains and external cavity occupied by the Anabaena azollae symbiont. This is the species known to be symbiotically associated with Azolla and belongs to phylum cyanophyta, order Nostocales and nonrandom distribution suggest the possibility of recognition site in the Azolla. Anabaena azollae increases from apex to base in the successive leaves concomitantly the nitrogen fixation capacity increases. Alga free Azolla has rarely been found in nature.

Azolla plants are triangular and polygonal in shape consists of floating branched rhizomes bearing small alternative overlapping leaves and simple roots. The bilobed leaves consist of green or purple chlorophyllous dorsal lobe which contains the algal symbiont (BGA) within the internal cavity and achlorophyllous boat shaped ventral lobes which contain spongy tissues, symbiont BGA and *Azolla*. The symbiotic BGA occurs as unbranched filamentous consists of vegetative cells and heterocysts. The heterocysts possess a complex wall sheltered and multienzyme system called nitrogenase fixes atmospheric nitrogen and supply to the *Azolla* and in return gets the photosynthetic products from *Azolla*.

Role of *Azolla*

Azolla has a unique role as biofertiliser:

1. In replacing/supplementing chemical fertilisers to meet nitrogen for increasing rice yield.
2. As organic input in rice cultivation.
3. It has the potential to create self employment among educated rural youth through its scientific production and distribution system in India.

Due to continued research in India and abroad the Blue Green Algae (BGA) and *Azolla* biofertiliser has been identified for rice cultivation.

It has been reported that single incorporation of *Azolla* in rice field is comparable to 30 kg N/ha. *Azolla* incorporation significantly increases total soil nitrogen, organic carbon, available phosphorus. Moreover it improves the soil physical condition, *i.e.*, decreases specific gravity and increases porosity. Its weed suppressing potential and beneficial residual effects to succeeding crops make it more popular. Under favourable environmental condition like stagnant water, temperature (20–30°C), light energy at visible spectrum, all macro and micro nutrients except nitrogen, *Azolla* found to grow better. It can tolerate the salt concentration of 90 to 15 mg salt/lit.

Inoculum Production of *Azolla*

In order to exploit the multiple benefits of *azolla*, particularly as a biofertiliser in rice production, the primary need is to produce large quantity of inoculum in the vicinity of the rice fields with least application of inputs and without involving productive crop land. A simple and low-cost technique has been developed to produce large quantity of both vegetative and spore inocula. As a result, 2.5–5.5 tonnes/ha/day biomass can be produced in hypereutrophic water bodies under the partial canopy of suitable trees during October through March. Simultaneously, biomass-based spore inoculum can be produced in open area of the water bodies during January through March. The spore production becomes of maximum when the plant turns brownish in open area. The spores germinate after 3 months of storage and remain viable for more than 12 months, however, germinability is reduced by 20–30 per cent after 9 months of storage. Several institutions has started producing vegetative inocula for the local farmers and spore inocula for the distant farmers. Though the vegetative inoculum has the ability to produce larger biomass within a short time, spore has also disadvantages. It can be used directly in the rice field or as a means of producing vegetative inoculum. It can tide over the unfavorable seasons of winter and summer and it could be transported over a long distance. The cost of production of vegetative inoculum is Rs. 25 per 250 kg packet, that of one kg biomass-based spore inoculum packet is Rs. 1 while that of spore only is Rs. 25 per 100 gm packet.

There are plenty hyper-eutropic water bodies in different parts of India, particularly in eastern India. So rural people, particularly women can utilise both the biomass production processes either for vegetative inoculum or spore inoculum to develop village industries at a very low cost without involving additional land, water, fertilisers etc. but recycling a considerable amount of nutrient from the aquatic system to the crop land wherefrom the nutrients are mostly lost. However, one has to take care while utilising the water bodies. The oxygen concentration and phytoplankton density are not reduced with half coverage of the *azolla* mat. So the *azolla* biomass should be harvested when the mat develops half coverage to avoid any deleterious consequences on the aquatic animals. Even the farmers can produce small quantity inoculum in the water bodies by fencing a small part of the water bodies through a floating chord.

Factors Affecting Successful *Azolla* Production

The most important factor limiting the growth of *Azolla* is temperature as according to FAO report the optimum temperature for proper growth of *Azolla* stands in between 20 to 30°C. Out of six sp. of *Azolla–A. carotiniana, A. nilotica, A. filicldoides, A. mexicana, A. microphylla* and *A. pinnata,* last sp. has been found most tolerant sp. to high temperature and competent under Indian conditions. As pH of the soil is another limiting factor as pH range of 6–7 is most suitable but according to work done at the CRRI, Cuttack, ideal pH for this purpose should be around 8.0. Standing water of 5–10 cm deep and application of super-phosphate at the rate of 5–10 kg/ha seemed essential for rapid growth of *Azolla.* To avoid *Azolla* nurseries from wind erosion, it is preferable to use plots of size 50 × 100 sq.m. than bigger plots. Use of carbofuran at the rate of 1–2 kg/ha is equally important to prevent the spread of insect parasites and the consequent destruction of *Azolla* is further strengthened either because of poor quality of water or because of unavailability of water.

Azolla Nursery

A simple rural oriented low cost method for raising *Azolla* inoculum for rice has been recommended. The details of multiplication from the starter culture include:

1. Inoculating 1.0 kg of fresh *Azolla* in 10 sq.m. plot having 10–30 cm standing water, apply 30 gms of single super phosphate and 1 gm a.i. carbofuran.
2. On 5th day apply 30 gm single super phosphate as foliar spray (1 per cent solution).
3. On 10th day harvest *Azolla* (10–15 kg) and inoculate in 100^2 in size plot having 10–30 cm standing water, apply 300 gms of single super phosphate and 10 gms a.i. carbofuran.
4. On 15th day apply 300 gms of single super phosphate as foliar spray (1 per cent solution).

5. On 20th day harvest 100–400 kg of *Azolla* and inoculate it to a field of 1000 m^2 size (0.1 ha) having 10 cm of standing water. Apply 3 kgs of single super phosphate and 100 gms a.i. carbofuran.
6. On 25th day apply 3 kg of single super phosphate as foliar spray on *Azolla.*
7. On 30th day 1 to 1.5 tonnes fresh *Azolla* is ready to harvest, which is sufficient to 1 ha of rice field.

Recently a simple technique has been developed to produce dried sporocarp inoculum for *Azolla* which has a shelf life of two years. Pre-treatment of dried sporocarp with growth regulators increases germination and enhances the growth of young sporophytes. Only 200 kg fresh weight of *Azolla* fronds is sufficient per hectare.

Field studies carried out under the auspices of All India Coordinated Agronomic Research Project (AICARP) of ICAR at different locations, have revealed that incorporation of 5 t/ha of *Azolla* gave *Azolla* yield equivalent to about 30 kg N/h at Bhubneshwar, Khaiagpur.

Azolla Nursery method has been evolved for its large-scale multiplication in the field forestry adoption by the cultivators. The field selected for its nursery must be thoroughly prepared and levelled uniformly and then divided into 20 m × 2 m plots by providing suitable bunds and irrigation channels. Water is maintained at a depth of 10 cm. Ten kg fresh cow dung mixed in 20 litres of water is sprinkled and *Azolla* inoculum at 8 kg is dispensed to each plot. Single super phosphate (100 g) is applied in three split doses at 4-day interval as a top dressing for *Azolla.* For insect pest control, corbofuran, granules at 100 g/plot is applied 7 days after inoculating field as source of primary inoculum. From one harvest, 40–55 kg fresh *Azolla* is obtained from each plot. Again, it is inoculated in the same plot and cow dung single superphosphate and carbofuran are applied. Four cents nursery plots are required for raising *Azolla* for inoculation to one hectare main field. Adding cowdung in flooded rice soil system is reported to increase its growth and nitrogenase activity. The use of cowdung and cattle-shed water increases its growth during winter.

Constraints

The biggest constraint in the large scale use of *Azolla* is inadequate availability of its location specific strains. Also its multiplication is a encumbersome process as it has to be grown and multiplied in separate ponds as an independent crop, requiring lot of assured supply of water and its growth is much influenced by temperature fluctuations. In fact, it has to be raised as a crop, requiring fertilisation and proper care. Because of these limitations, *Azolla* seems to have limited potential and raising of a crop of *Azolla* often turns out to be a specialised job, especially in regions with high temperature. The growth and efficiency of *Azolla* in general is high in hilly or semi-hilly humid areas or in coastal areas than in regions with high temperature.

Conclusions and Future Outlook

Azolla biofertiliser has a tremendous significance on the future scenario of fertilizer market. It is one of the best alternative source of Nitrogen and organic matter. Moreover, this technology may be manipulated to develop "Dry *Azolla* Technology" and to evolve stains which can sustain high temperature and low phosphorus. On weight basis, the yearly *Azolla* requirement of our country is approximately 400 million tonnes. This formidable figure needs a massive extension efforts, training and participation of farmers and entrepreneurs, adequate investment from Government and Non-Government agencies to achieve this target.

China and Vietnam are the leading countries in world where *azolla* application in rice cultivation has been a common practice since-long. Ze Jiang University at Hong Zhou (China) is the leading centre of *azolla* research. Realizing the role of *azolla* in Agriculture University of Hawaii has recently initiated an International *Azolla* Project being coordinated by Dr. Tom Lumpkin, a leading International *Azolla* Research Scientist. The positive effect of *azolla*, an experiment on use of *azolla* as an organ nitrogenous fertiliser for rice was chalked out in All India Coordinated Research Projects and results revealed that incorporation of *azolla* @ 5 to 10 t/ha was equivalent to the application of 300 N/ha as inorganic nitrogen fertiliser.

However, there is a great need to conduct trials on large scale both at research centres and farmers fields by the Agricultural Universities/Institutes of the rice growing areas to explore possibilities of practical use of *azolla* in rice cultivation in the country.

Though *Azolla* is known to be a very efficient green manure for rice, its use is restricted areas where water is available in off-seasons. Further, high temperature and severe winter in some regions prohibit its rapid multiplication and adoption technique. Identification of tolerant strains development of cheap and efficient techniques for their multiplication are important requirements promoting the technology. In addition, an effective network for *Azolla* production and distribution has to be initiated to facilitate its large scale adoption, in rice-growing regions.

Integrated Approach for Increasing Microbial Inputs

In bio-farming biotic agents play very important role in integrated soil fertility manganese (ISFM). These soil biota (biotic agents) have the ability to carry out biochemical transformations as serve as a source and sink for mineral nutrients. A range soil biota are involved in various beneficial transformation processes such as BNF, mineralization of nutrients, organic matter decomposition increased nutrient uptake and use efficiency etc. This is being optioned by several microbiologist that the microorganisms are not being exploited efficiently in agriculture for increasing production and productivity to a expected level.

Microorganisms in agriculture use and their activities governed by several facts such as microorganisms per se, their mechanisms of benefits, host controlled factors and lastly appropriate management practices and each factor is depended on other, hence it is suggested all have to integrate achieve maximum benefits. In this connection it is appropriate to use *Rhizobium/Bradyrhizobium* an example to study the ways for increasing the benefits in biofarming. Several attempts have been made to select the most efficient strains of the biofertilising agents. Biological nitrogen fixation (BNF) for example involve host, bacterium and environment. When more than one organisms are involved it becomes necessary to study the system involving major groups of beneficial organisms, so synergistically deliver maximum benefits.

The fundamental knowledge on the mechanisms by which the introduced microorganisms benefit the crop is very critical. Besides, the major activities of the biofertilising agents, studies have shown that other functions such as production of siderophores, hormones or antibiotics or increases productivity through increased root growth also help plants to increase yield. Further there is need to test strains of one group of organisms such as nitrogen fixers, VAM, PSM, etc. which can work in synergistic way in the soil. It is of prime importance that the activities of microorganisms is to be improve in field conditions. The survival and competitive ability of the strains to be introduced by improved. There is also need for identifying biological, chemical and physical factors that influenced establishment of the inoculated agents in the rhizosphere and full expression of its trait for benefiting the crop. The host factors such as genes, physiology, specificity and selectivity are also important.

Further there is need to identify type of legumes and also genotypes a given legume which yield more and also derive larger benefit cost the proportion of their nitrogen requirement through BNF in a particular cropping system. As low nodulating and nonnodulating plants were found in the existing cultivars and/or segregating populations, efforts are required to minimise the occurrence of such plants. Suitable management practices need to be provided to ensure maximum contribution from the inoculants. It is generally observed that VAM provide maximum benefit in low phosphorus situations, nitrogen fixers perform better under low soil nitrogen situation, etc. Nitrogen application should be manipulated in such a way that in intercropped situations slow released nitrogenous fertilisers should be used and these to be placed at the root zone of cereal crops rather than broadcasted, Application of organic matter improve the efficiency of the inoculants, Stagnation of water should be avoided and also water conservation practices should also be followed.

Economics of Biofertiliser Use

Inconsistency in crop response to biofertilisers is a rule rather than an exception. However, the cost of biofertilisers is so low that even small increases in crop yields, though not significant, will be in excess of likely cost of inoculation which would make this technology attractive to the farmers. Monetary gains from the use of biofertiliser was reported. The investment of one Rupee in *Azolla* dual cropping gave a net return of Rs. 21–24 as against a net return of Rs. 5 with fertiliser N. A net income of 2,000/- from 0.2 ha land by using BGA biofertiliser technology has been reported. Based on the field trials by ICAR and SAU in different agro-climates estimation of the biofertiliser equivalent of important biofertilisers has been made. According to that, fertiliser equivalent of 19–22 kg N/ha from *Rhizobium*, 20 kg N/ha from *Azotobacter* and *Azospirillum*, 20–30 kg N/ha from BGA and 3–4 kg N/ha of *Azolla* could be realised. If one takes into consideration the cost of fertiliser N equivalents of biofertilisers and the cost of fertiliser, it will certainly be very economical to use biofertiliser.

Obviously, even if half of the indicated potential of each biofertiliser could be realised under normal field conditions, their use would be highly cost effective.

Besides their direct effect on current crop, use of a biofertiliser also leaves considerable beneficial effect on soil fertility observed that yield of various Rabi legume fodders increased by 10–35 per cent with the aid of *Rhizobium* inoculation and the yield of subsequent Kharif sorghum fodder raised on residual fertility was also 17–33 per cent higher than the control; non-inoculated plots. The residual effect of a leguminous crop inoculated with a biofertiliser on the succeeding cereal crop is believed to range between 20 to 123 kg N/ha.

Rhizobium possesses better tolerance to salt and pH (5 to 10) under various ecological conditions, therefore, possesses better adaptability to different agro-climatic situations. Biofertilizers may exert favorable effect on root growth and crop stand by affecting general growth and development of plant and sustained greenness. They may hasten flowering and crop maturity to a certain extent. Soil application of 2.5 kg/ha of *Azospirillum* hastened flowering and maturity of sunflower by 15 days. The crop was also observed to withstand favorably the impact of dry spell towards later stage of the 1 crop. *Azospirillum* and Phosphobacterin produce growth promoting substances which have been reported to promote number and length of roots, besides shoot growth.

Use of biofertilisers is economical with a high cost, benefit ratio, without risk. Soil application of 800 gm of *Azospirillum* was found to give, on an average, yield increase of rice by 1500 kg/ha resulting in net profit of 3250/- per hectare.

If the strain of bio-inoculant is properly evolved, it might work simultaneously as a "biofertiliser" and a "biopesticide", as was observed in the case of *Azotobacter* strain 1006 by Shah and Joshi. The said strain was claimed to have shown potential in inhibiting seed borne pathogens in some cereals.

Biofertilisers are renewable sources of plant nutrients and supplements fertiliser requirement in crop production. The available data indicate that biofertiliser can supply upto 20 per cent of the nitrogen requirement of crops, Research findings have established that upto 2 kg of biofertilisers applied to one ha will fix 25 kg of atmospheric nitrogen in the soil. In term of cost per acre, biofertilisers worth Rs. 20/- applied to the crop reduces the requirement on urea worth Rs. 175/- even at the current prices. Hence, there is a very good economic incentive for the farmer to supplement chemical fertilisers with biofertiliser.

Outlook

This Is conclusively proved that *Rhizobium* inoculation of leguminous plants is highly effective in exploiting and harnessing atmospheric nitrogen for improving the nitrogen status of the soils. Symbiotic nitrogen fixation could occur only within the cross inoculation group. Nitrogen fixation by biological agents as distinguished from chemical N fixation in a fertilizer factory. BNF is mediated by several organisms. The ratio between chemically fixed N and BNF ranges approximately from 1 : 4 to 1 : 2 : 5. But Rhizobium fixes half of the total BNF to that of industrial fixation. However, the symbiotic N_2-fixation is adversely affected by adverse soil conditions. Seed pelleting is effective in protecting the inoculated seeds against adverse conditions like acidity, sodicity and pesticides. The quantity of nitrogen varies from 40–200 Kg/ha/year. The nitrogen fixed symbiotically not only benefits the host but also has a residual effect to benefit the subsequent crops.

Chapter 5

Nitrogen Fixing Microorganisms: Asymbiotic

Introduction

There is a range of non-symbiotic nitrogen fixer bacteria and fungi. However only few of them are exploited commercially for inoculating other than leguminous plants. Among them *Azospirillum* and *Azotobacter* are most common. Among algae Anabeana or blue green algae (BGA) is intensively used in paddy fields.

Genus–*Azospirillum*

The genus *Azospirillum* belongs to family Spirillaceae with four commonly available species *viz.*, *A. brasilence, A. lipoferum, A. amazonese* and *A. halopraeferens/seropedicae.* The bacterium is gram negative with cell size 1 micrometer in diameter and shape is curved rod and accumulates poly-β-hydroxy butyric acid. The members of *Azospirillum* are aerobic and utilize carbon sources from organic acids and sugars such as L-arbinose, D-fructose, D-glucose, sucrose and pectin and nitrogen sources through the fixation of nitrogen, ammonia, amino acids and nitrate.

Azospirillum is an micro-aerophilic nitrogen fixer commonly found in association with the roots of cereals and grasses of interest. It colonises the root mass and fixes N in loose association with plants. It fixes nitrogen in an environment of low oxygen tension. These bacteria induce the plant roots to secrete a mucilage which creates low oxygen environment and helps to fix atmospheric nitrogen. These found to be specially effective for N fixation in lowland rice. High N fixation capacity, low energy requirement and abundant establishment in the roots of cereals and tolerance to high soil temperature (30°C–40°C) makes these most suitable for tropical conditions. Use of *Azospirillum* inoculum under saline alkali conditions is also possible to maintain high nitrogenase activity under such stress conditions. Response of different crops to *Azospirillum* inoculation is highlighted in Table 5.1.

Table 5.1: Response to *Azospirillum* Inoculation in Different Crops

Crop	*N Applied (Kg/ha)*	*No. of Isolations*	*Increase in Yield Over Control (%)*
Rice	0	5	21.40
	40	10.57	
Wheat	0	3	4.41
	40	13.80	
Barley	0	6	16.70
	40	12.95	
Oats	0	6	42.23
	40	11.44	
Pearlmillet	0	2	6.31

Beijerinck in 1925 reported a nitrogen fixing bacterium under the name *Spirillum lipoferum*. The ability to fix nitrogen by certain *Spirilla* was first recorded by him. Although *Spirillum* fixed nitrogen in semi-pure cultures, its ability to fix nitrogen in pure culture remained doubtful. Schoreder could not only establish the ability to fix N_2 by *S. lipoferum* but also found that peptone or NH_3 was necessary for its growth. He isolated several strains of *S. lipoferum* exhibiting their characteristic morphology and movement but possessing only a single polar flagellum which led him to suspect that the genus could be either Vibrio or Spirillum.

Aerobic N_2-fixing *Spirilla* were reported by some Russian in 1956 and 1965, though no confirmatory studies with labelled N were carried out. Such a confirmation was given from African soils. The strain incorporated labelled N in pure cultures when grown with yeast extract supplement. Later, N_2 fixation by *S. lipoferum* has been confirmed using conventional reduction technique and the isotopic enrichment method involving labelled N. Before the revision of its nomenclature, the bacterium was referred to as *Spirillum*. A new orientation to the study of this bacterium has come with the observations that *Azospirillum* could be isolated from the roots of tropical grass *Digitaria decumbens* using a semi-solid N_2-free sodium malate enrichment medium. Surface sterilization of roots by 70 per cent alcohol and creation of micro-aerophilic (low oxygen requirements) conditions in the medium are the two essential steps for the isolation of the organism. *Azospirillum* is recognised as a very ubiquitous soil organism capable of colonizing effectively not only the roots of a widely variety of plants but also their above ground portions forming apparently an associative symbiosis.

Taxonomy

Azospirillum produces white dense and undulating fine pellicles (a thin skin or film specially one that reflects a part of the light falling upon it and limits the rest of the light through it and that is used for dividing a beam of light) is a very characteristic on the semi solid malate medium. Microscopic examination reveals polymorphism but the dominant forms on solid medium are characteristic curved rods of varying sizes with prominent refractive fat droplets. *Azospirillum* is gram negative, motile, generally vibroid in shape and contains poly-β-hydroxybutyrate granules. Thick pellicles are usually contaminated but the thin pellicles are *Azotobacter*, actinomycetes. It is essential that for obtaining pure culture serial transfers are necessary from thin pellicles into fresh semi-solid medium. Selective medium and technique is now available for *A. brasilence*. The technique essentially consists of growing on successive liquid enrichments in N-free semi-solid medium supplemented with streptomycin followed by the most probable number counting method and N-free medium supplemented with cycloheximide (250 mg/L), streptomycin sulphate (200 mg/L), sodium deoxycholate (200 mg/L), 2,3,5,tri-phenyl tetrazolium chloride (15 mg/L) and congo red (1000 mg/L). The size of the cells are variable even among isolates from the same roots. Electron micrographs show lateral and single polar flagellum. When grown in nitrogen-free media it behaves as micro-aerophillic and when supplied with fixed nitrogen it grows as an aerobe. The salts of organic acids such as malate, succinate, lactate and pyruvate have been found to be satisfactorily oxidizable carbon and energy sources. Certain strains were found to require low levels of yeast extract for growth in mineral media and to grow in glucose as C source, whereas strains that did not require yeast extract failed to grow on glucose. On these and several other observations the existence of two or possibly three groups of organisms within the species *S. lipoferum* was established.

A taxonomic study of *S. lipoferum* group was conducted based on the DNA homology and other tests. DNA homology experiments indicated the occurrence of two distinct but related homology groups. Group II strains were distinguished from group I by their ability to use glucose as a sole carbon source for growth in nitrogen-free medium, by production of acidic reactions in a peptone based glucose medium, by their requirement for solid, nitrogen-free malate medium. Their results indicate the existence of two species based on their characteristics. It is proposed that this may be assigned to a new genus, *Azospirillum*. Strains belonging to group II are named *A. lipoferum*, while those belonging to group I are named *A. brasilense*. A new species, *Azospirillum amazonense*, was described which differs from the other two species by the formation of white rather pink colonies on potato agar (BMS agar), the absence of lateral flagella and the ability to use sucrose as a sole carbon source for nitrogen fixation. The fourth species, *Azospirillum seropedicae* was described. This species differed from other *Azospirilla* by having a smaller cell diameter, by having 1-3 flagella at one or both the poles rather than a single polar flagellum, and by forming small, moist, green-centred colonies on N-free-based medium (NFB medium) containing yeast extract rather than the dry white colonies formed by other *Azospirilla*.

A large number of *Spirillum lipoferum* strains were examined for their denitification ability. It was presumed that *A. brasilense* contains two sub-groups, nir^+ and nir^-. Both the groups possess assimilatory nitrate reductase, but *A. brasilense* nir does not dissimilate nitrite to establish the affinity of *Azospirillum* to plant roots. Studies were carried out under aseptic conditions by growing plants on N-free seeding agar and inoculating them with a 48 hour-old culture of *Azospirillum*. The first sign of their adaptability to the root system was the proliferation and clustering of the bacteria around the root hairs, followed by the entry into the cortical layers often extending into the xylem. *In vitro A. lipoferum* produces siderophores when grown in iron deficient medium. Production of sideropliores by the bacterium may improve iron nutrition and offer protection from pathogens. This bacterium does not grow with nitrogen as the sole source under aerobic conditions.

The optimum temperature is 32 to 40°C and pH is 6.8 to 7.8. The occurrence of *Azospirillum* depends in soil on pH and high temperature. The bacterium preferred ammonia for N_2-fixation.

Isolation, Maintenance and Cultivation

Azospirillum has been isolated from the roots of a variety of grasses, legumes, grain crops and soil. *Azospirillum* is also found on rhizosphere root surface, root hairs, epidermal cells and vascular tissue. Subsequently these were reported in association with roots and rhizosphere of several cultivated and uncultivated plant species. *Azospirillum* are mesophilic and reported in association of crops grown in acidic to alkaline pH range. *Azospirilla* are metabolically versatile and can grow vigorously in presence of nitrogenous compounds present in soil but as soon the external combined nitrogen supply is exhausted the bacteria switch on to diazotrophy. The ability to fix nitrogen is unaffected by the presence of combined nitrogen sources and may account for the beneficial response of *Azospirillum* inoculation in fields receiving mineral fertilisers. Use of *Azospirillum* inoculation in saline alkaline conditions is possible because strains adapted to these stress condition maintained high nitrogenase activity, but no such observation was recorded with respect to high temperature. There are numerous reports indicating beneficial effect of *Azospirillum* inoculation of crops under Indian conditions.

Half centimetre long root pieces are surface sterilized in 70 per cent alcohol for 3–5 seconds and repeatedly washed in phosphate buffer (pH 7.0) and then they are plated in semi-solid, N-free sodium malate medium. The plates are incubated at 35°C for 48 hours. Characteristic growth of *Azospirillum* is indicated by the formation of white pellicles 2-4 mm below the surface of the medium. The cultures are purified in two steps. The isolates are enriched by repeated transfer into semi-solid malate medium in small screw-cap tubes. These cultures are then subjected to serial dilution upto 10^{10}. From each dilution which gives good growth is found to be pure and used for further studies.

The bacterium can be maintained on ammonium chloride containing agar medium formulated by Oken and coworkers. On this medium bacterium does fix nitrogen but proliferates profusely under aerobic conditions. The number of cells in the broth can be counted by plating dilutions on ammonium chloride agar medium. In large scale multiplication, the bacterium can be grown in ammonium chloride medium in flask on the rotary shaker at temperature of 35°. The best carriers for this bacterium is powdered sterilized farm yard manure and soil or FYM alone and charcoal and on this carrier the bacterium survives for a period of six months.

Physiology

Azospirillum lipoferum (AL) and *A. brasilence* (AB) can grow vigorously in salts of malic, succinic, lactic and pyruvic acid. AL is able to utilize glucose and sucrose as sole carbon sources for growth and N_2. whereas these sugars are poor substrate for AB. The correlation between organic acids as potentially good substrate for this bacterium and their accumulation in the roots of grasses has been emphasised in the support of the suggestions of a symbiotic association between these bacteria and grasses. Further these bacteria require low oxygen for the expression of nitrogenase activity. Experiments indicated that tricarboxylic acid pathway is operative in these organisms while glycolysis and pentose pathway is of minor importance. *Azotobacter* fixes N_2 in air because of respiratory and confrontational protection of its nitrogenese and because of the location this enzyme within the cell. Micro-aerophily studies have been reported that there is another mechanisms by which *Azospirillum* fixes N_2.

Genus–*Azotobacter*

The genus belongs to family Azotobacteriaceae and has six species *viz., A. chroococcum* most commonly used and *A. beijerinckii, A. paspali, A. vinelandii, A. insignis* and *A. macrocytogenes*. The cell size is usually large ovoid and the size ranging from 2.0–7.0 × 1.0–2.5 micron and are polymorphic. The bacterium is gram negative and accumulate poly-β-hydroxybutyrate granules. The physiological studies indicated that the bacterium uses a variety of carbon sources (mono, di and certain polysacchrides) and organic acids whereas the nitrogen sources comes from fixation of amino acids, NH_4 and NO^-_3. *Azotobacter* is a heterotrophic, aerobic and grows on Ashby and Jensen's medium. *Azotobacter chroococcum, A. beijerinckii* and *A. vinelandii* have the unique characteristics of producing extracellular slime, although its formation is related to nutritional conditions such as type of carbon and nitrogen sources. It is above all a genetically controlled character which is specific for different species. The various *Azotobacter* species produce colour pigments in the medium for example *A. chroococcum* produces black to brown (insoluble), *A. beijerinckii* and *A. insignis* produce light brown (insoluble): *A. vinelendii, A. agilis* and *A. insignis* green (soluble). It fixes on an average 20 Kg N/ha/ year and sometimes produces growth protein substances which has shown to be antagonistic to pathogens. It has shown increase in growth and yield of a wide variety of cereals, millets, pulses, vegetables, fruit crops etc.

Introduction

Azotobacter a free living heterotrophic nitrogen fixing bacterium encountered in neutral to alkaline soil conditions not only provides the nitrogen but produce a variety of growth promoting substances. Some of these growth promoting substances are indole acetic acid, gibberellins, B vitamins, and antifungal substances. Another important characteristics of *Azotobacter* associated with crop improvement is excretion of ammonia in the rhizosphere in presence of root exudates and helps in modification of nutrient uptake by the plant. These strains were better competitors than (survived better in soils compared to) the non excreting strains. The genus *Azotobacter* is highly versatile in utilising carbon sources therefore application of organic carbon containing sources to the soil improves the asymbiotic nitrogen fixation capacity of the diazotroph solubilisation of different inorganic phosphates by *A. chroococcum*, have been reported.

The benefits of *Azotobacter* inoculation on crop production are well established. Crop receiving *Azotobacter* inoculation along with moderate levels of N gave similar grain yields as the crops receiving higher doses of mineral fertilizers but uninoculated. This indicated the possibility of reducing application by inoculating the crop with appropriate strains. This inoculation is useful for cereals and non-leguminous crop plants. The benefits acrue from their inoculation in the form of increased grain yield, biomass and N uptake. These may be ascribed to N addition through biological N fixation, development and branching of roots, production of plant growth hormones, like indole acetic acid, gibberlins etc. enhancement in the uptake of NO^-_3, NH_4^+, $H_2PO_4^-$, K, Fe etc. and improved water status of plants, increased nitrate reductase activity and production of antifungal compounds.

While these are believed to be as good as *Rhizobium* under irrigated conditions, these are often relatively less effective in rainfed soils. This is because non-symbiotic organisms essentially depend on soil matter (OM) reserves for their energy requirement, and most of rainfed soils are poor in OM. Therefore, only low to moderate contribution of N through these in dry land conditions could be expected, unless suitable application of an organic source is made to the soil. Thus the beneficial effects of *Azotobacter* inoculation appear to be qualitatively real and are attributable to the multiple action of *Azotobacter*, namely, the fixation of atmospheric nitrogen, synthesis of growth-promoting substances and anti fungal antibiotics. However, considering the cheapness of the process the small increments obtained may be of advantage when fertilizers are in scarce supply and there is need to conserve them by maximizing their utilization.

There is controversy in the literature as to the survival and multiplication of *Azotobacter* in the rhizosphere of crops. Larger populations of *Azotobacter* according to some investigations are found to increase vitamin C content in tomatoes. Increased biological activity in terms of the total nitrogen fixation in the mixed culture of *Azotobacter chroococcum, Thodopseudomonas capulatus* and *Cylindrospermum musciola* than in pure cultures of each organism showed that high counts of *Azotobacter* were established in the rhizosphere of wheat when a mixed culture of *A. chorococcum* and capsulants was used.

A course of events may be visualised as occurring in an ecosystem involving three organisms, namely, blue green algae, *Azotobacter* and photosynthetic bacteria. Both BGA and photosynthetic bacteria can carry out their metabolic activities in presence of light. In the special case of the photosynthetic organism, the bond is provided by photochemical reactions that convert radiant energy into chemical bond energy, which is then used to drive the machinery of the cell. The photosynthetic evolution of oxygen by BGA can provide an aerobic environment to *Azotobacter.* The utilization of oxygen by *Azotobacter* may result in partial or complete anaerobiosis in the immediate vicinity, where the anaerobic *Rhodopseudomonas* can thrive well. *Rhodopseudomonas* can use lower fatty acids and excrete carbohydrates. In this course, the bacteria can fix nitrogen and assimilate carbon dioxide. On the other hand, heterotrophic bacteria can use the carbohydrates and can excrete lower fatty acids. This carbon dioxide released in the heterotrophic dark course can be assimilated by the photosynthetic bacteria.

Cultures of this bacterium form considerable quantities of biologically active substances (biosynthesis of growth promoters: BGP). The major BGP are Vitamin B-complex such as nictonic acid and panthothenic acids, biotin, heteroauxin and gibberllins. *Azotobacter* has ability to produce antifungal antibiotics and fungicide compounds against pathogens like *Fusarium, Alternaria* and *Trichoderma.*

Seeds whose surface has been sterilised and treated with a thick suspension of *Azotobacter* culture, development of the shoots was much better than that of control seeds treated with water. Thus the increased germination and vigour of young plants lead to an improved stand in crops inoculated with *Azotobacter* cultures and prompted its use. Inoculation with *Azotobacter* accelerated the stem and leaf growth of tomato and shortened the time between the bud appearance and petal fall. The onset of flowering of wheat was also hastened similar effects of gibberellin treatment and *Azotobacter* inoculation on tomato seeds was also observed.

Distribution

Azotobacters are found in rhizospheres of many plants that fix nitrogen and produce growth promoting hormones. They are parasitized by Bdellovibrio and Azotophages are distributed widely. The population in soil ranges from negligible to 10^4/gram of soil depending on organic content, microbial interactions and temperatures. *Azotobacters* are active N_2 fixers under lab conditions, ubiquitous in geographic distribution and can use a wide variety of carbon and energy sources for their growth. A survey of sorghum and maize growing areas showed that the most probable number (MPN) of N_2 fixers varied 1000 fold, from 100 to 100,000/g soil. Isolates obtained from these soils, 42 per cent showed nitroganase activity *in vitro.* In another study, out of 546 isolates obtained from the rhizosphere of pearl millet grown at ICRISAT Center only 17 per cent isolates showed nitroganase activity, *in vitro.* Pearl millet rhizosphere was dominated by azospirilla (72 per cent of total N_2 fixing isolates) followed by enterobacters (12 per cent), azotobacters (11 per cent) and 5 per cent pseudomonads. Similar information is lacking for other crops.

Azotobacter chroococcum strains is isolated by Subramoney from red loamy soils in India. Studies regarding the distribution of *Azotobacter* species indicated that they are widely spread and in top layers and the population decreases with depth. *Azotobacter chrocooccum* and *A. beijerinckii* are deemed to be the most commonly occurring species. The later appeared in acid soils while *A. chroococcum* in neutral and alkaline soils.

Several *Azotobacters* have the unique characteristic of producing extracellular slime. Alt its formation is related to nutritional conditions such as type of carbon and nitrogen sources, it is above all genetically controlled character which is specific for different species.

Classification

Various suggestions have come across for classifying bacteria of the genus *Azotobacter.* Based on morphological and physiological features *Azotobacter chroococcum, A. beijerinckii, A. vinelandii* and *A. paspali* were categorised under the genus *Azotobacter,* while *A. agilis* was categorised under the genus Azomonas.

Morphology and Taxonomy

The bacterium is heterotroph, aerobic organism, fix dinitrogen as nonsymbionts and besides occurs in the rhizosphere of many plants. The bacterium is rod shaped gram negative, polymorphic, size ranging from 2–10 × 1–2.5 micrometers. Young cells have peritrichous flagella and are used as locomotory organs. The old bacteria produces encaspulated forms and are heat resistant. Like *Azospirillum, Azotobacter* also contains poly β-hydroxy butyrate and produces polysacchrrides and melanin. The ideal pH is 6.8 to 7.8 and temperature 35 to 40°C. *Azotobacter* are found in rhizosphere of many plants and fixes nitrogen and produces hormones like growth substances.

Cells occur singly, in pairs and sometimes in chains with large capsules. As cultures ages the cells become coccoid, form cysts and turn yellow or cinpamon with water insoluble pigment. The cell size and shape vary considerably with species, strains, age of culture and growth conditions; *A. chroococcum* cells have accuminated ends, while *A. beijerinckii* are oval with rounded ends. However, the cells of *A. beijerinckii* are larger than that of *A. chroococcum.* Young cells of *A. vinelandii* closely resemble to *A. chroococcum* type, being comparatively smaller.

Cells of the family Azotobacteriaceae are usually variable in morphology. The vegetative cells give rise to specialised spherical resting cells known as cysts. Each cyst is produced from 3 a single vegetative cell. There is a distinct morphological difference exists between vegetative cells and cysts. As compared to the rod-shaped structure of vegetative cells, the cysts under a light microscope appear as spherical cells with contracted cytoplasm and a double contoured thick wall. During cyst germination, a young vegetative cell emerges from the cyst structure leaving behind a residual portion of the cyst coat. Encysted cells have the ability to tolerate adverse conditions of desiccation, sonic treatment and ultraviolet radiations which are hardly tolerated by the vegetative cells. Motility in most of the cells is carried out by means of petrichous flagella. Among the three different typical species, *A. beijerinckii* are less motile when compared with *A. vinelandii* and *A. chroococcum* has the highest motility of the three. This serves to differentiate among the main species of *Azotobacter.* The pigment is commonly attributed to the presence of melanin. The melanin is formed as a result of oxidation of tyrosine by tyrosinase, a copper-containing enzyme.

Isolation

Azotobacter spp. can be isolated from soil and water by dilution plating method. Ten grams of soil sample are transferred to 100 ml sterile distilled water and mixed thoroughly by shaking the flask for 5 minutes. Serial dilution of the suspension are made using sterile distilled water. Ashby/Jensen's Nitrogen free agar media specific for *Azotobacter* prepared and poured into sterile petri-plates. One ml samples from the appropriate dilutions are spread evenly over cooled agar medium in petri-plates. The plates are incubated at 30°C for 3–4 days. *Azotobacter* colonies appear as translucent dew-drop, soft, milky and mucoid on agar plates. Direct isolation of *Azotobacter* from soil has been followed by Russian scientists. Lumps of soil are spread on N-free agar medium and the plates are incubated at 28°C. *Azotobacter* colonies develop on the agar plates after 3 days of incubation.

Crop Responses

The application of *Azotobacter* resulted in beneficial effect in several crops such as rice, wheat, sorghum, maize, pearl millet and cotton. Wani reported an increase in gram yield of pearl millet from 0–27 per cent by *Azotobacter* application. In sorghum, Kavimandan observed higher population of *A. chroococcum* in rhizosphere compared to non-rhizosphere. A yield increase of 9 to 38 per cent has been reported when inoculated by this organism. Shinde reported that *Azotobacter* population was observed in high count in rice and grain yield of rice increased to 5–31 per cent under different N, P and K combinations. Palarwar reported an improved germination of wheat by 30 per cent with *Azotobacter* and 16–30 per cent increase in yield of wheat was recorded.

The results (Table 5.2) of the trials conducted on mustard yield increased by about one quintal over application of 20 Kg N alone. Thus, by spending Rs. 12.50 only per hectare on biofertilizers, there is net saving of Rs. 800/- per hectare. Application of 20 Kg N + *Azotobacter* seed inoculation resulted in 56.6 and 10.8 per cent higher gross income over control and application of 20 kg N alone respectively. Thus cost benefit ratio by using low cost benefit input (biofertilizer-*Azotobacter*) in mustard is approximately 1:65. Although application of 40 Kg N/ha (recommended dose in mustard) results in higher net income but the FVM treatment (organic manure) also enriches the soil with other nutrients, thereby, improving the overall fertility of the soil. In comparison with 20 Kg N + *Azotobacter* seed inoculation also, the results are less beneficial but it reduces the dependence on inorganic fertiliser which may be identified specially to the farmers with limiting resources. Because both the 20 Kg N + *Azotobacter* as well as FYM + *Azotobacter* treatments are definitely superior over control as they result in net benefit of nearly Rs. 3000/ha over control. Therefore, farmers are advised to inoculate the seeds of mustard with *Azotobacter* at the time of sowing. For direct application in the field, it has been observed that if *Azotobacter* is applied alongwith FYM, the response is more as compared to its application alone.

Table 5.2: Effect of Organic, Inorganic and Biofertilizer on Grain Yield (q/ha) of Mustard and Net Income

Treatment	*Grain Yield*	*Gross Income (Rs./ha)*	*Expenditure on Nutrient (Rs./ha)*	*Net Income (Rs./ha)*
Control	5.8	5220	–	5220
20 Kg N/ha	8.3	7370	118.7	7351.3
Azotobacter seed inoculation	9.3	8280	131.2	8148.8
4 tonnes of FYM + *Azotobacter* seed inoculation	9.5	8550	492.5	8057.5
40 Kg N/ha	9.9	8910	237.4	8672.6

Blue Green Algae

Blue green algae (BGA) or cyanobacteria are free photosynthetic nitrogen fixer and distributed worldwide and are believed to contribute to soil fertility in many agricultural ecosystems. Its potential in low land rice is well documented and it could contribute 25–30 kg N/ha. However, like *Azolla* commercial production of BGA is set with scores of problems which limits its large scale use.

Morphology

They are universal in distribution. They are either single celled or consists of branched or unbranched filaments. Some of them possess a peculiar structure known as heterocyst and heterocystous forms can fix nitrogen from air. Some BGA can fix N without heterocysts have also been found to fix N under special conditions like low oxygen tension. A range of BGA has been isolated in the all Indian survey which fix N under different ecosystem. The important BGA which are found in Indian soil are given in the Table 5.3.

Table 5.3: Some Important Algae Genera

Aulosira
Anabaena
Calothrix
Nostoc
Plectonema
Scytonema
Tolypothrix

Paddy fields posses an efficient non-symbiotic N-fixing ecosystem, since a daily integrated N-fixing is of the order of magnitude as that of a symbiotic system. At IRRI, Philippines, using ^{13}N radio active, values ranging from 40 to 100 kg N/ha/year and biomass 50 tonnes/ha have been recorded in paddy fields. BGA are responsible for BNF in the paddy fields is clearly discernable by the marked influence of light under simulated day-night conditions on the total N fixing capacity. Ammonical N below 40 ppm or presence of pesticides have no adverse effect on BNF.

Constraints

Some of the problems of its commercial productions are:

1. Freshly inoculated BGA are often overwhelmed by the indigenous algal population.
2. Use of large scale fermenters of factory level production of BGA is very costly.
3. The alternative of having small BGA nurseries in farmers fields, although appears to be feasible, but it is difficult to develop and maintain such nurseries because of technical difficulties and the extra care required.
4. Like *Azolla*, growth and development of BGA is sensitive to temperature fluctuations and water level in the nursery. However, in this regard it is relatively less sensitive than *Azolla*.
5. Technology is yet to be developed to have location specific thermo insensitive strains and formulation of the inoculant in dry form for better feasibility in their handling and transportation to different locations.

Field studies at Central Rice Research Institute (CRRI), Cuttack revealed that BGA increased grain yield of rice to the extent of 6 to 35 per cent. On the other hand, experiments conducted at Farmer's fields in Tamil Nadu, revealed much less utility (generally 2.5 to 5 per cent yield increase) of BGA (Table 5.4).

Table 5.4: Performance of BGA on Farmer's Fields in Tamil Nadu

Treatment	Grain Yield of Rice (kh/ha) at Different Location		
	Vellore (Av. of 11 Trials)	Trichy (Av. of 8 Trials)	Cuddalor (Av. of 10 Trials)
$N_{60}P_{25}K_{25}$	4235	4557	3789
$N_{60}P_{25}K_{25}$+BGA	4455	4735	3884
$N_{75}P_{27.5}K_{27.5}$	4400	4485	3953
$N_{75}P_{27.5}K_{27.5}$+BGA	4565	5117	4156

Future Outlook

The field results of BGA inoculation are found to be errotic and inconsistent. The occurrence of ineffective algal strains and grazing by microcrustaceans are some of the problems encountered in large scale multiplication of BGA. Simple techniques for rapid multiplication of BGA have to be made available to the farmers and extension workers.

Chapter 6
Crop Response to Biofertilizers

Biofertilizers have an important role to play in improving the nutrient supplies in our search for alternate sources of soil fertility build-up through renewable sources. They can help in the biologically fixed atmosphere N or increase the native P availability to crop plants. Among Biofertilizers for increasing N supply, N-fixing bacteria (*Rhizobium, Azotobacter* and *Azospirillum*), blue green algae (BGA) and *Azolla* are important. Among those associated with increased P availability, different P solubilising bacteria (PSB) and mycorrhiza are of significance.

The critical role of nitrogen and organic matter in improving plant production is well recognised, particularly through non-symbiotic sources, offers a great potential by augmenting nitrogen supplies. There are evidences of biological nitrogen fixation in rhizosphere and rhizoplane of crops by non-symbiotic diazotrophs. Non-symbiotic nitrogen fixation seems to have magnitudes of agronomic significance. Nitrogen fixation in the rhizosphere of the grasses has been suggested as a mechanism to explain the substantial nitrogen gain observed under plant fallows. Nitrogen fixing bacteria of many diverse genera occur in high numbers in rhizosphere.

High productivity in irrigated areas leads to higher nutrient turnover and a larger gap between nutrient supply and removal. Therefore, there is a strong need to have complementary use of available source of plant nutrients including biofertilizers along with mineral fertilisers for maintenance of soil productivity. Results from various cropping systems and ecologies illustrate that positive interactions results from the integrated use of mineral fertilisers and organic/biological sources of plant nutrients within the framework of integrated plant nutrition. Benefits from Biofertilizers in terms of increased yields and economy in fertiliser use of are not consistent in irrigated areas (Table 6.1).

Dryland agriculture stretches over 100 m ha in India and produces 44 per cent food and 40 per cent population. Cultivation of more than 90 per cent of the coarse cereals, 80 per cent groundnut, 85 per cent pulses. 70 per cent cotton, 55 per cent rice and 22 per cent wheat is concentrated in rainfed areas. Average productivity of food-grains (coarse cereals plus pulses) in rainfed areas is around 680 kg/ha. The comparable yields of irrigated crops are two to three times higher. Low and unstable yields of dryland crops are due to unpredictable swings in rainfall, a host of dryland crops are due to unpredictable swings in rainfall, a host of biotic and abiotic stresses and adherence to age old farms and crop management practices. Among the abiotic constraints, poor fertility of rainfed soils is a major factor contributing to dismally low yields of dryland crops. Current addition of about 25 kg nutrients (N+P+K) per hactare through fertilisers to coarse cereals is nowhere close to crop related removals. Possibility on narrowing the nutrient addition-removal gap through chemical fertilisers seems bleak because vast majority of the dryland farmers are economically poor to invest on costly fertilisers and to bear risk if the crop failure happen. Apparently, the situation demands that a dryland farmer is burdened minimum with expensive inputs. Since abridgement of negative balance arising from nutrient depletion by crops and its replenishment through external sources is compulsory to sustain productivity goals, harnessing natural nutrient reservoir becomes an important aim. Productivity of drylands will be required to be raised to 2 tonnes/ha to keep pace with the food needs of population which is burgeoning at a rate of about 2.1 per cent.

Table 6.1: Response Under Irrigated and Rainfed Conditions of Biofertilizers to Several Crops

Crop	*Biofertilizer*	*Range or Increase in Crop Yield (%)*
Irrigated Conditions		
Potato	PSB	30–60
Gram	PSB	10–30
Chickpea	PSB	9–16
Paddy	BGA	2.5–12.4
Paddy	PSB	10–20
Wheat	PSB	7–42
Buffel grass	Ar	18–35
Sugarcane	Ar	9–24
Groundnut	R	10–12
Cowpea	R	12–18
Urd bean	R	2–8
Mung bean	R	0–10
Non-leguminous	Ar	7–17
Fodder	As	4–25
Rainfed Conditions		
Chickpea	R	9–76
Pigeonpea	R	10–46
Green gram	R	9–51
Cluster bean	R	4–17
Red gram	R	40–47
Soyabean	R	0–5
Sorghum bean	Ar	0–15
Pearl millet	Ar	0–19

PSB: Phosphate solubilizing bacteria; Ar: Azotobacter, AS: Azospirillum; R: Rhizobium.

Organic manures and Biofertilizers are a vital natural resource which can be utilised to boost the overall nutrient turnovers. This explains the ever-increasing emphasis on plant nutrient supplementation through indigenous sources. Alleged involvement of chemical fertilisers in environment degradation is another factor reviving vigorously the value of organic manures and biofertilizers in soil fertility management appraises the role of niofertilizers (BF) in augmenting the natural nutrient transfers to irrigated rainfed and forage crops. Biofertilizers, in this regard, means generation of certain plant nutrient in response to introduction of appropriate microorganisms.

The forages are the main stay of animal production are often considered a low value and a low priority crop in human food chain. India has about 15 per cent of world's livestock population with only 2 per cent of world's geographical area. This is indicative of animal pressure on limited resources. Low productivity of our livestock is mainly due to poor nourishment. The fodder produced in the country is not sufficient to meet the requirement of the present livestock population. Present area under cultivated fodder crops is about 8.3 million hactare (m ha). For meeting the projected requirement of green fodders (990 mt) by 2000 AD, this area should ideally be 15 m ha, but this appears to be a difficult task to achieve, in addition, an area of about 12 m ha is under pastures and grasslands. From these resources, the present availability of forages (450 mt) are to be tune of 50 per cent of the desired needs. A summary of work carried out on response of biofertilizer to crop has been presented in this chapter.

Symbiotic Nitrogen Fixation

Rhizobium

Irrigated Crops

Contribution of BNF to agriculture is derived from the symbiosis between legumes and *Rhizobium.* Legume inoculation is a long-established and successful practice to ensure adequate N nutrition in place of fertiliser N in most agricultural soils throughout the world. The *Rhizobium-legume* association can fix upto 100–300 kg N/ha in one crop season and in certain situations leave substantial N for the following crop. Inoculation with efficient *Rhizobium* strain specific to each crop is very essential for the N gains and better crop yields.

The field response to *Rhizobium* inoculation under irrigated conditions is quite variable. For example, with chickpea, significant improvement in grain yield was reported, predominantly in Central and Western India with yield increases varying from 14 to 30 per cent over control yields. Studies carried out in the All India Coordinated Agronomic Research project, have indicated significant response to inoculation in mungbean and urdbean during Kharif. Soyabean showed good response only at Jabalpur while cowpea responded both at Hisar and Bichpuri. Groundnut responded significantly at Akola. The residual effect of *Rhizobium* inoculation of different Kharif legumes on the succeeding cereal crop could be noticed. Significant response to inoculation in summer mungbean was also observed at Delhi. However, combining *Azotobacter* or *Azospirillum* with *Rhizobium* had no significant additional advantage. Urdbean in spring season and field peas in rabi season also responded significantly to *Rhizobium* inoculation under irrigated conditions at Delhi. Typically, the responses are substantial when indigeneous ineffective *rhizobia* are absent and soil nitrate is low. Application of starter N generally reduces responses to *Rhizobium* inoculation. This is food legumes normally does not leave any significant residual effect. This is understandable as the N_2 fixed by food legumes is often less than the N removed with the grain. In most situations, food legumes probably do not increase the soil N pool.

Before legume inoculation is recommended as a practice, the most important point is do we need this in a region where these crops have been grown over long periods. It is generally recognised that the need to inoculate is not universal and should be carefully determined for each individual situation before investing in inoculant production and use. Legumes like groundnut, mungbean, urdbean, soyabean and chickpea are increasingly cultivated in irrigated areas and all possibilities of integrating *Rhizobium* as a part of nutrient supply systems in the cropping systems involving these should be explored.

Dryland Crops

During the last two decades, multilocational inoculation experiments. Involvement *Rhizobium*, have been conducted across contrasting rainfed conditions under the aegis of All India Coordinated Research Projects on Dryland Agriculture (AICRPDA). Pulse Improvement (AICPIP) and Oilseeds (AICORPO). In addition, several inoculation studies on pulses and oilseeds were conducted at ICAR Institutes, ICRISAT and State Agricultural Universities.

A summary of the trials conducted under network at different research centres on *Rhizobium* indicate that only some locations, significant response was obtained to inoculation. At other sites, the yields increase ranged from zero to 13 per cent. A critical appraisal of the data obtained over the variable crop growing environments of the coordinating centres did not provide any significant clues to explain the positive response, or the lack of it. Rainfall, soil type, nutrient applied or crop species, were hardly useful in delinating the response pattern. Among the crops, clusterbean was benefited by inoculation at Agra (sandy loam) and Jhansi (loamy sand) but not at Hisar and Jodhpur (Sierozemic soils). Pigeonpea responded at Kovipatti (deep black soils), but not at other centres. Green gram and black gram were non-responsive to inoculation at most of the centres. During the first year of experimentation, inoculation did not work for chickpea at Rewa and Varanasi. During the second year, however, the yield increased by 36 per cent at Varanasi.

Consistency in grains yield from inoculation with *Rhizobium* gave positive response of more number of locations. The overall increase in yield over uninoculated control varied from 9 per cent to 70 per cent with pulses like chickpea, pigeonpea and green gram. The response data showed high degree to abnormalities both in time (interannual variability) and space (locational variability).

The Coordinated experiment groundnut across different locations in the country also presented a variable response picture. Significant and positive response were obtained with Kharif groundnut involving Robut-33-1 cultivar and NC-92 strain, while with several other varieties the results were non-significant. With Rabi/summer groundnut *Rhizobium* strains IGR-6 and IGR-40 failed to increase the yield that could reach the level of statistical significance.

An analysis of the data on *Rhizobium* inoculation trials conducted under AICRPDA from 1974 to 1985 revealed that the commercial "Nitragin" was used as an inoculant. The life of these experiments seldom exceeded 2 years. Crop growth, nutrient uptake and yield levels being intimately linked to the seasonal rainfall and its distribution, it is not surprising that definite conclusions have often evaded the BF research in drylands. This deficiency was partly made up in the experiments conducted under the auspices of AICPIP and AICORPO, which have (*i*) longevity of more than 5 years at many locations, and (*ii*) used locally isolated efficient strains, superior inoculation methods and compatible cultivars. With the format, relatively greater response to inoculation was obtained in these experiments. However, the problem of inconsistency in response continues to haunt chiefly due to dependence of legume-Rhizobium symbiosis on several unpredictable external variables.

Rhizobium inoculation has not become popular with the rainfed farmers to the extent commensurable to the advantage from its adoption. While timely availability of quality cultures is one bottle-neck, the problem of convincingly demonstrating yield improvements still remains a challenge. A few trials on *Azospirillum* inoculation showed promise, but the response seems to depend largely on rainfall, soil type and mineral-N. More vigorous testing and consistent data are required before these organisms can be considered as dependable sources of N supplementation in drylands.

Dryland Legumes

The symbolic action of *Rhizobium* in siphoning atmospheric N has attracted research on quantifying N gain that may follow on incorporation of a legume component in inter- and sequence-cropping systems in drylands. The hypothesis that legume N may be transferred directly to the non-legume component crop in the intercropping system has been found not to hold ground. Consequently, response to applied N in legume-non-legume intercropping system was generally poorer than, the non-legume sole crop. Incorporation of legume in crop rotation/sequence has, however, been found to give a usual effect measured in terms of N gain by the crop in system. Total soil N increased by 56 to 70 kg/ha by including pigeonpea and greengram in crop rotation with castor in a semi-arid Alfisol. Nitrate-N content in the top-soil of semi-arid Vertisol was 20-25 kg/ha following cowpea in contrast to 18 kg/ha after sorghum. Sub-surface soil contained significantly more nitrate-N in rotation with a legume as compared to the system without it. Short duration legumes like cowpea and green-gram were more effective in increasing yields of following crops in rotation through higher N build up in the soil. The beneficial effect of N gained through the presence of a legume in crop rotation/sequence in attributed to the evaluation of N from the breakdown of enriched nodules that formed with *Rhizobium* symbiosis and root mass of legume crop.

Fodder Crops

Fixation of nitrogen by legume-bacterial symbiosis depends upon the presence of an effective strain on *Rhizobium* that infects the legumes root, availability of soil and plant nutrients, environmental conditions, competition of associated vegetation and intensity of defoliation/grazing. A majority of tropical legumes are readily infected with wide range of rhizobial types commonly present in the soils of tropics and subtropics. Some tropical legumes require a specific types of *Rhizobium* for effective nodulation and nitrogen fixation (Table 6.2). Cultivation of leguminous plants provides the most economical means of adding N to the soil-plant animal system. Studies showed a wide range in quantities of fixed, but the amount were generally comparable under tropical and temperate conditions.

Table 6.2: Important Fodder Crops and their Rhizobia

Rhizobium	*Fodder Crop*
Rhizobium sp (Cowpea Group)	Cowpea, cluster bean, velvet, bean, stylos and siratro
Rhizobium meliloti	Lucerne, clover, fenagreen
Rhizobium trifolli	Berseern
Rhizobium phaseoli	Lablab bean

The annual production of 567 kg/ha of N by *Leucaena leucocephala* obtained in Australia approaches that of 604 kg produces by *Trifolium repens* in New Zealand. Such quantities exceed those commonly obtained under experimental conditions.

Inoculation with *Rhizobium* to all important cultivated legume fodders, pasture legumes and shrubs was studied over the years. The increased fodder yields due to inoculation were observed to be 10–28 per cent in case of cultivated fodders and 11–31 per cent in pasture legumes and shrubs over control (non-inoculated). High response was observed in case multi-cut fodders like lucerne, berseem and Persian clover, and in pasture legume like butterfly pea. *Rhizobium* inoculation to some important fodder legumes were also studied in different soils. It was observed that the crops gave much higher response in less fertile red soil than in high fertile black soil. The benefit of left over nitrogen from has been reported by many workers. The studies indicated that left over nitrogen from berseem and lucerne on cultivated fields was to the extent of 40 to 30 kg N/ha which was reflected on the yield of subsequent rice crop, 5 q/ha after lucerne and 7 q/ha after berseem. In another study, *Rhizobium* inoculation to cultivated rabi legume fodder crop indicated that the increase in yields of different crops were 10 to 35 per cent and the subsequent Kharif fodder sorghum yield was also increased by 17–33 per cent over control.

Azolla

Irrigated Crop

Aquatic fern *Azolla*, is distributed in both temperate and tropical rice growing regions and fixes atmospheric N in symbiotic association with a heterocystous blue green algae (BGA), *Anabaena azollae*. This association is a live, floating nitrogen factory using energy from photosynthesis to fix atmospheric N amounting to 100–150 kg/ha annually from about 40–60 tonnes of biomass. *Azolla* can be used both as a green manure before transplanting and as a dual crop after transplanting of rice. Both practices are feasible in India, but dual cropping is more practicable, although growth and N fixation of *Azolla* as well as yield of rice are generally higher in green manuring than the dual cropping. Adoption of *Azolla* green manuring is restricted to areas having water available before transplanting.

The studies carried out under AICARP have clearly indicated that incorporation of 5 t/ha of *Azolla* can give yield equivalent to about 30 kg N/ha at several locations such as Kharagpur, Bhubneshwar and Titabar both in Kharif and Rabi seasons. However, in regions with high temperature like Mangalore, Nandyal, Maruteru, Thanjavur and Kraiyiruppu, the benefit from incorporation of 5 t Azolla/ha was less than 30 kg N/ha.

Incorporation on one crop of *Azolla* as a green manure increased the grain yield of rice as obtained with 25–30 kg N/ha as urea whereas two to three crops of *Azolla* increased the yield of rice similar to 60–90 kg N/ha applied as urea. *Azolla* besides supplying N to the current crop also leaves substantial residual effect on the succeeding crop.

The available data indicate that incorporation of *Azolla* can be a very useful supplement to N only under specific soil and environmental conditions where it is possible to multiply *Azolla* without much additional efforts. In areas with very high temperature during Kharif season as in Punjab and Haryana. *Azolla* cannot be multiplied for incorporation at the time of transplanting of rice.

Non-Symbiotic Nitrogen Fixation

Blue Green Algae (BGA)

Irrigated Crops

Blue green algae is a free living N fixer which is distributed worldwide and contributes to the soil fertility in many agricultural ecosystems. The agronomic potential of BGA in rice is established beyond doubt. Judicious use of BGA could provide to the country's entire rice hectarage as much nitrogen as obtained from 15–17 lakh tonnes of urea. Field experiments conducted in India have reported about 14 per cent yield increase over control, which corresponds to 450 kg grain/ha. Studies at CRRI, Cuttack indicated that BGA increased grain yield by 6 to 35 per cent. Field trials conducted at different locations on farmers' fields in Tamil Nadu indicated beneficial effect of BGA inoculation on rice productivity. It is estimated that at farm level. BGA can contribute 20–30 kg N/ha.

Nitrogen from BGA becomes available to the rice crop after their decomposition. Only apart of BGA-N is absorbed by the current rice crop and rest remains in the soil as residual N and repeated algalisation produces a cumulative residual effect. BGA can withstand high temperature and variable moisture regimes better than *Azolla* and hence has better adaptability. Soil and climatic conditions of a region must be primary conditions for selection of BGA for rice crop.

Azotobacter

Irrigated Crops

This is group bacteria which are free living N fixers. The mechanisms by which the plants inoculated with *Azotobacter* derive positive benefits in terms of increased grain, plant biomass and N uptake are attributed to small increase in N input from BNF, development and branching of roots, production of plant growth hormones, enhancement in uptake of NO_3, NH_4, H_2PO_4, K, P and Fe, improved water status of the plants, increased nitrate reductase activity and antifungal compounds.

Plant responses in inoculations with *Azotobacter* are varying. In irrigated wheat, significant response to inoculation was observed in 342 out of 411 trials carried out on farmers' fields and the response ranged from 34 to 247 kg/ha. Application of 5 kg/ha of *Azotobacter* culture by root band method at planting of sugarcane in medium black soil increased the yield of sugarcane at Padegaon in Maharashtra. The culture with 250 kg N/ha gave an economy of 100 kg N/ha by producing cane more or less equal to that with 350 kg N/ha. In Bihar, an economy of 30–35 kg N/ha is obtained with *Azotobacter* application. Higher yields of sugarcane ranging from 9 to 24 per cent were obtained by various *Azotobacter* treatments over control in spaced transplanting technique. Inconsistencies in response to *Azotobacter* are very common and the responses depend on crops, cultivars, locations, seasons, agronomic practices, bacterial strains, level of soil fertility and interaction with native soil microflora. Use of FYM, green manures and other organic amendments enhance the benefits from inoculation.

Treatment of paddy seedlings with *Beijerinckia* along with soil application of 40 kg N/ha in the form as urea resulted in significantly increase in yield over the control, while urea at 40 kg/ha did not give significant increase in yield over the control. They observed that the yield with urea of 40 kg/ha + *Beijerinckia* did not significantly differ from that obtained with urea at 80 kg/ha.

Field experiments are in progress with crops like wheat, rice and vegetable crops onion, brinjal, tomato and cabbage and sugarbeet responded better to *Azotobacter* inoculation than other crops.

Inoculation of sorghum with *Azotobacter* and *Azospirillum* resulted in a marked decline of shootfly (*Atherigona soccata* Rond) damage as compared to control through increased levels of phenol content in the shoots of *Azospirillum* inoculated plants has been implicated.

Responses with inoculations with *Azotobacters* in cereals and non-cereals are often reported in terms of increased yield, and yield contributing factors and disease infestation. Recent reviews have evaluated the worldwide crop responses to inoculation with *Azotobacters* and *Azospirilla*. The results indicated that in many cases inoculations increase plant yields but such increases are variable. Responses were varied with crops, cultivars, locations, seasons, agronomic practices, bacterial strains, level of soil fertility, and interaction with native soil microflora. In an evaluation of the reported worldwide success of *Azotobacter*, it was concluded that statistically significant yield increases were obtained in approximately 60 per cent of the trials (Table 6.3).

Table 6.3: Effect of Seed Inoculation with Selected Strain of *Azotobacter chroococcum* on Different Crops Grown Without Fertilizer Application

Crop/Yield	*Location*	*Control No Nitrogen*	*Azotobacter Alone*	*C.D. at 5%*	*% Increase in Yield*
Sorghum (q/ha)	Pali	12.8	14.0	1.22	9.3
	Dharwar	23.6	32.6	10.56	38.1
Maize (q/ha)	IARI	7.8	13.4	4.80	71.7
	Dharwar	32.0	.43.7	9.90	36.5
Cotton (q/ha)	Surat	1254	1339	241	6.7
	Indore	366	401	104	9.5
	Khandwa	559	708	165	26.6

Multi-locational trials showed that seed inoculation with *A. brasilense* increased the grain yields of pearl millet and sorghum significantly. The yield increase with pearl millet varied from 10 to 17 per cent and with sorghum from 7–31 per cent. In inoculation trials with pearl millet increases in grain yield with inoculation were higher 11 per cent in case of *A. lipoferum* and 8 per cent with *A. chroococcum* over the inoculated controls.

Dryland Crops

Very few experiments have been conducted with *Azotobacter* inoculation in the dryland crops. Nevertheless at other research centres on dryland crops like sorghum, pearl millet and rainfed cotton, *Azotobacter* inoculation led to beneficial effects; response ranging from 15 per cent to 20 per cent. With assured moisture availability. contribution of *Azotobacter* tended to be superior. This contention supports the higher response obtained more often with crops grown in high rainfall areas or under irrigated conditions.

Field trials were conducted (Table 6.4) on pearl millet, jowar, cotton, etc. using *A. chroococcum* cultures. Although phenomenal increase was recorded but there were also failures at some locations. Previously its acceptance was very slow on account of its inconsistent performance. Those failures were due to single factor that one culture was utilized for various crops and it was assumed that it should serve the purpose of every crop at different parts of the country.

Table 6.4: Benefits of *Azotobacter* Inoculation in Some Dryland Crops

Crop	*% Yield Increase Over Control*
Sorghum	15–20
Pearl millet	0–27
Cotton	11–16

Azospirillum

Irrigated Crops

The bacterium colonises root mass and fixes N in loose association with plants. In recent years, there has been considerable interest in the use of *Azospirillum* species as N-fixing bacteria in association with the roots of monocotyledonous plants, particularly grasses and grain crops. Their inoculation has shown positive interaction with applied N in several field crops with an average response equivalent to 15–20 kh/ha of applied N reported that the study carried out at Delhi and Dehradun, inoculation with *Azospirillum* increased the yield of irrigated barely significantly at low or no application of N.

Large increase in ground nut yield due to *Azospirillum* inoculation was reported by Raverker. Significant field responses to *Azospirillum* were also reported in sugarcane, sorghum and wheat. In *Kharif* rice, out of 108 trials carried out on farmers' fields in Tamil Nadu, in 48 trials it was possible to reduce 25 to 50 per cent N from fertiliser source amounting to 25 to 62.5 kg N/ha by inoculation with *Azospirillum.* In Rabi season, such reduction in fertiliser N was possible in 31 out or 55 trials.

The inconsistency in yield responses to *Azospirillum* inoculation is probably the result of ecological and environmental factors. The success of inoculation obviously depends on many factors, including the appropriate choice of carrier and inoculation, the ability of the bacterium to establish itself and to compare with the native microflora, favorable soil chemical and physical conditions, climatic conditions and agricultural practices.

Dryland Crops

Compared to *Rhizobium,* the agronomic significance of non-symboitic N_2-fixing organisms in drylands is less understood. Fewer number of studies conducted with them is a major factor. Since, non-symbiotic organisms essentially depend on soil organic matter serves for energy supply, and majority of the rainfed soils are poor in organic matter, it is only logical to expect low to moderate contribution of N through them. Favourable response was obtained with non-symbiotic bacteria. Inoculation trials with *Azospirillum* showed highly incoherent results. According to Das sorghum and pearl millet responded positively, but during the latter years, the response was not significant (Table 6.5). *Azospirillum* failed to serve pearl millet during 1984–86 and cotton during 1985–87. Method of inoculcation (seed vs. soil) could not alter the response pattern. Beneficial effect of *Azospirillum* on the straw (20 per cent) and grain (15 per cent) yields of pearl millet was found. The bacterium promoted plant growth essentially by phytohormone production rather than via N_2-fixation.

Table 6.5: Effect of *Azospirillum* Inoculation on Grain Yield of Sorghum and Pearl Millet

Treatment	*Grain Yield Per Hectare*	
	Sorghum (730 mm)	*Pearl Millet (590 mm)*
Control	26.7	13.5
Inoculated	31.5	16.6
C.D. (0.05)	4.07	1.38

Coordinated trials under sorghum and millet improvement projects demonstrated beneficial effects of *Azospirillum* inoculation in terms of straw and grain yield at many centres across the country. About one third of the total N application for pearl millet could not be substituted by *Azospirillum.*

Fodder Crops

Wide variation occurs among grasses in nitrogen fixed by the associative organisms, but estimates of about 1.5 g/ha of N per day have been made for *Paspalum notatum* and *Digitatia decumbens. Azospirillum* species have been isolated from soils and roots of wide variety of plants from temperate as well as from tropical regions. Specific associations or preferential stimulations of various nitrogen fixing bacteria in the rhizospheres of different genera/species come to light *viz.*, *Azotobacter paspali* in notatum and *Enterobacter* in maize. The organism *A. lipoferum* was found concentrated in the cortex cells of Pangolagrass. Important nitrogen fixing bacteria observed in the rhizosphere of grasses are listed (Table 6.6). In signal grass (*Bracharia brizantha*), *A. lipoferum* increased the forage yield by 15 per cent over noninoculated situation and a saving of nitrogen fertilisation equivalent to 30 kg N/ha. In a pot culture experiment on red soil (pH 8.1), Pahwa studied the efficacy of *A. lipoferum* strain-1 and *Azotobacter chroococcum* in influencing the herbage yield of different pasture grasses. The yield increases were ranged between 18.4–34.4 per cent with seedling inoculation and 11.4–19.5 per cent with seed inoculation over uninoculated situation. *Azospirillum* was found to have increased efficacy than *Azotobacter.* He further reported that in other field experiment different strains of *Azospirillum* were tried as seedling inoculants of buffel grass (*Cenchrus ciliaris*) and the results indicated that *A. lipoferum* was much more effective in influencing forage yield than *A. brasilense.* Amongst the strains, *A. lepoferum* was the most efficient nitrogen fixer giving a yield increase of 35 per cent over control and crop nitrogen gain by 37 kg.

Table 6.6: Important Nitrogen Fixing Bacteria Observed in the Rhizosphere of Grasses

Bacterium	*Plant*
Azotobacter paspali	*Paspalum noratum*
Azotobacter sp.	*Cynodon dactylon, Zea mays*
Beijerinckia sp.	*Cynodon dactylon*
	Digiraria decumbens, Cyperus and other forage grasses
Derxia sp.	Tropical grasses
Azospirillum sp.	Most tropical forage grasses including maize
Bacillus macerans	*Agrostis renuis, Festuce* sp.
Enterobacter cloacea	
E. aerogenes	*Zea mays, Panicum maximum,*
	Juncus balticus, Agnostis tenuis
	Andropogon geradil, Panicum virgatum
Klebsiella pneumonaiae	*Panicum maximum, Chloris divariecata*
Clostridium	*Digiteria smutzii*

Dr. Pahwa from IGFRI and coworkers reported that pelleting is practised in grass seeds for proper germination and its establishment. Subsequently in a study on red sandy soils at Jhansi, pelleting of grass seeds was done in relation to inoculation with *A. lipoferum*. The bacterial inoculation has helped in obtaining increased herbage yield by 23 to 35 per cent in buffel grass and by 27 to 29 per cent in marvel grass. *A. lipoferum* was found much more effective than *A. brasilense* in its influence on herbage yield of buffel grass. The former increased the forage yield by 33 per cent over control (uninoculation), with a yield advantage equivalent to 40 kg N/ha; whereas, the later one gave a yield advantage of 20 per cent and yield contribution equivalent to 20 kg N/ha. The effect of different non-symbiotic nitrogen fixing bacteria like *Azotobacter* and *Azospirillum* as inoculants was studied in array of forage crops under the aegis of All India Coordinated Project for Research on Forage Crops (AICPRFC) over the years. It was observed that inoculation improved the forage yield in almost all crops. The increase in yields were varied between 6–17 per cent with *Azotobacter* and 4–21 per cent with *Azospirillum* in case of cultivated annual fodder crops; whereas, the increased ranged from 7–15 per cent and 12–17 per cent respectively, in case of perennial crops. In case of pasture grasses, increased yields were ranged from 13–16 per cent and 6–25 per cent with *Azotobacter* and *Azospirillum* inoculation, respectively. The benefit from both these non-symbiotic nitrogen fixers in increased forage yields (10–30 per cent) and saving of nitrogen (20–40 kg/ha) have been reported by many workers.

Phosphate Solubilizers and Fixers

Mycorrihza

Irrigated Crops

Mycorrihza, mutualistic symbioses between plant roots and certain soil fungi, play an unquestionable role in p cycling and in the uptake of phosphate by the plant. Because the known world resources of p could be depleted in a few decades, the contribution of this symbiosis to the reduction of fertiliser requirements is of increasing interest.

Among mycorrihizae, VAM (Vesicular-Arbuscular Mycorrihizae) has good potential in improving nutrients cycling. Field experiments with VAM inoculation are fewer as there is a major constraint due to inability to produce clear, pure inoculum on a large scale as the fungi are obligate symbionts and have to be maintained and multiplied on live plants. Significant yield increases with VAM under field conditions in irrigated environment were reported in chillies and groundnut. In many other studies, percentage increase over control was observed by inoculating VAM as in Soybean (Table 6.7).

Table 6.7: Effect of VAM Inoculation on the Yield of Soybean

Sl.No.	Treatment	Percentage Increase Over Control
1.	Control	–
2.	VAM Inoculation	5.2
3.	Rhizobium Inoculation	11.0
4.	Combined application of Tr. 2 and 3	18.7
5.	VAM + 30 kg P_2O_5/ha	20.6
6.	VAM + 60 kg P_2O_5/ha	25.2

Crop response to VAM inoculation is governed by soil type, host variety, and VAM strains in addition to the biotic and also abiotic factors mentioned earlier. In general, field experiments with VAM inoculation are fewer than those reported for *Rhizobium*, *Azospirillum*, to *Azotobacter*. The major constraint for field trials with VAM has been the inability to produce 'clean pure' inoculum on a large scale as the fungi are obligate symbionts and have to be maintained and multiplied on living host plants.

A recent survey revealed moderate to high mycorrhizal colonisation in grain legumes like pigeonpea and chickpea. Observations in other soils confirmed heavy infections of roots with VAM of many field grown crops. However, field experiments involving VAM inoculation of rainfed crops are fewer in number. Limited studies with sorghum, pigeonpea and groundnut revealed that dual inoculation of VAM + *Rhizobium* or VAM + *Azospirillum* was more advantageous than single inoculation with either organism.

Bhagyaraj reported that VAM fungi can save fertiliser to the tune pf 25–30 per cent besides improving plant yield. Until suitable methods are evolved to multiply VAM on a large scale for inoculation of crops directly sown in the field, the best strategy to utilize VAM fungi for crop normally grown in nursery beds where they can be easily inoculated with selected strains and then transplanted.

Dryland Crops

Vesicular arbuscular mycorrhiza offers promise for augmenting supply of P and micronutrients like Zn for a variety of arable crops. Currently, the field scale application of VAM is limited because of inability to produce this biofertilizer on a large scale. Since, dryland crops as such are widely infected with native VAM, only a few inoculation trials have been conducted thus far to assess its potential under field conditions. With the emphasis of shifting on agroforestry system, VAM better scope in establishment and survival of silvan species. Its value in rehabilitating soils with depth limitations will presumably be far greater because fungus filaments are able to increase the absorbing surface of roots. P solubilisers are yet another group of heterotrophic organisms which may have applicability to rainfed agriculture, particularly that covered by Vertisols having problems of poor P availability.

Fodder Crops

A considerable evidence is available that mycorrhizal associations are of importance to plant nutrition specially on nutrient deficient, soil. Inoculating the host plant with VAM fungi specific for a compatible host soil fungus combination can be led to more economical use of costly phosphatic fertilizer and better exploitation of inexpensive fertilizer sources.

The results of various studies revealed that the mycorrhizal plant possess a greater ability to absorb phosphorus from soil than the control (non mycorrhizal) plant. A number of fodder crops were tested with VAM inoculation which revealed that an increase in fodder yield between 5 to 18 per cent in annual cultivated legumes. 3–5 per cent in annual cultivated cereal fodders, 13–33 per cent in perennial grasses were observed over the non-mycorrhizal plants.

VAM inoculation to perennial grass such as guinea grass was tried at three distinct locations of humid region *viz.*, Jorhat in east, Palampur in north and Vellyani in south which clearly indicated that VAM inoculation had a distinct influence on forage yields, specially in southern region YAM fungi was also compared against non-symbiotic nitrogen fixing bacteria on guinea grass in low fertility (0.3 per cent organic carbon) acid soils of Palarnpur at subtropical region. The data revealed that VAM fungi was much more effective in influencing forage yield to that both of *Azotobacter* and *Azospirillum*. VAM fungi increased green forage by 13 per cent and dry forage yield by 24 per cent as compared to non-symbiotic bacteria, 5–7, and 15–17 per cent for the corresponding yields over control (uninoculated) plots. This indicated a possible saving of 25 kg N/ha due to VAM inoculation, similar observation was also noted on an experiment conducted at Jhansi in low fertile red sandy loam soils. Guinea grass gave 13 per cent green forage yield increase with VAM inoculation over uninoculated situation. In another study at Jhansi VAM inoculation in conjunciton with 40 kg N/ha to forage sorghum gave similar yield level to that obtained with 80 kg N/ha and about 17 per cent yield increase over plots fertilised with 40 kg N/ha. In a trial in an alluvial soil at Ludhiana, VAM (*Glomus facsiculatum*) inoculation gave an increased yields of 32 per cent in multi-cut sorghum (3 cuts). 28 per cent in guinea grass (6 cuts) and 7 per cent in pearl millet (4 cuts); whereas. 8 per cent in single cut maize over respective untreated controls.

The results of field trials conducted in India indicate that VAM inoculations increased yields significantly in around 50 per cent trials and the response varied with soil type, soil fertility and VAM cultures. This scenario is similar to that for *Rhizobium* and *Azospirillum*. In such a situation, until suitable methods are evolved to multiply VAM on a large scale for field inoculation of crops directly sown in the-field, the best strategy to utilize VA mycorrizal fungi for crop production is to concentrate on crops normally grown in nursery beds where they can be easily inoculated with selected strains and then transplanted.

Phosphate Solubilizing Microorganisms

Irrigated Crops

The problem of phosphorus supply is more acute as it is mainly depended on imported raw materials. However, there are substantial deposits of low grade rock phosphate in the country which cannot be made use of for direct-application in neutral to alkaline soils. Integrated use of suitable microbial cultures along with low grade rock phosphate can add about 30–35 kg P_2O_5/ha. To achieve this, it is imperative that efficient PSM are isolated and congenial conditions created for their establishment by devising better inoculation techniques and modifying hostile environment.

Soil bacteria and fungi possess the ability to bring sparingly soluble/insoluble inorganic and/or organic phosphates into soluble forms by secreting organic acids, lowering of soil pH by immobile forms of phosphate. Some of the hydroxyacids may chelate with Ca, Al, Fe and Mg resulting in effective solubilisation of soil P and thereby its higher utilization. Field studies carried out under AICARP have indicated significant effect of inoculation with *Bacillus megaterium* and *B. circulans* in rice and wheat, respectively, which was equivalent to the yield increases obtained with upto 50 kg P_2O_5/ha. Field trials conducted during the 60's, showed significant increase in yield in case of rice, wheat, chickpea, pigeonpea, soyabean, groundnut and berseem. In multilocational trials conducted with different crops, increased yields ranging from 0–50 per cent were obtained due to inoculation with a culture of PSB with or without rock phosphate addition. Potato yields were increased dramatically (50–60 per cent) due to inoculation with *Aspergillus awamori, P. striata* and *B. polymyxa* cultures. The residual effect of inoculation was also observed on the succeeding crops. Soil inoculation of *P. striata* besides increasing grain yield of wheat and pea showed residual effect by increasing the grain yield of maize crop in alluvial soil.

Factors Affecting Crop Response to Biofertilizers

There are several factors which influence the performance of biofertilizer to crop. Some of the important ones are enumerated here.

Interaction Effect of Microbial Strains

There seems to occur a synergistic effect of one bacterial strain on the other leading to better modulation, higher yield and better crude protein content of the host plant. While studying the effect of multi-microbial inoculation on shrubs and forage trees favourable response was observed. Inoculant containing multi-strains of *Rhizobium loti* (Tal 82, Tal-582 and LL 28-2) gave better performance in terms of herbage yield and crude protein content than any of these pure strains. Similarly, multi-microbial inoculation involving either *Rhizobium* and VAM, *Rhizobium* and *Azotobacter chroococcum* or *Rhizobium* and *Azospirillum*, gave superior performance than their pure strains. Several other studies lend support to such a positive interaction between different strains leading to crop yield.

Effect of Nutrient Interactions

Inoculants being living organisms do require some nutrition for their normal growth and development and effective activity. The symbiotic as well as non-symbiotic N fixing bacteria, in general, perform well if they are provided with nominal initial N fertilisation through the host crop (20–40 kg N/ha), which serves as the starter dose. Similarly, there seems to be a positive interaction of zinc, molybdenum or sulphur fertility on the growth and performance of *Rhizobium* and *Azotobacter.* Application of 20 kg S/ha had a stimulating effect on nodulation and forage yield of berseem (26 per cent) and peas (31 per cent). Application of 10 kg ZnSO/ha exhibited favorable effect on the growth and activity of *Azotobacter* leading to significant increase in the yield of pearl millet, than their sole application. Similarly, an increased forage yield by 75 per cent and enhanced nodulation following the foliar application of 1.5 MoO_4/ha.

Beneficial effect of application of *Azotobacter* @ 5 kg/ha in association with nitrogen application was reported on the yield of sugarcane in Maharashtra. Application of *Azotobacter* by root band method of planting of sugarcane in medium black soil along with 250 kg N/ha resulted in economy of 100 kg N/ha by producing sugarcane more than that obtained with 350 kg N/ha.

Dryland Crops

Performance of P solubilising bacteria has not been investigated exhaustively in drylands. Being heterotrophic in nature, their dependence on organic matter for energy supply is a likely constraint for their optimum activity in rainfed soils.

Fodder Crops

Another important area which has been less researched upon in the exploitation of beneficial role of soil microorganisms responsible for transformation of soil phosphorus and solubilisation of less soluble applied phosphatic fertiliser like rock phosphate. Most frequently the isolated species are *Pseudomonas* and *Bacillus megaterium var phosphaticum* are widely used in many other crop plants as bacterial inoculant known as phosphobacterin. Limited work has been done at Indian Grassland and Fodder Research Institute (IGFR) in this direction. Inoculation with PSB improved the efficacy of less soluble rock phosphate like Mussoorie, Udaipur, Laccadive and Vishakhapatnam. Phosphobacterised seeds resulted in an increased yield of oats by 31 per cent. Similarly, in another study, it was observed that bacterised seeds with *Bacillus polymyxa* H6 gave higher yield than non-bacterised seed. Mussoorie rockphosphate-phosphate solubilising culture (*B. polymyxa*) gave highest forage yield.

Methods of Inoculation

The generally followed seed inoculation methods has often been found wanting to deliver adequate numbers of rhizobia into the rhizosphere. In soils where surface layer dries rapidly, this practice could lead to poor survival of the bacterium. Soil inoculation in liquid or slurry seed treatment of groundnut which is commonly dressed with fungicides. Using pelleted or granular inocula has also been suggested for rapidly drying soils of the tropics. However, these methods have not been widely evaluated in India. As an indigenous alternative, another method has been proposed in that the use of baked clay granules soaked with rhlzobial cells as inocula for rainfed crops. Issues concerning inoculation techniques are highly relevant for drylands when survival and multiplication of the inocula is always at stake. The details of the application techniques have been discussed later.

Other Factors

Besides the foregoing specific variables, the competitive ability of the introduced strain to establish against the native rhizobia is an important factor that influences the performance of BF under all situations including drylands. Very few studies have been conducted in the past on this aspect using marker strains. Experiments on chickpea and groundnut revealed that only 20 per cent to 30 per cent of the nodules were formed by the introduced strain. Long-term basic research is necessary to understand the host bacterial relationship and achieving higher nodule occupacy by the introduced strain. This may, perhaps, be achieved by developing specific strain-host combinations that exclude all other organisms.

Under field conditions the above variables interact in a complex mode which makes it often difficult to isolate individual constraints that regulate the response level. Recently, several models have been proposed. A modelling approach using soil rhizobial population and mineral N levels to predict the response of a crop to inoculation at a given location. Such approaches can be practical for delinating the areas that might respond to inoculation. Further majority of the inoculation trials conducted thus far were on the research stations where soil fertility characteristics and native rhizobia population have hardly any resemblance to those encountered in farmers fields which suffer from far wider constraints. Therefore, to clearly spot out the agents that favour superior inoculation response, emphasis of research should shift to farmers' fields.

With regard to the non-symbiotic organisms, soil organic matter levels and the energy availability in the rhizosphere through root exudates remain the major limiting factor. Therefore, an integrated approach where soil organic matter reserves are improved through organics would be an essential strategy in order to derive optimum benefit from the non-symbiotic organisms.

Host Response to Biofertilizers

Plant is a major controlling factor in inducing and sustaining microbial activity through over changing quality and quantity of root exudates. The earlier concept har nitrogen fixing power of rhizosphere is limited, is fast changing; it may be capable of allocating large amounts of polysaccharides with a high C/N ratio to the nitrogen fixing bacteria. This is further corrobrated by the increasing numbers of observations on substantial mucigel formation on roots of various plants. Balandreau and Fares Hamad estimated energy allocation by rice roots to fix atmospheric nitrogen to be around 23 per cent of the carbon fixed due to photosynthesis which includes carbohydrates only amino acids and sloughed off cells. Similarly a single wheat plant releases over a hundred organic acids and compounds excluding in the root zone. The wheat plants excrete about 19 to 25 per cent of its photosynthetes into the soil through roots. Glucose and fructose are the main exudates. Samtsevish estimated root cap cell secretions of about 700 m^3 per hectare per annum in winter wheat weighing at least 7 metric tonnes. The mucigel at root tip cells largely containing highly hydrated carbohydrates. With such a large amount of carbohydrates released by wheat plants, it is reasonable to conceive that the threshold concentration of NH_4-N at which sphere, may be much higher. The large amounts and a variety of carbohydrates released by wheat roots thus have potential to support large number of bacteria which can fix nitrogen and solubilise phosphorus and their other activity.

The host also plays an important role in depression of nif genes in bacteria. Wheat and sorghum were shown to induce nitrogenase activity both in *Azospirillum* and *Rhizobium*. The activity obtained by both plant species with *Azospirillum*, however, was 100 times higher than with *Rhizobium* as observed by these workers. They and Kavimandan accepted that in presence of more than one readily respirable substances-pentose sugar, Na glutamate and low PO_2, many strains of *Rhizobium* can reduce nitrogen to ammonia in free living conditions. The nitrogen fixed in the root region of crop plants may not be readily available to plants alone-the non-nitrogen fixing organism may also compete for such fixed N excreted by or released due to cell autolysis into the environment. Amongst the free living diazotrophs *Azotobacter* and *Azospirillum* may excrete about 15 per cent of the fixed nitrogen into the medium. *Rhizobium* on the other hand, can release upto 94 per cent of the fixed nitrogen; this is possible because with derepression of nif genes, there is considerable decrease in NH_3 assimilating enzyme activity in *Rhizobium.*

Isotope dilution technique was used to quantify the amount of N derived from atmosphere in wheat-Rhizobium system. About 10–50 Kg N per hectare was found to derive from atmosphere in wheat cultivar HD 2285 inoculated with *R. leguminosarum* and *R. trifolii*. He has further confirmed rhizobial contribution of atmospheric N by growing wheat HD 2270 in sterile sand and sonalika in unsterile soil, supplemented with 15 N urea.

Organic amendments and crop residue management can possibly be used profitably in conjunction with biofertilizers particularly with assured irrigation facilities. Such usage of organo-biofertilizers can reduce amount of inorganic N to be applied as well as problems of nitrate pollution of underground water. In rice-wheat sequence, 5 tonnes wheat/rice straw/FYM were incorporated prior to transplanting/sowing so that the period of maximum immobilization of nutrients does not coincide with increasing needs of mineral nutrients of plants. Algal inoculation for rice and *Azotobacter* inoculation for wheat were part of these treatments. Application of crop residues, biofertilizers and reduced levels of N had favourable effect on yield of these two important crops. Frequent additions of decomposable organic matter (About 5 tonnes per ha.) appears to be useful in enhancing heterotropic N_2 fixation in root regions of crops. In order to harness the maximum benefit from biofertilizers use of inexpensive carbonaceous materials as amendment to carrier material with a view to provide additional energy source to bacteria prior to seed inoculation and enrichment of carrier with root extract/exudates to enhance compatibility of diazotorpoh with crop cultivar. IAA, GA, 6-Benzyl-adenine the last one has resulted in 10 fold increase in nitrogenase activity in Petunia *Rhizobium* system and use of biofertilizer in conjunction with plant hormones may prove to be useful.

Interaction of Inoculants with Other Nutrients

The inoculants as a living organism require some amount of nutrition of their effective growth and performance. The non-symbiotic nitrogen fixing bacteria in general, performed well under low fertile soils or through nominal initial nitrogen fertilisation to crops (20–40 kg N/ha.). The affectivity of these organisms were less conspicuous in high fertility soils or higher level or fertilisation. This is the reason that the real benefit of the application of non-symbiotic nitrogen fixing bacteria has a well deserved place to pasture grasses and to grasslands where less or no fertilisation is practised. However, this is not a situation with some other Biofertilizers like *Rhizobium* and VAM and also to some extent to phosphate solubilising bacteria.

Shree Kumar in a significant study reported that a combined application of *Azospirillum brasilense* and VAM alongwith 50 kg N/ha gave 102 per cent higher forage yield over absolute control and even marginally higher yield (99 per cent) than with similar combination plus an additional application of 50 kg P_2O_5/ha (Table 6.8). Application of 50 kg N+50 kg P_2O_5/ha gave only 87 per cent yield increase over control. This signifies that a combined application of *A. brasilense* + VAM alongwith 50 kg N/ha could effected in a saving of 50 kg P_2O_5/ha from application to forage sorghum crop. In an experiment to find out optimum level of Zn, Mo and S under rhizobial inoculated crops in red soil (pH 8.2). Pahawa observed that an application of 20 kg S/ha to berseem had a stimulating effect on nodulation and an increased forage yield of berseem by 26 and 31 per cent in pea over non-inoculated situation. Application of 20 kg $ZnSO_4$/ha had resulted in an increased yield of lucme. Hazara reported that a combined application of $ZnSO_4$ @ 10 kg/ha and *Azotobacter* significantly increased forage yield of pearl millet than their sole application. Similarly, *Azotobacter* inoculation alongwith 20 kg $ZnSO_4$ha increased forage yield of oats (14 per cent) over their independent applications. Application of 2 kg

MoO_4 ha was required in enhancing nodulation forage legumes which reflected in enhanced yields (20–30 per cent). In another study, Hazara recorded and increased forage yield by 75 per cent over control due to an application of 1.5 kg MoO_4/ha, specially foliar application. Pahwa indicated that rock pyrite, which was otherwise less effective in red soils, had shown increased efficacy and forage yield increase in lucerne with *Rhizobium* inoculation. Rock pyrites application @ 5 t/ha inoculated with *Rhizobium* gave highest forage yield.

Table 6.8: Effect of Combined Application of *Azospirillum brasiliense* and Vesicular-Arbuscular Mycorrhizal (VAM) Fungus (*Glomus fasciculatum*) on the Forage Yield of Sorghum

Treatment	*Forage Yield (q/ha)*	*Per cent Increase in Yield Control*
Control	65.7	
50 kg N/ha	118.5	80
50 kg P_2O_5/ha	78.5	19
50 kg N and 50 kg P_2O_5/ha	122.7	87
A. brasilense alone	80.5	22
VAM alone	75.8	15
A. brasilense + VAM	92.2	41
50 kg N/ha + *A. brasilense* + VAM	132.7	102
50 kg P_2O_5/ha + *A. brasilense* + VAM	98.7	50
50 kg N and 50 kg P_2O_5/ha + *A. brasilense* + VAM	130.8	99

It is thus evident from the foregoing studies that an appropriate and adequate nutrition is required for the effective functioning of the inoculants being used as bio-fertiliser to various forages. Addition of these nutrients, especially micronutrients, not only help as a plant nutrition but their interaction effects were also observed when applied in conjunction with microbial inoculants.

Multi Microbial Inoclation

In a study on various inoculants on shrubs and trees of forage value (Pahwa 1990), studied the effect of different strains of *Rhizobium loti viz.*, LL-28-2, TAL-82 and TAL-582; VAM fungi (*Glomus fasciculatum*); phosphate solubilising bacterium (*Bacillus polymyxa* H5); phosphate solubilising fungi (*Aspergillus awamoori*) and non-symbiotic nitrogen fixing organism like *Azotobacter chroococcuin* strain ICM-2001 on subabool (*Leucaena leucocephala*), seedling (var. Silvi-4). It was observed that the associative effect of *B. polymyxxa* H5, *A. awamoori*, *R. loti*, LL-28-2. VAM fungi were found to be more effective in terms of nodulation, herbage yield and nitrogen content than *Rhizobium* alone. Inoculant containing multi-strains of *Rhizobium* (TAL-82, TAL-582 and LL-28-2) gave superior performance in terms of herbage yield and crude protein content than any of these pure strains. Inoculation of *Rhizobium* in presence of VAM was found to produce higher nodulation of Rhizobium in presence of VAM was found to produce higher nodulation and forage yields than *Rhizobium* or VAM alone. It was also interesting to note that inoculation with *Azotobacter chroococcum* alongwith *Rhizobium* did have added advantage in terms of forage yield and nitrogen content of the plants than pure *Rhizobium* inoculation. He further reported that in red soils of Jhansi, it was observed that dual inoculation of *Rhizobium* + VAM gave an increased average yield of seven forages to the extent of 31 per cent over uninoculated control against 15 per cent from VAM and 19 per cent from *Rhizobium* alone. In another study in acid soils of Vellayani, combined application of *Rhizobium* + VAM in stylos (*Stylosanthes hamata*) grown as an intercrop in tapioca gave 59 per cent increased green forage and 77 per cent dry forage yields as compared to yield advantages of 33 and 37 per cent with *Rhizobium* and 34 and 53 per cent with VAM respectively, over uninoculated control (Table 6.9).

Table 6.9: Effect of Inoculation of *Rhizobium* and VAM Fungi to *Stylosanthes hamata* Grown as Intercrop in Tapioca

Inoculation	*Yield (q/ha)*		*Per cent Increase in Yield Over No Inoculation*	
	Green Forage	*Dry Forage*	*Green Forage*	*Dry Forage*
VAM	52.6	23.5	34	53
Rhizobium	52.3	21.1	33	37
VAM + *Rhizobium*	62.4	27.2	59	77
Control (no inoculation)	39.2	15.4	–	–
CD 5 per cent	6.35	2.57	–	–

Tilak while working on *Rhizobium* inoculation to Rabi legume forages *viz.*, Indian clover and lucerne tried to find out the possible saving of initial nitrogen application to the crops with the help of combined application of *Rhizobium* and *Azospirillum brasilense* (Table 6.10). The study indicated that a combined application of *Rhizobium* and *A. brasilense* could possibly save 25 kg/ha of initial nitrogen application to the crops.

Table 6.10: Effect of *Rhizobium, Azospirillum* and its Combined Application to Legume Forages

Treatment	*White Clover*		*Lucerne*	
	Green Forage Yield (q/ha)	*Per cent Increase in Yield Over Control*	*Green Forage Yield (q/ha)*	*Per cent Increase in Yield Over Control*
Uninoculated control	308	–	255	–
25 kg N/ha	385	25	292	14
Rhizobium	370	20	325	27
Azospirillum brasilense	302	2	248	3
Rhizobium + 20 kg N/ha	450	46	350	37
A. brasilense + 25 kg N/ha	380	23	290	14
Rhizobium + *A. brasilense*	405	31	355	39
Rhizobium + *A. brasilense* + 25 kg N/ha	455	48	385	51
CD at 5 per cent	52		38	

Compatibility Between Biofertilizers and Chemical Fertilizers

During the last three decades world agricultural witness tremendous growth and high production. In India the yearly production of grains increased from 5 crore tonnes in 1950 to 20 crore tonnes in 1995. The credit goes to agro chemicals particularly chemical fertilizers. This period is also called chemical dominance period. However, the intensive use of chemical fertilizers caused underground water pollution and soil and environmental pollution. The microbial population decreased significantly. The soil became sick and yields have gone down. Besides the cost of agrochemicals increased tremendous and also unavailability in time of need. However, the production level has to be maintained and also increased to feed our growing population. The scientists engaged in agriculture diverted their attention and started exploring the biofertilizing about such as *Rhizobium, Azotobacter, Azospirillum* and phosphate solubilizing microorganisms (PSM). Their role in crop fertility and productivity was known earlier but their utilization in crop production was limited. This has to be borne in mind that neither biofertilizing agents nor chemical fertilizers alone can solve the food problem. They have to integrate/blend each others to achieve higher productivity without polluting water, soil and environment. Experiments are being conducted in J.N. Agricultural University, Indore Campus to find out best and economic ratios of biofertilizers and chemical fertilizers which give a significant increase in yields.

Experiments in cotton seed inoculation indicate that 75 gm *Azotobacter* culture was used for 15 kg of seed; the culture was dissolved in about 2 litre of water and paste is formed. The seeds were mixed in this paste by hands. The care should be taken that a thin film is formed of *Azotobacter* paste on the seed and then dried in shade and planted. Before inoculating the seeds, they were treated with fungicidals either with Thiram or Banstr or Derosal for protection against seed borne fungal pathogens. The seed inoculated plots gave an average yield of 1700 kg as compared to control (1000 kg). In these experiments nitrogen and phosphorus were applied at half the recommended. *Azotobacter* besides fixation of nitrogen also releases antibiotics and hormones which protects plants from diseases and increases growth respectively.

Studies carried out at College of Agriculture, Indore on the blending of FYM with Biofertilizers indicate synergistic effect of income in soyabean (Table 6.11).

Table 6.11: Economics of Blending FYM with Biofertilizer in Soyabean

Treatment	*Yield in Quintal/ha*	*Net Return/ha in Rs.*
Check	10	8000
FYM[2]	13	104000
Biofertilizer[1]	12	9600
FYM + Biofertilizer	15	72000

[1] Biofertilizer was *Rhizobium* (59/kg of seed treatment) and phosphate solubilizing bacteria (5 kg/ha soil applications)

[2] FYM at the rate of Rs. 25 per ha.

The advantages of arbuscular mycorrhizal fungi (AMP) in crop productivity raises concern for their survival and preservation in soils. The conservation in soils covers many aspects of agricultural crop production and management programmes. Some of these concerns include preventing reduction of infection through excessive use of fertilizers, high levels of phosphorus and nitrogen, inhibition of infection and reduced population caused by pesticides and agrochemicals. Some chemical treatments are of particular interest because of their highly toxic effects on non-target organisms such as AMF, but other can actually enhance colonization and sporulation. Vyas have reviewed the subject on the influence of pesticides on the AMF and their formation.

The cultures of *Glomus* spp. were maintained on maize (*Zea mays*) for the studies. These consist of spores and chlamydospores in root fragments and the predominant species were *Glomus fasciculatum*, *Gigaspora marginata*, and *Acaulospore* sp. The composite culture of these AM fungi was used. Garden soil containing 25 to 30 per cent sand and 5 per cent organic matter, low in phosphorus was steam-pasteurized and mixed with a pot culture inoculum of 2500 spores per 20 cm clay pot for the experiment. Root fragments of maize containing spores and chlamydo-spores were mixed with soil and then pesticides with their recommended rates for the control of insects, weeds, and fungi were applied as powder during the mixing of soil. These treated and inoculated soils were put in pots: three replications maintained, and three plants were kept per pot. All the pesticides were applied once to the potting soil. Respective controls were maintained for the comparison of the treatments. Soyabean seeds of variety JS-72-44, treated with thiram (0.3 per cent) and inoculated with *Rhizobium japonicum* (10g per kg of seed), were planted in the clay pots and fertilized. At harvest, plant highest and combined root and shoot weight were recorded. The chlamydospores number was determined by wet sieving 25g of soil from each pot at the termination of the tests and was done as recommended by Gerdemann. The mycorrhizal infection of the root was checked as per Phillips and Hayman. Root samples for each treatment were pleached with 25 ml of 10 per cent (w/v) aqueous KOH at 80°C for one hour. Then the root pieces were stained for three hours with 0.05 (w/v) acid fuschin dissolved in lactophenol solution. The data were statistically analysed using Duncan's multiple range tests.

The results of the study are presented in Tables 6.12–6.14. The data on interaction between *Glomus* spp. and various systemic fungicides-carbendazim, metalaxyl and fosetyl-Al, and non-systemic fungicides-thiram and mancozeb, are presented in Table 6.17. Carbendazim, fosetyl-Al, and mancozeb possess a significant growth promoting effect, which is evident from the data.

Table 6.12: Growth of Soyabean, Mycorrhizal Infection, and Chlamydospore Production in Inoculated Soil with *Glomus* spp. and Treated with Fungicides

Fungicides	*Rate/ha(kg)*	*Plant*		*Mycorrhizal Infection*	*Chlamydospore/ 25 g*
		Height (cm)	*Weight (g)*		
Thiram (Thiride)	15.50	40.00d	25.00c	50c	177b
Carbendazim (Bavistin)	10.00	45.00a	35.00a	65a	180b
Triadimefon (Bayieton)	10.00	28.00d	30.00b	10d	62c
Metalaxyl (Ridomil)	25.00	35.00c	28.00b	65b	185b
Fosetyl-Al (Aliette)	25.00	40.00b	32.00a	80a	251a
Mancozeb	25.00	38.00b	26.00b	53c	170b
Control 1	–	24.00d	15.00d	80a	185b
Control 2	–	20.00e	10.00e	–	

Note: The data are means of three replicates, the application rates of fungicides are field rates used for control of soil borne plant pathogens; numbers not followed by the same letter are significantly different at P–0.05, in Duncan's multiple range tests; Control 1 is inoculated but not treated and Control 2 is uninoculated and untreated.

Similarly they significantly enhanced both mycorrhizal infection (MI) and the chlamydospores number. Hence, there is some relationship between the healthy growth of plants and higher mycorrhizal activity. Triadimefon, a fungicide, inhibits sterol biosynthesis in plant and fungi and thus results in stunting plant growth. This also adversely affects the growth of mycorrhiza. By and large, the application of fungicides enhanced the growth of MI and chlamydospore number. The results also indicate that metalaxyl and fosetyl-Al have stimulatory effect on mycorrhizal development after soil application. This should be exploited for harvesting more mycorrhizal spores.

Among the pesticides, insecticides were least toxic to MI and chlamydospores production. The data indicate that a definite trend has been established based on their toxicity toward AMF. The descending order of their toxicity toward mycorrhiza is: thimet > carbaryl > methyl-parathion > Chlorophyriphos > quinolphos (Table 6.13). Herbicides were found to be more toxic than the other pesticides. The MI and chlamydospores number is significantly reduced in various treatments. This cautions us their use should be judicious to harness the benefits of mycorrhiza in crop productivity (Table 6.14).

Table 6.13: Growth of Soyabean, Mycorrhizal Infection, and Chlamydospore Production in Inoculated Soil with *Glomus* spp. and Treated with Insecticides

Insecticides	*Rate/ha(kg)*	*Plant*		*Mycorrhizal Infection*	*Chlamydospore/ 25 g*
		Height (cm)	*Weight (g)*		
Quintolphos Gr (Ekalux)	12	35.00b	25.00d	80a	86b
Carbaryl (Sevin)	30	45.00a	36.00a	68b	90b
Thimet (Phorate Gr)	12	30.00c	29.00c	65b	85b
Methyl Parthion (Folidol)	30	35.00b	32.00b	70b	90b
Chlorophyriphos (Ruben)	1.5lt	40.00a	30.00b	75a	90b
Control 1	–	25.00d	18.00e	80a	170b
Control 2	–	20.00	15.00e	–	–

Note: The data are means of three replicates, the application rates of Insecticides are field rates used for control of soil borne plant pathogens; numbers not followed by the same letter are significantly different at P–0.05, in Duncan's multiple range tests; Control 1 is inoculated but not treated and Control 2 is uninoculated and untreated.

Table 6.14: Growth of Soyabean, Mycorrhizal Infection, and Chlamydospore Production in Inoculated Soil with *Glomus* spp. and Treated with Herbicides

Herbicides	*Rate/ha(kg)*	*Plant*		*Mycorrhizal Infection*	*Chlamydospore/ 25 g*
		Height (cm)	*Weight (g)*		
Fuchloralin (Basalin)	Pre-pl(1 Kg.)	40.00a	25.00a	30b	50c
Metolachlor(Dual)	Pre-E(1 Kg.)	35.00b	20.00b	20e	40d
Metribuzin (Sencor)	Pre-E(0.5 Kg.)	38.00a	21.00b	20e	60b
Alachlor(Lasso)	Pre-E(2 Kg.)	40.00a	27.00a	50b	50c
Nitrofen (Tok)	Pre-E(2 Kg.)	30.50b	20.00b	25c	40d
Control 1	–	22.00c	72.00c	80a	175a
Control 2	–	20.00c	10.00c	–	–

Note: The data are means of three replicates, the application rates of herbicides are field rates used for control of weeds; numbers not followed by the same letter are significantly different at P–0.05, in Duncan's multiple range tests; Control 1 is inoculated but not treated and Control 2 is uninoculated and untreated.

This study concludes that as the pesticides and mycorrhizal interactions studied through this technique showed varied effects, the compatibility between these should be pre-determined. Among the pesticides, herbicides are more toxic towards MI. Hence, their use should be restricted and judicious. A similar study should also be conducted under different soil, variety cropping pattern and climatic conditions to understand the pattern of toxicity, as pesticides and biofertilizing agents such as AMF are indispensable in the present day commercial farming system. Those that show toxicity towards AMF should be either discarded or should be used judiciously.

Several studies indicate that VAM and ectomycorrhizal (ETM) fungi influence plant disease caused by root rotting pathogens. *S. rolfsii* and *R. bataticola* are important soil-borne plant pathogens affecting a wide range of host plants. Soyabean (*Glycine max*) is being cultivated in Malwa since last 15 years intensively as such the pathogen, *S. rolfsii* is so well established that the successful cultivation of this crop will not be possible in near future. Similarly, *R. bataticola* is very serious pathogen of *Leucaelw* sp. Hence, study was undertaken to study the interaction of mycorrhizal fungi and plant pathogens in soyabean and Leucaena in the University. The soil used in the experiments was a phosphorus deficient sandy loam of pH 6.0. Soil was autoclaved at 15 lb pressure for 1 hr. The pots (25 cm) were filled with the autoclaved soil and five seeds of soyabean and Leucaena were planted; the pot soil was inoculated with either pathogen and mycorrhizal fungi, or both. *Glomus* spp. maintained on maize roots which were used as the inoculum. A 25-ml inoculum containing about 250 chlamydospores and infected root pieces was added around the seeds. *S. rolfsii* was grown separately in 30 ml Erlenmeyer flasks for 15 days at 30°C on a shaker in potato dextrose agar. One ml of the culture containing mycelia and young developing sclerotia were used for inoculation around the seeds. There were five replications for each inoculation treatment sm., (*i*) Control (no inoculation); (*ii*) VAM alone; (*iii*) pathogen alone and (*iv*) VAM and pathogen. Plants were harvested after 100 days. Root and shoot dry weight were recorded. Mycorrhizal infection of the root and spore count in root zone soil were estimated.

Studies on the ETM role in the control of dry root rot (DRR) in *Leucaena leucocephale* was carried out in a similar manner. Ectomycorrhizal fungi can be grown in the laboratory in culture whereas VAM are obligate parasite and they cannot be cultivated in the laboratory on artificial medium; hence the culture of ETM was substituted in the second experiment instead of VAM and root pieces. Mortality growth, and ETM development of Leucaena was studied.

The data presented in Table 6.15 indicate that the maximum mycorrhizal infection occurred in plants inoculated with the *Glomus* spp. The presence of *Sclerotium* reduced the mycorrhizal infection of the root. Addition of mycorrhizae were added and not the pathogenic culture, the mortality was nil. Plant treated with only mycorrhizal fungus produced the highest root and shoot dry weight (Table 6.16). Plant inoculated with only pathogen made poor growth and such plant produced the least root and shoot weight.

The study demonstrated that infection by the pathogen, *Sclerotium rolfsii* reduces growth and addition of mycorrhizal fungus together with the pathogens in soyabean nullified the effect of pathogen. Similar observations were made by Davis. It is well known that mycorrhizal fungus augment P nutrition of plants and this may playa role in reduction in severity of the disease. Inoculation with mycorrhizal fungus brings about anatomical and physiological changes in the plants. Histochemical studies have shown that mycorrhizal roots contain more phenols that comparable non-mycorrhizal roots. Phenol deposition in mycorrhizal tomatoes has been reported to impede the infection and spread of *Fusarium* reported that collar rot caused by *Sclerotium rolfsii* was controlled in peanut by the VAM fungus

Glomus fasciculatum. The present study establish that VAM could reduce the severity of disease caused by the fungi in crop plants.

Table 6.15: Percentage Mycorrhizal Infection, Number of Extramatrical Spores of VAM Fungi in Soil and Sclerotial Bodies of the Pathogen as Influenced by Interaction Between *Glomus* spp. and *Sclerotium rolfsii* in Soyabean

Treatment	*Percentage Mycorrhizal Root Infection*	*Mycorrhizal Spores per 100 ml of Soil*	*Sclerotial bodies per 100 ml of Soil*
Control	Oc	Oc	Oc
Sclerotium rolfsii (*SR*)	Oc	Oc	56a
SR + *Glomus* spp.	80b	250b	26b
Glomus spp.	91a	289a	Oc

Note: The data are mean of five replications; Means followed by the same letters in rows are not significantly different (P = 0.05) from each other.

Table 6.16: Root and Shoot Dry Weight or Soyabean Plants as Influenced by Interaction of *Glomus* sp. and *Sclerotium rolfsii* and *Rhizoctionia bataticola*

Treatment	*Dry Weight per Plant (g)*	
	Root	*Shoot*
Control	60c	20b
Sclerotium rolfsii (*SR*)	30d	10c
Glomus spp.	69b	25c
Glomus spp. + *SR*	90a	40a

Note: Plants were harvested after 100 days for root and shoot weights after inoculation; are the means of 10 seedlings. Means followed by the same letter within a column for a particular parameters are not significantly different (P = 0.05) from each other.

Table 6.17: Seedling Mortality, Growth and Ectomycorrhizal Development of *Leucaena leuco cephale* Inoculated with *Paxillus involutus* and *Rhizoctonia bataticola* in Pot Culture

Treatment	*Mortality (Percentage)*	*Shoot Height (cm)*	*Root Length (cm)*	*Total Dry wt (mg)*	*Short Root*	
					Mycol[1]	*Non Mycol*[2]
Control	0	30b	25b	25b	0.0	55a
Paxillus involutus (PI)	0	30b	27b	30b	6a	40b
Rhizoctnoia bataticola + *Sclerotium rolfsii* (SR)	20	10c	9c	10c	0.0	15d
PI + RB + SR	5	40a	30a	60a	4a	33c

Note: Seedlings were harvested eight weeks of inoculation: values are the means of 10 seedlings. Means followed by the same letters within a column for a particular parameter are not significantly different (P = 0.05) from each other. 1 = Mycorrhizal and 2 = Nonmycorrhiza.

The results of the studies with leuacena and ETM *P. tinctorius* and *Paxillus involutus* and pathogen *Rhizoctonia bataticola* indicate that the application of ETM reduced the DRR significantly when the soil inoculated with the fungus and ETM. Shoot height, and root and shoot dry weight were significantly higher as compared to control (Tables 6.17 and 6.18). Mycorrhizal colonization was significantly higher as compared to control. The possible use of ETM fungi in the biological control of pathogenic fungi has been discussed.

Table 6.18: Seedling Mortality, Growth and Ectomycorrhizal Development of *Leucaena leucocephale* Inoculated with *Picolithus tinctorius* and *Rhizoctonia bataticola* and *Sclerotium rolfsii* in Pot Culture

Treatment	*Mortality (Percentage)*	*Shoot Height (cm)*	*Root Length (cm)*	*Total Dry wt (mg)*	*Short Root*	
					Mycol[1]	*Non Mycol*[2]
Control	0	37b	26b	30b	0.0	60a
Picolithus tinctorius (PT)	0	35b	29c	30b	8a	40
Rhizoctnia bataticola (*RB*) + *Sclerotium rolfsii* (SR)	25	20c	18	10	0.0	16d
PT + RB + SR	5	45a	40a	67a	6a	36c

Note: Seedlings were harvested eight weeks of inoculation: values are the means of 10 seedlings. Means followed by the same letters within a column for a particular parameter are not significantly different (P = 0.05) from each other. 1 = Mycorrhizal and 2 = Nonmycorrhiza.

Studies were also conducted on the effect of seed treatment with fungicide metalaxy fosetyl-Al indicated that they are compatible with *Rhizobium* and *Glomus* sp. (Tables 6.19 and 6.20). These treatments increased various growth promoting parameter, such as dry weight, seed weight and yield. The perusal of data indicated that mycorrhizal infection, nodule number and nodule weight were also increased. The yield increase was significant over treatments.

Table 6.19: Effect of Metalaxyl Seed Treatment and Foliar Application of Mycorrhizal Infection, Root Nodulation and Plant Growth of Soyabean after dual Inoculation with *Golmus* spp. and *Rhizobium japonicum*

Treatment	*Germination*	*Mycorrhizal Infection (%)*	*Nodule Number*	*Nodule Weight (g)*	*Plant Height (cm)*	*Plant Dry Weight*
						Root
Uninoculated	53c	70c	5c	1.30c	24.73c	3.25b
Rhizobium japorncum	60b	14c	22c	4.15b	27.05b	3.11b
Glomus spp.	61b	60b	7c	1.61c	26.55b	5.11a
R. japonicum + *Glomus* spp.	68a	72a	36a	5.00a	28.85a	5.76a

*: Data are mean of three replications; nodule count was made after 21 days of germination whereas mycorrhizal infection was recorded after 55 days.

**: Number not followed by the same letter are significantly different at the P = 0.05 in Duncan's multiple range tests.

***: Seed treatment at the rate of 0.2 mg a.i. per kg was given at the time of sowing after inoculation whereas foliar application was given in 45 days old seedlings.

Table 6.20: Effect of Seed Treatment and Foliar Application of Fosetyl-Al on Mycorrhizal Infection, Root Nodulation and Plant Growth of Soyabean after dual Inoculation with *Golmus* spp. and *Rhizobium japonicum*

Treatment	Germination	Mycorrhizal Infection (%)	Nodule Number	Nodule Weight (g)	Plant Height (cm)	Plant Dry Weight
						Root
Uninoculated	61c	10c	8c	0.9c	25.75c	3.25b
Rhizobium japonicum	68	12c	21c	1.9b	28.05b	4.50b
Glomus spp.	71b	68b	19c	1.7b	29.00b	5.11a
R. japonicum + *Glomus* spp.	78a	76a	25a	2.2a	33.00a	5.70a

*: Data are mean of three replications; nodule count was made after 21 days of germination whereas mycorrhizal infection was recorded after 55 days.

**: Number not followed by the same letter are significantly different at the P = 0.05 in Duncan's multiple range tests.

***: Seed treatment at the rate of 0.2 mg a.i. per kg was given at the time of sowing after inoculation whereas foliar application was givon in 45 days old seedlings.

Adaptive Trials

Experiments conducted by the research institutes and results obtained have to be demonstrated to the cultivators, the end users. Hence trials on farmers field were conducted and results are presented in tables (Table 6.21 and 6.22).

The data in the table indicate that half dose of fertilizer with PSB+Rh gave equivalent yield to full do. e of fertilizer in adaptive trials in 5 districts of Malwa region where gram is grown predominantly under rainfed conditions.

Table 6.21: Field Trial to Study the Effect of Fertilizers alone and Fertilizer Blended with Biofertilizers (PSB or VAM and Rh.) in Soyabean, Pigeonpea and Wheat

Crop	Fertilizer	Treatment and Yield				Yield (q/ha)
		PSB	VAM	Rh.	Azo.	
Soyabean	Full dose (20:60:20)	–	–	–	–	24.4
	Half dose (10:30:10)	–	+	+	–	25.14
	–do–	+	–	+	–	26.51
Pigeonpea	Half dose (20:30:10)	–	+	+	–	14.25
	–do–	+	–	+	–	14.34
	–do–	–	–	–	–	11.40
Wheat	Half dose (60:40:20)	–	+	–	+	30.29
	–do–	+	–	–	+	35.14
	–do–	–	–	–	–	31.81

Table 6.22: Adaptive Trials of Biofertilizers in Gram on Farmers in Five District of Malwa Region

Fertilizer	*Rh*	*PSB*	*Indore*	*Ratlam*	*Shajapur*	*Mandsaur*	*Ujjain*
Full Dose (20:60:20)	–	–	13.08	9.56	9.36	11.02	11.16
Half dose (10:30:10)	–	–	–	10.38	9.28	11.22	11.20
Half dose	+	+	13.14	10.50	9.58	12.52	11.22

Outlook

The perusal of literature indicates that, there is a positive response of crop plants to biofertiliser application. Nevertheless, there appears to be a degree of inconsistency of results depending upon the type of inoculant used, agro-climatic conditions, methods of inoculation/application, interaction between strains, nutrient interactions etc. It is this inconsistency of crop response which poses a problem in the popular use of Biofertilizers which needs to be tackled and their use perfected for their routine application.

Chapter 7
Soils and Plant Growth

Soil, in its traditional meaning, is the natural medium for the growth of land plants, whether or not it has developed discernible soil horizons, in this sense soil has a thickness that is determined by the depth of rooting of plants. Soil is defined as the unconsolidated mineral material on the immediate surface of the earth that serves as a natural medium for the growth of land plants. Soils are that portion of the earth's crust in which land plants can grow, if water and temperature are adequate, at least the minimum nutrients are available, and toxic substances are in low concentration.

Soils are often deposits of weathered, loosened, and transported particles. These materials are called parent materials. They are deposited by water, glaciers (ice), wind, or gravity into land areas called landforms. Many landforms are named for their method of transport and final shape: Alluvial is from flowing water; moraines are from glacier movement and melting; dunes are from wind movement of sands; and colluvium is from gravity's action. All of these various deposits, and others, becomes soils when exposed to weathering at the earth's surface. The soils, and their landform materials, vary enormously in the sizes of particles, slopes, uniformity of properties, and extensiveness across the landscape. The soils vary widely in quality and productivity.

Sand dunes do not ordinarily support plant life, but with stabilization (retarded mobility) and water, if needed, they could do so, and so can be considered soils. Organic deposits, such as are found as peat or muck bogs, are very different from the majority of soils because they sometimes contain almost no mineral materials, which are the major components of most soils. nevertheless, they are soils.

Mineral vs Organic Soils

Because soils range from almost no organic matter to nearly 100 per cent organic substances, an arbitrary division of soils into mineral soils and organic soils is made. Organic soils have more than half of the upper 80 cm (32 in.) as organic material. Mineral soils are all other soils.

Soil Factors for Plant Growth

Soils are the medium in which crops grow and provide food and cloth to the mankind. The basic needs of crop production are to maintain soil fertility and soil productivity.

Soil Fertility

It is the inherent capacity of soil to provide essential chemical elements for plant growth. Soil fertility is vital to a productive soil. But a fertile soil is not necessarily a productive soil. These are so many factors *e.g.* poor drainage, insects, drought etc. that can limit production, even when fertility is adequate. Soil fertility in modern-day agriculture is a part of a dynamic system. Nutrients are constantly being 'exported' in the forms of plant and animal products. Unfortunately, others can be lost by leaching or erosion. Still others, like phosphorus (P) and Potassium (K), can be tied up by certain soil clays. Organic matter and soil organisms immobilize, then release, nutrients throughout time. If production agriculture were a closed system, nutrient balance might be relatively stable.

Soil Productivity

Productivity emphasizes the capacity of soil to produce crops and it is expressed in terms of yields. To understand soil productivity, one must recognize existing soil plant relationships. Certain external factors control, plant growth: air, heat (temperature), light, mechanical support, nutrients and water. The plant depends on the soil (at least partly) for all these factors, except light. Each directly affects plant growth. Each is linked to the others. Since water and air occupy the pore spaces in the soil, factors that affect water relations necessarily influence soil air. In turn, moisture changes affect soil temperature. Nutrient availability is influenced by soil and water balance as well as by soil temperatures.

Essential Plant Nutrients

Sixteen chemical elements are known to be essential for plant growth. They are divided into two main groups: non-mineral and mineral.

Non-Mineral Nutrients are carbon (C), hydrogen (H), and oxygen (O). These nutrients are found in the atmosphere and water. They are used in photosynthesis. Products of photosynthesis account for most of the increased plant growth.

The 13 mineral nutrients, which are available in soil are divided into three groups: major, secondary and micronutrients:

Major Nutrients

Nitrogen (N), Phosphorus (P), Potassium (K). The major nutrients usually become deficient in the soil first, because plants use them relatively in large amounts.

Secondary Nutrients

Calcium (Ca), Magnesium (Mg), Sulfur (S).

Micronutrients

Boron (B), Chloride (Cl), Copper (Cu), Iron (Fe), Manganese (Mn), Molybednum (Mo), Zinc (Zn).

Five additional nutrients-sodium (Na), cobalt (Co), vanadium (V), nickel (Ni), and silicon (Si) have also been established as essential nutrients in some plants. These five micronutrients are almost never deficient in soils. The secondary nutrients and Micronutrients are usually deficient less often and smaller amounts are used. But they are just as important as the major nutrients. Plants must have them when and where they need them.

Soil and Root Environment

The growing tips of the root wriggles through the growing medium following the line of least resistance. Roots are able to enter cracks, which are, or can be readily opened up to about 0.2 mm in diameter. Compacted soils severely restrict root exploration but once into these narrow channels the root is able to overcome great resistance to increase its diameter. Anything, which reduces root, exploration and activity, can limit plant growth. When this happens action must be taken to remove the obstruction to root growth or to supply adequate air, water and nutrients through the restricted root volume.

In order to grow and take up water and nutrients the root must have an energy supply. A constant supply of energy is only possible so long as oxygen is brought to the site of uptake. Consequently the soil spaces around the root must contain air as well as water. A lack of oxygen or a build-up of carbon dioxide will reduce the root's activity. Furthermore, in these conditions anaerobic bacteria will proliferate and many produce toxins such as ethylene. In warm summer conditions roots can be killed back after one or two days in waterlogged soils.

Soil Physical Properties

The physical properties of soils-texture, structure, density, porsity, water content, strength (consistency), temperature, and colour are dominant factors affecting the use of a soil. These properties determine the availability of oxygen in soil, the mobility of water into or through soils, and the ease of root penetration.

Soil Texture

Natural soils are comprised of soil particles of varying sizes. The soil-size groups, called soil separates, are sands (the coarsest), silts, and clays (the smallest). The relative proportions of soil separates in a particular soil determine its soil texture.

Texture is an important soil characteristic because it will, in part, determine water intake rates (infiltration), water storage in the soil, the ease of tilling the soil, the amount of aeration (vital to root growth), and will influence soil fertility. For instance, a coarse sandy soil is ease it till, has rapidly and easily loses plant nutrients, which are drained away in the rapidly lost water. High-clay soils (over 30 per cent clay) have very small particles that fit tight together, leaving little open pore space, which means there is little room for water to flow into the soil. This makes high-clay soils difficult to wet, difficult to drain, and difficult to till.

Soil Separate Sizes

The U.S. Department of Agriculture (USDA) has established limits of variation for the soil separates and has assigned a name to each size class (Table 7.1). This system has been approved by the Soil Society of America and is the one used in this book. Other particle-size classification systems are used in the United States and throughout the world.

Table 7.1: Soil Separates and their Diameter Ranges

Soil Separate Name	*Diameter Range (mm)*	*Visual Size Comparison of Maximum Size*
Very coarse sand	2.0–1.0	House key thickness
Coarse sand	1.0–0.5	Small pinhead
Medium sand	0.5–0.25	Sugar or salt crystals
Fine sand	0.10–0.10	Thickness of book page
Very fine sand	0.10–0.05	Invisible to the eye
Silt	0.05–0.002	Visible under microscope
Clay	Less than 0.002	Most are not visible even with a microscope

Soil Textural Classes

Textural names are given to soils based on the relative proportions of each of the three soil separates are sand, silt, and clay. Soils that are preponderantly clay are called clay (textural class); those with high silt content are silt (textural class); those with a high sand percentage are sand (textural class). A soil that does not exhibit the dominant physical properties of any of these three groups (such as a soil with 40 per cent sand, 40 per cent silt, and 20 per cent clay) is called loam. Note that loam does not contain equal percentages of sand, silt, and clay. It does, however, exhibit approximately equal properties of sand, silt, and clay.

The textural triangle is used to determine the soil textural name after the percentages of sand, silt, and clay are determined from a laboratory analysis. Since the soil's textural classification includes only mineral particles and those of less than 2 mm diameter, the sand plus silt plus clay percentages equal to 100 per cent (organic matter is not included).

Table 7.2: Textural Classes of Soil

Textural Class	*% Sand*	*% Silt*	*% Clay*
Clay (see center)	20	20	60
Silt loam	20	70	10
Sandy loam	65	25	10
Loam	40	40	20

Separation of sand into various sand sizes and the U.S. standard sieve numbers used for these separations. A 10-mesh sieve has 10 wire divisions per inch in the two directions making the screen.

Soil Structure

Soil texture is the relative proportion of sands, silt, and clay in a soil; how these particles are grouped together into stable collections or aggregates is soil structure. Natural aggregates are called peds. They vary in their water stability; the word clod is used for a coherent mass of soil broken in any shape by artificial means such as by tillage.

The formation and stability of soil aggregates is a very important phenomenon. Although there is still some uncertainty about the exact mechanism, which lead to formation of granules, several factors are responsible for their geneses. These include:

1. Soil tillage
2. Wetting and drying
3. Freezing and thawing
4. Activities of plant roots and soil fauna
5. Decomposition of organic matter
6. The modifying effects of adsorbed cations.

The stability of aggregates is of great physical importance. Three major factors influence aggregate stability:

1. Binding action of soil microorganisms.
2. Cementing action of intermediate products of microbial synthesis and decay, such as gums and certain polysaccharides, and

3. Cementing action of the more resistant stable humus components, which contributes to long term stability of the soil aggregates.

Aggregate stability is not entirely an organic phenomenon as interaction does take place between organic and inorganic components. Polyvalent inorganic cations (*e.g.*, Ca^{++}, Mg^{++}, Fe^{++}, and Al^{3+}) are thought to act as bridges between the organic matter and soil-clays, thus encouraging development of clay-organic matter complexes. Soils rich in kaotinite and hydrons oxide clays have aggregates with highest stability. The larger the aggregates in a soil, the lower is its stability.

Soil Structural Classes

Soil structural units (peds) are described by three characteristics: type (shape), class (size), and grade (strength of cohesion). Types of structure describe the ped shape with the terms angular blocky, subangular blocky, columnar, granular, platy, and prismatic.

Structureless

Soils have not noticeable peds. It might be an unconsolidated mass such as noncoherent sand and is called single grain, or it might be a cohesive mass, such as could occur in some loams or clayey soils, and is called massive. Soil structure influences the rate of infiltration of water in soil.

Soil stability-resistance to erosion by wind or water-is determined partly by the particle sizes in the soil (soil texture). More often, soil structure-the bonding of individual mineral particles into larger stable configurations called aggregates-resists erosion even further.

Natural or machinery-induced hardened soil layers (hardpans) are common to many soils. Such hardpans may inhibit root penetration and thereby reduce the effective soil volume from which plants can absorb nutrients and water. Subsoiling, a process of ripping the hardpan with tillage equipment, may be needed. Water may be trapped above a hardpan, causing a waterlogged layer of soil. Oxygen needed by the roots cannot pass through these water filled pores even for short distances of a few centimeters (an inch or so).

The magnitude of plant growth reflects a composite of many favorable and unfavorable factors. Favorable growth factors include adequate aeration, water, nutrients, adequate soil depth, and proper soil temperature. Unfavorable growth factors are many, among which are toxic levels of certain elements, diseases, harmful insects, adverse temperatures, and inadequate or excess sunlight. Any accurate estimate of the plant productivity of a soil must include all favorable and unfavorable factors.

Soil Porosity and Permeability

Pore spaces (also called voids) in a soil consists of that portion of the soil volume not occupied by solids, either mineral or organic. Pores in soil are the result of irregular shapes of primary particles and their aggregation; the forces of penetrating roots, worms, and insects; and of expanding gases entrapped by water. Under field conditions, pore spaces are occupied at all times by air and water.

Pores are officially described according to their average diameter in millimeters are follows:

Very fine < 0.5; Fine 0.5–2; Medium 2–5; Coarse >5

Water drains by gravitational force from pores larger than about 30-60 microns (æm) (0.0012–0.0024 in.). In comparison, root hairs, the smallest plant roots, are between 8 and 12 æm (0.00032–0.00048 in.) in diameter. In soils where most pores are smaller than 30 μm (0.0012 in.), attraction forces in the soil retain the water within the fine pores, which results in a waterlogged soil and poor aeration. Thus, to the growing plant, pore sizes are of more importance than total pore space. The best balance of water retention (smaller pores) plus adequate air and water movement (larger pores) is an medium-textured soils such as loams. Aggregation, or the lack of it, can modify this balance of large and small pores, which results from the soil texture. Soluble bicarbonates, silicates, and iron during wet periods will move in the soil and precipitate when drying occurs. Over many decades and centuries, pores become filled with the precipitates and often are cemented into hard layers.

The relative amounts of air and water in the pore space fluctuate continuously. During a rain, water drives air from the pores, but as soon as soil water disappears by deep percolation (downward movement) evaporation, and transpiration (evaporation from plant leaf openings), air gradually replaces the water as it is lost from the pore spaces.

Soil Air

To survive, all living organisms require exchanges. The most important of these in soils are for respiration by plant roots and for organic matter decomposition by microorganisms. The most desired condition is well-aerated soil, a condition in which oxygen exchange between soil air and atmospheric air is rapid. The factors influencing the rate of gaseous exchange include soil pore sizes and continuity, temperature, depth in soil, wetting and drying of soil, and coverings (mulches) on the soil surface.

Composition of Soil Air

Atmospheric air has about the following composition of the gases important in soils:

Dinitrogen (N_2)	79 per cent	Oxygen (O_2)	20.9 per cent
Carbon dioxide (CO_2)	0.03 per cent	Water vapor (relative humidity)	20–90 per cent

The most important of these of plant growth is the oxygen content. Oxygen is required in respiration and organic material oxidation (decomposition). As roots and microbes use oxygen, they give off carbon dioxide into the soil air.

Rates of Oxygen Exchange

The rate at which in the soil exchanges with oxygen in the atmosphere is called the oxygen diffusion rate (ODR). Many large soil pores would speed air exchange (diffusion). Small pores, with bottleneck portions filled with water, would decrease exchange rates.

The Oxidation-Reduction Potential (Eh or Redox Potential)

In most soils, plants grow best oxidized soil (rice and marsh plants are exceptions). The major reason is the need for oxygen. Oxygen is the primary acceptor of electrons released in the oxidation of carbon in organic materials. If the oxygen concentration is too low, certain elements or irons become the electron acceptors and are, in turn, reduced. Mostly undesirable things happen when the ODR is too low and redox value drops; toxic organic acids form, nitrate is denitrified and nitrogen is lost to the atmosphere, available sulfate is reduced to sulfide, and most plant roots cannot respire. Thus, plant root growth slows or stops.

Aeration and Plant Growth

Reduced aeration causing deficient oxygen concentrations can be caused by various factors. Some of these are: (1) waterlogging (ponding); (2) soil compaction of loams and clays (3) very water high contents of clay that swell when wet, thereby closing large pores; and (4) organic matter decomposition by soil microorganisms that uses oxygen in soils already low in ODR.

Soil Strength

Soil strength is the degree of resistance of a soil mass to crushing or breaking when force is applied. A non-cemented soil mass is evaluated under two moisture conditions; air-dry and field moisture capacity. When a soil mass is cemented, soil strength is evaluated when air dry and again after it has soaked in water for 1 hour.

Soil Color

Soil color indicates many soil features. A change in soil color from the adjacent soils indicates a difference in the soil's mineral origin (parent material) or in soil development. White colors are common when salts or carbonate (lime) deposits exist in the soil.

Dark soil absorbs more than do light-colored ones. Some black coal mining wastes and dark-colored oil-shale residues reach temperatures of 65.6–70.0°C (150–158°F), which are lethal to many plants that could otherwise grow in those soils. Although black soils having high humus content absorb more than light-colored soils, they also frequently hold more water. The water requires relative larger amounts of heat than the soil minerals to raise its temperature; water also requires considerable heat to evaporate it.

Within geographic regions, darker colors usually indicate higher organic matter contents. Soil color determination is standardized and determined by the comparison of the soil color to Munsell color charts. These color charts are similar to books of color chips found in paint stores where many gradations of different color groups are shown.

Soil Temperature

Soil temperature has important role in soil biological activities, nutrients recycling, water movement etc. major regulatory factors of soil temperature are soil texture, structure, depth, organic matter content and vegetation cover. Soil temperatures resulting from the sun's radiation change with depth and with the time of day. For example, the maximum daily temperature at deep soil depths is delayed, even by several hours, after the time when the air temperature reaches a maximum.

Soil Water

Water is required for all life. In soils, water is supplied to plants through the roots, it lubricates the soil allowing root penetration, it is necessary for microbial mobility and action, and it allows nutrient mobility. In a dry portion of soil, water uptake stops, nutrient absorption essentially ceases, and the root growth practically stops. A typical wetted loam soil will contain water in about 25 per cent of its volume, with only about half of that water being available to plants.

The properties of soil water are usually most favorable in the medium-textured soils (loams, sandy loams, silty clay loams). Water entry into soils (infiltration) is too fast in sands and too slow in clays for optimum crop production. Inadequate air exchange will occur in the small and water filled pores in clayey soils. Sandy soils hold low plant available amounts of water following wetting. Thus, sands will require frequent rains or irrigations to supply adequate water. Clayey soils will have slow drainage of excess water sandy soils will be well drained, if drainage is possible.

Water Retention Forces

Soil serves as a water reservoir but a leaky one. When too much water is added, the excess runs off over the surface or in deeper layers. Water is held in soils because of the attraction between unlike charges–a positive ion attracted to a negatively charged ion. The positive charged hydrogens of water are attracted to nearby negatively charged ions such as oxygen, even to the oxygen of another adjacent water molecule.

Water potential is defined as the work the water can do when it moves from its present state to a pool of water (at a specified elevation, temperature soluble salt content, etc.). Because it refers to work, it is an energy term, not a pressure term. The soil water potential is a combination of the effects of the surface area of soil particles and small soil pores that adsorb water (matric potential), the effects of dissolved substances (solute or osmotic potential), and the atmospheric or gas pressure effects (pressure potential), it nonsalty well-drained soil, the matric potential is almost equal to the water potential.

Soil Water Classification for Water Management

The most used classifications of water are those relating water plants can use and to water which they cannot use. Terms include gravitational (drainage) water, plant available water, field capacity, and permanent wilting point.

Ecological Classification of Soil Water

Plant-available water is the portion of stored soil water that can be absorbed fast enough by plant root sustain life. It is defined as the weight percentage of total soil water with a water potential between –33 and –1500 kPa (for sands, –10 to –150 LD). The water held within these potentials makes up most of the stored water used by plants, (Gravitational water) also is equally available as it flows past the plant roots if adequate is maintained. Since gravitational water is present only for short periods of time permeable soils, it cannot be depended upon in calculating plant water requirements.

The *wilting point* (sometimes called permanent wilting point percentage) is the percentage of soil water held with water potential less (more negative) than –1500 kPa; it is held so strongly that plants are not able to absorb it fast enough for their needs.

Field capacity is the percentage of soil water that is held with water potential less (more negative) than –33 kPa and is a measure of the greatest amount of water that a soil can, hold, or store, under conditions of complete wetting followed by free drainage. Field capacity values are used to determine the amount of irrigation water needed and the amount of stored soil water available to plants. Plant-available water is equal to the difference of water percentage at field capacity and at permanent wilting point.

Capillary water moves through the soil pores because of a water potential gradient. Water potential gradient is the difference in total water potential between two locations in the soil.

The saturation percentage is the water content of soil when all pores are filled with water. It is approximately double the field capacity value. The saturation percentage is approximated by the water in a soil.

Soil Water Reservoirs

Water in soil is held as films on particle surfaces and in small pores. Large pores (as in sands and in between large aggregates) allow water to drain by gravitational flow. Small pores (clays) retain water by capillary force. Generally, the more clayey the soil and the higher the humus content, the larger the amount of water the soil can store not only does the quantity of water retained in clay soils remain large, but much of it is highly held to the large clay surface area. This means that the clay soil will hold a large amount of water at the permanent wilting point percentage.

When Plants Most Need Water?

Plants are in need of water when they begin to wilt, but by the time wilting is visible, plant growth has already been reduced significantly, Wilting usually is due to insufficient soil moisture. Most plants have critical periods of growth during which a lack of water is most damaging. For plants producing seed, the period from flowering to seed set (fertilization) is the most critical. If drought occurs during this period, grain or seed yield will be greatly reduced.

Other Soil Physical Properties

Additional soil physical properties include plasticity, stickiness, smeariness and fluidity.

Plasticity is the degree to which soil is permanently deformed, without rupture, by a force applied continuously in any direction. Plasticity is determined in the field by rolling wet soil between the hands until a 3-mm wire will break when the soil has dried to approximately the plastic limit.

Stickiness means the property of wet soil to adhere to another object.

Smeariness is a field term for clays that are thixotropic. Thixotropy is the property exhibited by certain fine-textured clay gels of being solid upon standing but turning liquid when shaken or mixed.

Fluidity refers to non-thixotropic clay soils that flow under hand pressure.

Tilth

The soil surface or seedbed should be carefully managed to produce the required tilth: the structure of the top 50 mm of the soil. Sandy soils are easily broken down to the right size with cultivation equipment. Heavier soils are less easy to cultivate and benefit from weathering to produce a frost mould.

The fineness of a seedbed should be related to the size of seeds but ideally consists of aggregate or crumbs between 0.5 and 5 mm in diameter. Cloddy surfaces lead to poor germination as well as poor results from soil herbicide treatments; Tilth is broken down by the rain on the soil surface. As soil crumbs break up, the particles fill in the gaps; this reduces infiltration rates. As the surface dries, a cap or pan of formed. Thus fine "dusty" tilths should be avoided and the soil crumbs should be stable be that they can withstand the effect of rain until the crop is established. This is particularly important on fine sandly and silty soils which tends to have poor structural stability.

Soil Colloids and Chemical Properties

The chemical properties of soils are determined by the colloidal portion of soils. A colloid is any solid substance whose particles are very small; thus, its surface properties are relatively more important than its mass (weight). Most colloids are smaller than a few micrometers (microns) in diameter. Because of their large surface-to-mass ratio, soil colloids settle slowly from suspension and have many reactions at their surfaces that determine soil properties. The predominant soil colloids are soil clays and humus.

Humus is a mixture of residues left after partial decay of organic substances in and on top of soils. Chemical properties of soils are more influenced by the clays and humus than by equal weights of the larger silt and sand particles.

The chemical properties of soils include mineral solubility, nutrient availability, soil reaction (pH), cation exchange, and buffering action.

Clays, the major mineral cation exchange particles, have negatively charged sites in their lattices and attract and hold positively charged ions (cations) at the clay surface. The quantity of these cations that can be held, or exchanged, by a given amount of soil is the cation exchange capacity of that soil. The exchangeable cations are not easily lost by leaching action until they are replaced (exchanged) by other cations. Plants often excrete hydrogen ions (H^+) to the soil solution, which can replace nutrient cations from their exchange sites, thus allowing the nutrients to move to the plant root and be absorbed.

If the adsorbed ions on the colloids are mostly $Al(OH)^{2+}$ and some H^+, the soils are acidic. The pH is a measure of acidity or alkalinity: soil pH ranges from about pH 3.5 (extremely acidic) to pH 7 (neutral) to about pH 11 [extremely alkaline (or basic)]. Plants grow well between pH 5 and pH 8.5. Strongly acidic soils are undesirable because soluble aluminium and manganese can reach toxic levels and microbial activity is greatly reduced. Strongly alkaline (basic) soils have lower micronutrient availability, except for molybdenum; iron, zinc, manganese, and macronutrient phosphorus may be deficient. Truly, the colloids in soils are very important to the properties of soils. In fact, the colloids have dominant influences on the soil's physical and chemical properties.

Organic Colloids (Humus)

Humus is a temporary intermediate product left after considerable decomposition of plant and animal remain-temporary because the organic substances remaining continue to decompose slowly. They humus is often referred to as an organic colloids and consists of various chains and loops of linked carbon atoms.

Humus is amorphous, dark brown to black, nearly insoluble in water, but mostly soluble in dilute alkali (NaOH or KOH) solutions. It contains about 30 per cent each of the nitrogen-rich proteins, the slow-to-decompose lignis, and complex sugars (polyuronides). The polyuronides comprise much of the varied organic substances that cement soil aggregates together. Humus has about 50 per cent carbon, some oxygen, 5 per cent nitrogen, and lesser amounts of sulfur, phosphorus, and other elements. It also has a cation exchange capacity on a dry weight basis many times greater than that of clay colloids. On a weight basis, a few per cent humus exerts much greater influence than does several times more per cent clay, especially for properties such as the cation exchange capacity.

Cation Exchange

Plant growth is altered by modification of chemical properties, such as soil acidity, availability of nutrients, and toxicities. These concerns involve cation exchange, soil acidification, and nutrient balance. Any electric charged colloidal surface area will attract mobile substances of opposite charge in the soil water. The dominant residual charge on most soil colloids is negative. These negative charged sites attract positively charged ions in the soil water. Positively charged ions are called cations. Soil acts as a cation exchanger.

Cation exchange is an important reaction in soil fertility, in causing and correcting soil acidity and basicity, in changes altering soil physical properties, and as a mechanism in purifying or altering percolating waters. The plant nutrients calcium, magnesium, and potassium are supplied to plants in large measure from exchangeable forms. Cation exchange is very important in soils because of the following relationships:

1. The exchangeable K is a major source of plant K.
2. The exchangeable Mg is often a major source of plant Mg.
3. The amount of lime required to raise the pH of an acidic soil is greater as the CEC is greater.
4. Cation exchange sites hold Ca^{2+}, Mg^{2+}, K^+, Na^+, and NH^4 ions and slow their losses by leaching.

5. Cation exchange sites adsorb many metals (Cd^{2+}, Zn^{2+}, Ni^{2+}, Pb^{2+}) that might be present in waste waters. Adsorption removes them from the percolating water, thereby cleansing the water that drains into groundwaters of surface waters.

The amounts of cations in the soil solution are intimately related to the exchangeable ions. Any change in concentration of a cation in the solution (fertilizer or lime) forces a change in proportions of all exchangeable ions.

The cation exchange sites, immobilize cations but also keep them exchangeable, and, thus, available to plant roots. Any water moving through the soil (exchanger) will lose many of its soluble cations to the soil (exchanger) and pick up those cations that have been replaced from exchange sites.

Cation Exchange Capacity (CEC)

Cation exchange capacity (CEC) is the amount of exchangeable cations per unit weight of dry soil. It is measured in centimoles (+) of cations per kilogram of soil [cmol(+)kg]. The term centimoles (+) is used because the number of negative sites in a given soil sample does not change, but the weight of the cations that may be adsorbed to those sites at one time do change. One centimole (+) of cation X occupies the same number of cation exchange sites as one centimole (+) of cation Y (a different cation).

Anion Exchange and Adsorption

Colloids that have an appreciable anion exchange are those that also have low cation exchange.

Multicharged positive ions, such as iron and aluminium, have hydroxyls (OH^-) that can be exchanged with sulfate (SO_4^{2-}), phosphate ($H_2PO_4^-$ or HPO_4^{2-}), molydbate (MOO_4^{2-}), and some other anions. The highest anion exchange capacities (AEC) occur in amorphous silicate clays (volcanic ash soils) and in iron and aluminium hydrous oxide clays; lesser exchange occurs in kaolinite. Phosphates particularly can be held firmly on anion exchange sites.

The Soil Reaction (pH)

The term pH is from the French *pouvoir hydrogene* or hydrogen power. Soil reaction (pH) is an indication of the acidity or basicity of the soil and is measured in pH units. The scale ranges from 0 to 14 with pH 7 as the neutral point. At pH 7, hydrogen ion concentration (H+) equals the hydroxyl ion concentration (OH), From pH 7 to 0 the soil is increasingly more acidic; from pH 7 to 14 the soil is increasingly more alkaline (basic). The H^+ concentration, which is the substance measured when determining pH, has a tenfold change between each whole pH number. Thus, a soil of pH 5 has 100 times more H^+ in solution than a soil solution with a pH of 7.

Importance of Soil pH

The soil pH is easily determined and provides clues about other soil properties. The soil pH greatly affects the solubility of minerals. Strongly acidic soils (pH 4–5) usually have high and toxic concentrations of soluble aluminium and manganese. Azaleas, tea, rhododendrons, cranberries, pineapple, blueberries, and several conifer timber species tolerate a strong acidity and grow well. In contrast, alfalfa, beans, barley, and sugar beets only do well in slightly acidic to moderately basic soils because of a high calcium demand or inability to tolerate soluble aluminium. Most minerals are more soluble in acid soils than in neutral or slightly basic solutions.

The soil pH also influences plant growth by the pH effect on activity of beneficial microorganisms. Most nitrogen-fixing legume bacteria are not very active in strongly acidic soils. Bacteria that decompose soil organic matter and thus release nitrogen and other nutrients for plant use are also hindered by strong acidity. Fungi usually tolerate acidity better than do other microbes.

On mineral soils most agricultural crops do best in slightly acidic soils (pH 6.5); on organic soils, about pH 5.5.

Soils become acidic, even from basic parent materials, by the leaching away by rainwater of the basic cations (Ca^{2+}, Mg^{2+}, K^+, Na^+) and replacement of many of them by H^+ from carbonic acid (H_2CO_3) formed from water and dissolved carbon dioxide. Eventually, $Al(OH)^{2+}$ replaces H^+.

Soil basically, although more difficult to alter than soil acidity, may be just as undesirable for plants. Nonleached soils or those high in calcium (low rainfall areas) may have pH values to 8.5. With increased exchangeable sodium, soils may reach values of over pH 10. Plants on soils of pH greater than about 9 usually have reduced growth, or even die. Some plants (halophytes) are tolerant of high salt or pH. Soluble $Al(OH)^{4-}$ may exist at these higher pH values.

Most micronutrient problems caused by high soil pH are solved by adding special fertilizers such as water-soluble chelates (key'-lates), which are "stable" but soluble complexers of the metal ions (although susceptible to microbial decomposition). If excess exchangeable sodium exists in the soil, it is usually removed by adding gypsum, followed by irrigation to leach the exchanged Na^+ ions.

Buffering in Soils

Most soils can resist appreciable pH changes when large amounts of a material either strongly acidic or basic are added, such as an acid-forming or base-forming fertilizer. This ability to resist a change in pH is the buffering capacity of the soil. The buffering capacity increases as the cation exchange capacity increases. To exhibit buffering, the soil must remove hydrogen ions (H^+) of added acids or neutralize the hydroxyls (OH) of added bases. The concept of buffering can be broadened to include a resistance to change in the concentration of any ion in the solution that is also adsorbed to the colloid.

Buffering action should be effective in controlling soluble (and sometimes toxic) concentrations of H^+, Al^{3+}, Ca^{2+}, Mg^{2+}, K^+, and Na^+. In all instances the larger the CEC of the soil, the greater should be the buffering capacity. Soils high in humans and/or clay, particularly montmorillonite or vermiculite clays, will have high buffering. Organic soils and clays have higher CEC values and are more strongly buffered than are sandy soils of the same geographic area.

Origin of Organic Soils

Organic soils are derived from plants growing in environments that decompose dead residues slowly. Usually, stagnant waters (swamps, marshes, moors, bogs) in temperate climates allow good growth of adapted plants, which die or shed leaves into the water. The decomposition of plant material in stagnant water lacks enough oxygen (is anaerobic) and so is very. Over centuries, such accumulations may reach depths of several meters. Examples of these are California's Stockton delta, Flonda's everglades, and the many ponded waters in depressions in northern Michigan, Minnesota and Canada.

Common organic soil parent materials include many mosses (such as sphagnum), pondweed, cattalils, sedges, reeds, grasses, and various "water-loving" deciduous and coniferous shrubs and trees. The term fibric means low extent of decomposition; more than two-thirds of the material can be recognized as to its plant part or source.

Soil Biology and Ecology

A small portion of the seemingly inert soil mass consists of live organisms-insects, rodents, worms, plants and their roots, and millions of microorganisms (microbes). The scientific study of these living organisms in and on soils is called soil biology. When interactions are included it is soil ecology. As detailed knowledge has grown, the classification of organisms has been modified. Previous classifications of all organisms into either animals (faunae) or plants (florae) is now expanded into new categories, all of them not always accepted by all scientists. In this book we use the system of five kingdom, which are Animalia (animals, worms, insects), Plantae (plants, algae), fungi (molds, yeasts, rusts), protista (protozoa, slime molds), and monera (bacteria, actinomycetes).

Some soil microbes and Animalia (macroorganisms) have little obvious effect on the soil one way or another. Some are absolutely essential to plant yields and by spreading animal infections and disease. Although the detrimental soil organisms are many, the beneficial ones exceed them in effect. Animalia aerate and aggregate soils. Some bacteria convert toxic carbon monoxide to carbon dioxide, fix inert atmospheric dinitrogen into useful forms, decompose organic materials, and reduces excess nitrates to relatively inert dinitrogen gas (N_2). Actinomycetes are a major source of antibiotics used for disease control and some also fix dinitrogen gas to usable forms. Even the lowly algae contribute to soil productivity by adding organic matter and fixing atmospheric dinitrogen. Microbes are major decomposers of organic materials; they continually work away at the resistant soil organic residues called humus.

Humus, the residual parts of one-living matter after initial rapid decomposition of organic materials, results primarily from the actions of bacteria, fungi, and actinomycetes and their enzymes. Fungi and bacteria are the most important organic matter decomposers. Both groups form symbiotic relationships in which two different organisms live in association, with both benefiting. Mycorrhizae, the symbiotic root-fungi association is found most dramatically on roots of trees but now is known to be possible on most plants. Mycorrhiza aid nutrient and water uptake. Rhizobia (N_2-fixing bacteria) form nitrogen-fixing nodules most Commonly on legumes; similar microbe-plant relationships exist on many other plants, even on corn and other grasses. Certain bacteria called autotrophs get their energy from oxidizing important soil elements, including nitrogen, sulfur, iron, manganese, and carbon monoxide. Some algae fix atmospheric nitrogen and become part of the organic matter. Actinomycetes fix N_2 and help decomposition of organic materials. Protozoa are also decomposers and act as predators on bacteria.

Soil Formation

Soil formation comprises two different processes. First, the changes from a consolidated mass (rock) not capable of growing plants to the development of an unconsolidated (loose) layer of material that can support plants if climate is suitable and water is available; Second, the changes occurring within the loose material as time passes. This latter process is also called soil development. Actually, the change from a solid mass to loose soil material and other changes within the soil profile occur simultaneously. Soil formation is used to mean both the production of unconsolidated material by weathering processes and soil profile development, which are the changes involved in development of horizons.

Horizons are soil layers which differ from adjacent layers. They tell much about the characteristics of a soil. They include information about hardpans, depth of organic matter accumulation, soil denseness from clay deposition; and the extent of leaching.

Changes During Soil Formation

Soil formation-the processes of weathering and of change within a soil mass is caused by many factors. Additions, losses, transformations, and translocations occur constantly.

Additions are made by water (rainfall, irrigation), nitrogen from bacterial fixation, energy as sunlight, sediment from wind and water, salts, organic residues, and fertilizers, and in other ways.

Losses result from chemicals soluble in soil water, eroded small-sized fractions, nutrients in grazed and harvested plants, water losses, carbon losses as carbon dioxide, and denitrification loss of N_2.

Transformations happen because of the many chemical and biological reactions that decompose organic matter, form insoluble materials from soluble substances, and alter or dissolve some minerals. As water and organisms move within the soil, many substances are translocated to different depths-for example, clays to deeper layers, and soluble salts into the groundwater or to the soil surface as water evaporates. Even localized sorting is active, such as forming deposits of carbonates or iron oxides in portions of the profile.

Physical Weathering

The processes of soil formation begin with the weathering of rocks and minerals. Physical abrasion of particles and chemical reactions constantly cause changes in the mineral and organic soil mass. Generally, the chemical processes are more dominant than physical processes in warm wet climates, but the centuries during which weathering is active allow physical weathering to exert even dominant effects, especially in cold arid zones.

Climate and Soil Formation

Climate is an increasingly dominant factor in soil formation with increased time, mainly because of the effects of precipitation and temperature. Some direct effects of climate on soil formation include the following:

1. A shallow accumulation or retention of lime (carbonates) in areas having low rainfall occurs because calcium bicarbonates (from dissolving carbon dioxide, minerals, and lime) are not leached if sufficient water is not present. Such soils are usually alkaline.
2. Acidic soils form in humid areas due to intense weathering and leaching out of basic cations (calcium, sodium, magnesium, potassium).
3. Erosion of soils on sloping lands constantly removes developing soil layers.
4. Deposition of soil materials downslope covers developing soils.
5. Weathering, leaching, and erosion are more intense and of longer duration in warm and humid regions, as in Hawaii, where the soil does not freeze. The reverse is true in cold climates, as in Alaska.

Moisture is important in soil formation. A soil is said to be developed when it has detectable layers (horizons), such as of accumulated clays, organic colloids, carbonates, or soluble salts that have been moved downward by water. The extent of colloid movement and the depth of deposition are determined partly by the amount and pattern of precipitation, which produce the leaching action.

Climate also influences soil formation indirectly through its action on vegetation. Semiarid climates encourage only scattered shrubs and grasses. Arid climates supply only enough moisture for spars, short grasses or shrubs, which may not be dense enough to protect the soil against wind and water erosion. Many arid soils exhibit only slight organic matter accumulation and small amounts profile development.

Soils resulting from year-around hot and humid weathering, such as occurs in some parts of Hawaii on old, and well-drained surfaces, are typically deep (often more than 3 m; 10 ft), reddish in color (due to the presence of oxidized iron), with well-decomposed organic matter, and low in essential elements because of the leaching by the high rainfall.

Biota and Soil Formation

The activity of living plants and animals and the decomposition of their organic wastes and residues (the living environment, the biota) have marked influences on soil development. Differences in soils that have resulted primarily from differences in vegetation are especially noticeable in the transition where trees and grasses meet. Minnesota, Illinois, Missouri, Oklahoma, Texas, and most western states are places where such differences can be observed readily.

Some soils beneath humid forest vegetation may develop many horizons, are leached (washed, eluviated) in the surface layers, and have slowly decomposing organic matter layers on the surface. In contrast, some grassland soils near the transition zone of forests are rich in well-decomposed organic matter, frequently to depths of 30 cm or more (1 ft or more) into the mineral soil.

Burrowing animals, such as moles, gophers, prairie dogs, earthworms, ants, and termites, are highly important in soil formation when they exist in large numbers. Soils that harbor many burrowing animals have fewer but deeper horizons because of the constant mixing within the profile, which nullifies the organic colloid and caly movements downward.

Microorganisms help, soil development by slowly decomposing organic matter and forming weak acids that dissolve minerals faster than pure water. Some of the first plants to grow on wathering rocks are crust like lichens, which are a veneficial (symbiotic) combination of algae and fungi.

Topography and Soil Formation

The earth's surface contour is called its topography (sometimes called relief). To graphy influences soil formation primarily through its associate water temperature relations. Soils within the same general climatic area developing from similar parent material and on steep hillsides typically have thin A and B horizons because less water moves down through the profile as a result of rapid surface runoff and because the surface erodes quite rapidly. Similar materials on gently sloping hillsides have more water passing vertically through them than do materials on steeper slopes. The profile on gentle slopes generally is deeper, the vegetation more luxuriant, and the organic matter level higher than in similar materials on steep topography.

Materials lying in landlocked depressions receive runoff waters from surrounding higher areas. Such conditions favor greater production of vegetation but slower decomposition of dead plant remains because of oxygen deficiency in waterlogged (saturated) soil; this results in soils with large amounts of organic accumulations. If the area above the soil surface is wet for many months of the year, organic (peat or muck) soils develop. If the accumulating waters dissolve salts from surrounding soils, the depression may become a salt marsh with unique tolerant plants growing, or it may develop toxic salt conditions with no plants at all. When soils on the watershed are strongly acid, iron may leach from them and be deposited in depressions to form bog iron (limonite). Alkaline soils on sloping topography in humid regions may result ill lime being eroded into depressions and lead to the formation of marl.

Time and Soil Formation

The length of time required for a soil to develop the distinct layers called genetic horizons depends upon many interrelated factors of climate, nature of the parent material, the organisms, and topography. Horizons tend to develop most rapidly under warm, humid, forested conditions where there is adequate water to move colloids (subvisible-sized clays, humus, and other materials). Acid sandy loams lying on gently sloping topography appear to be the soils most conducive to rapid soil profile development.

Under ideal conditions, a recognizable soil profile may develop within 200 years; under less favorable circumstances, the time may be extended to several thousand years.

Soil development proceeds at a rate that is a combination of the effects of time plus the intensities of climate and biota (organisms), further modified by the effect of land relief (topography) on which the soil is situated and the kind of parent material from which it is developing.

Interactions of Soil Forming Factors

The formation of soil is a diverse and complex process, with effects from the five major factors working in combination. For instance, soii with good drainage (topography or parent material) and mild temperature and high rainfall (climate) will probably support ample plant life (biota) because favorable drainage provides an aerated location for the plant roots. The plants, in turn, decompose, producing carbon dioxide (CO_2), which combines with water from rainfall to form carbonic acid (H^+ and HCO_3^-). Acidity increases the solubility of parent minerals; sodium, potassium, calcium, and magnesium are dissolved and leach away in the draining soil water, and the soil becomes ever more acidic. High rainfall also moves some clays and small organic matter particles deeper into the soil's B horizon.

The same climate (or clayey parent material) but different topography (a low-lying valley or depression instead of a gentle slope) might produce water-logged soil (which has poor drainage). Poor drainage results in stagnant water and non-aerated soil, often with resultant poor plant growth and decreased rates of organic matter decomposition. The accumulating drainage waters also mean the accumulation of dissolved salts. The horizons of these adjacent soils developed from common factors but with differing topography will become very different.

Soil Taxonomy

So far we have discussed soils as a product of continual natural change. Soil horizons are usually altered slowly, so in soil we see now few changes are expected during the next decades or even centuries. Thus, soils can be grouped (classified) into categories based on their present properties, and those classifications will remain unchanged for decades.

Taxonomy (derived from the Greek taxis, meaning arrangement or order) is the systematic grouping of similar things; soil taxonomy is the scientific grouping of similar soils. According to The U.S. Soil Classification System the most general soil category is called order. AU world soils are placed into 11 orders; in the United States these are subdivided into 47 suborders, 185 great groups, 970 subgroups, 4500 families, and 12,620 soil series. As a brief orientation, simple descriptions of the 11 orders.

The 11 soil order (which includes the tentative order called Andisols) are categorized by the nature of the developed pedogenic horizons they contain, by the extensive mineral weathering (Oxisols and to some extent Ultisols), by the importance of swelling clay contents (Vertisols), and by being organic soils (Histosols).

Table 7.3: Brief Characterizations of the 11 Soil Orders of the United States

Soil Orders	*General Features*
Entisols	Entisols have no profile development except perhaps a shallow marginal A. Many recent river floodplains, volcanic ash deposits, unconsolidated deposits with horizons eroded away, and sands are Entisols.
Inceptlsols	These soils, especially in humid regions, have weak to moderate horizon development. Horizon development has been retarded because of cold climates, waterlogged soils, or lack of time for stronger development.
Andisols	A tentative soil order. Andisols are soils with over 60 per cent volcanic ejecta (ash, cinders, pumice, basalt) with bulk densities below 900 kg/m^3. They have enough weathering to produce dark A horizons and early-stage secondary minerals (allophane, imogolite, ferrihydrite clays). Andisols have high adsorption and immobilization of phosphorus and very high cation exchange capacities.
Histosols	Histosols are organic soils (peats and mucks) consisting of variable depths of accumulated plant remains in bogs, marshes, and swaps.
Aridisols	Aridisols exist in dry climates and some have developed horizons of lime or gypsum accumulations, salty layers, and/or A and Bt horizons.
Mollisols	Mostly these are grassland, but with some broadleaf forest-covered soils with relatively deep dark A horizons; they may have B horizons and lime accumulation.
Vertisols	Vertisols have a high content of clays that swell when wetted. Vertisols require distinct wet and dry seasons to develop because deep wide cracks when the soil is dry are a necessary feature. Usually, Vertisols have only deep self-mixed A horizons (top soil falls into cracks seasonally, gradually mixing the soil to the depth of cracking). These soils exist in most temperate to tropical climates with distinct wet and dry seasons.
Alfisols	Alfilsols develop in humid and subhumid climates, have precipitation of 500–1300 mm (about 20–50 in.), and are frequently under forest vegetation. Clay accumulation in a Bt horizon and available water much of the growing season are characteristic features. A thick E horizon is also common. They are slightly to moderately acid.
Spodosols	Spodosols are typically the sandy, leached soils of cool coniferous forests. Usually, O horizons, strongly acidic profiles, and well-leached E's are expected. The most characteristic feature is a Bh or Bs or accumulated organic matter plus iron and aluminium oxides.
Ultisols	Ultisols are strongly acid, extensively weathered soils of tropical and subtropical climates. A thick E and clay accumulation in a Bt are the most characteristic features.
Oxisols	Oxisols are excessively weathered; few original minerals are left unweathered. Often Oxisols are over 3 m (10 ft) deep, have low fertility, have dominantly iron and aluminium oxide clays and are acid. Oxisols develop only in tropical and subtropical climates.

Conclusion

In the organic farming, soil is considered to be live and knowledge of origin (birth) and various factor affecting its behaviour and health, are necessary for holistic management of soil in sustainable organic production system.

Chapter 8
Organic Matter

Since ancient times, farmers have recognized significant benefits of soil organic matter to crop productivity. These benefits have been the subject of controversy for centuries and some are still debated today. The following list includes many of the recognized benefits of soil organic matter.

1. It serves as a slow-release source of N, P, and S for plant nutrition and microbial growth.
2. It possesses considerable water-holding capacity, and thereby helps to maintain the water regime of the soil.
3. It acts as a buffer against changes in pH of the soil.
4. Its dark colour contributes to absorption of energy from the sun and heating of the soil.

The term soil *organic matter* is generally used to represent the organic constituents in the soil, excluding undecayed plant and animal tissue, then partial decomposition products, and the soil biomass. Thus, this term includes:

1. Identifiable high-molecular-weight organic materials such as polysaccharides and proteins.
2. Simpler substances such as sugars, amino acids, and molecules, and
3. Humic substances.

Soil organic matter is frequently said to consist of humic substances and Nonhumic substances. *Nonhumic substances* are all those materials that can be placed in one of the categories of discrete compounds such as polysaccharides and sugars, proteins and amino acids, fats, simple organic acids, and so on; humic substances are the other, unidentifiable components.

The term *humic substance* refers to an operationally defined, heterogenous mixture of naturally occurring organic materials. These substances are ubiquitous in nature and arise from the decay of plant and animal residues in the environment. Humic substances are generally classified into humic acid, fulvic acid and humin on the basis of their solubility in water as a function of pH.

Interactions Between Soil Biota, Soil Organic Matter, and Soil Structure

Physical protection mechanisms are important determinants of the stability of organic matter in soils. It is necessary to integrate the different concepts of physical protection of soil. This is partly based on the observation that the turnover rates of easily decomposable compounds are much higher in liquid microbial cultures than in soils and on electron microscopy studies that have demonstrated sites of accumulation of organic residues of clearly cellular origin. The phenomenon that drying of soil samples and disruption of soil aggregates can increase C and N mineralization is indirect evidence of the existence of physical protection in soil. It has been established that there is more physical protection in fine-textured soils than in coarse-textured soils, leading to higher carbon (C) contents and larger amounts of microbial biomass.

Several following mechanisms have been suggested to explain the physical protection of organic matter in soil against decomposition:

1. The adsorption to or coating by clay and silt particles is the main mechanism of physical protection of organic matter and microbes in coarse-textured soils. The clay silt fraction remains highly enriched with organic C and microbial biomass. In fine-textured soils this mechanism does not seem to be the main mechanism of physical protection.
2. The clay and silt fraction does not enriched with organic C and microbial biomass. Here, the location in small pores in aggregates seems to be an important mechanism of protection. Physical disruption of fine-textured soils released some of the protected organic compounds.
3. The positive correlation between clay content and total organic C and microbial biomass indicates that the capacity to physically protect microbes and organics is greater in fine-textured soils than in coarse-textured soils.
4. The entrapment of organics in small pores in or between aggregates, which renders them inaccessible to the decomposing microbial community.
5. The biomass of nematodes and bacteria are regulated by pore size, distribution and clay content, leading to a higher biomass of nematodes in coarse-textured soils and a higher biomass of bacteria in fine-textured soils.
6. Soil structure may also control decomposition processes by its effect on the grazing intensity of the soil fauna on microbes. The soil fauna may stimulate microbial growth rates through grazing, and predation of microbes by protozoa and nematodes has been suggested as an important mechanism of nutrient turnover in soil. A large proportion of bacteria may occupy pores <3 μm, while protozoa and nematodes are restricted to larger pores. This means that a large part of the bacterial population will be physically separated from the protozoa and nematodes in soil.
7. Organic matter turnover and microbial activity have been studied in different soils. It is generally found that the turnover of organic matter and microbial activity is less in fine-textured soils than in coarse-textured soils.
8. The apparent higher grazing pressure on the bacteria in coarse textured soils in comparison with fine-textured soils, however, did not lead to a more active bacterial population.
9. Most of the bacteria were protected from bring grazed in both coarse- and fine-textured soils.

To understand how structure can affect soil organic matter turnover it is important to know how primary soil particles and organic matter interact. Tisdall and Oades presented a soil structure model describing the association of organic matter with free primary particles (*i.e.*, sand, silt, and clay), microaggregates, and macroaggregates. The basic structural units in soils are considered to be microaggregates (clay-polyvalent, metal-organic matter complexes, <250 μm in diameter) that are water stable and not affected by agricultural practices. These microaggregates are linked together to from macroaggregates, which are defined as being larger than 250 μm. Macro and microaggregates contain primary particles, organics, and pores of different size.

Fungal and Bacterial Pathways of Organic Matter Decomposition and Nitrogen Mineralization in Arable Soils

The development of sustainable agricultural practices depends largely, on promoting the long-term fertility and productivity of soils at economically viable levels. Efforts to achieve these goals have focused on:

1. Lowering fertilizer inputs in exchange for a higher dependence on biologically fixed and recycled nutrients.
2. Reducing pesticide uses while relying more on crop rotation and biocontrol agents.
3. Decreasing the frequency and intensity of tillage.
4. Increasing the return of crop residues and animal wastes to land.

The principal objectives of these approaches are to match the supply of soil nutrients with the fertility demands of the crops, to maintain acceptable pest tolerance levels, and to develop soil physical properties that optimize oxygen supply, water infiltration, and water-holding capacity at levels that minimize the losses of nutrients by leaching and gaseous export. Determining the suitability of these sustainable practices to a broad range of crops, soil types, and climatic regimes requires an understanding of their effects on the physical, chemical, and biological properties of soils.

The importance of soil biota as causal mechanisms for sustaining the fertility and productivity of soils has been the focus of several major programs on the Ecology of arable farming systems. These include, but are not restricted to:

1. The long-term conventional (CT) and no-tillage (NT) trails (Horseshoe Bend site) of the Georgia Agroecosystems Project in the United States.
2. The conventional and integrated farming trials (Lovinkhoeve site) of the Dutch Programme on soil Ecology of Arable Farming Systems.
3. The barley, grass ley, and lucerne ley trials (Kjettslinge site) of the Swedish project on the Ecology of Arable Lands.
4. The long-term stubble mulch and no-tillage trials (Akron site) at the Central Great Plains Research Station in the United States.
5. The cultivated barley trials (Ellerslie and Breton sites) of the University of Alberta, Canada.

In all the above programmes, much attention has been directed at under rstanding the contributions of fungal and bacterial-based food webs to the accumulation and loss of soil organic matter (SOM) and to nutrient cycling. The importance of distinguishing these two primary pathway is based on the theses that:

1. Bacteria have lower C assimilation efficiencies and faster turnover rates than fungi, factors that are likely to increase rate of nutrient mineralization and organic matter decomposition in bacterially dominated soils.

2. The mycelia growth form is more conservative of energy and nutrients, enhancing organic matter storage and nutrient retention in fungal dominated soils.

Table 8.1: Estimates of N Flux (kg N ha^{-1} 10 $cm^{-1}yr^{-1}$). Through bacteria and fungi and the n passing through microbial consumers and predators in bacterial and fungal channels

Dept (cm)	*Biomass N*		*Energy Channel*	
	CF	*IF*	*CF*	*IF*
		Bacteria		
0–10	32	49	30	55
10–25	36	40	30	25
		Fungi		
0–10	0.60	1.10	0.70	0.85
10–25	0.70	0.90	0.70	0.55

When bacterial production is greater, bacterial-feeding fauna are expected to dominate. The most common bacterial-feeding fauna are protozoa and many non-stylet-bearing nematodes, which require water films for locomotion and feeding. They are generally believed to increase organic matter loss and nutrient mineralization due to their relatively large biomass and high turnover rates and to their feeding on bacteria. The predominance of fungal- and bacterial-based food webs in NT and CT agroecosystems, respectively, could account for many of their observed differences in organic matter turnover and nutrient cycling.

Soil Structure and Soil Organic Matter Storage

Microbial Influence

Microbes (fungi and bacteria) indirectly influence the storage and release of nutrients through their effects on soil physical properties and the protection of soil organic matter (SOM). Mycelial fungi and bacteria contribute directly to the formation and stabilization of soil aggregates through their deposition of extracellular polysaccharides and hyphal entanglement. Where aggregates remain intact for relatively long periods of time (as in no-tillage soils), clay surfaces and micropores (<1 µm diameter) become occluded with the extra-cellular byproducts of microorganisms, restricting the access of microorganisms and extracellular enzymes to physically isolated SOM. As such, the formation and stabilization of soil aggregates represent potentially important mechanisms for the stroage and protection of SOM.

Polysaccharides and humic substances are the dominant-binding agents produced by several species of bacteria and streptomyces. In contrast, humic and lignin-like compounds are found to be the principal binding agents produced by most species of fungi. The aggregates bound by filamentous structures (and/or their associated binding agents) are more resistant to physical disruption than those bound by bacterial polysaccharides. The stabilization of macroaggregates (>250 µm) is most often attributed to fungi, while that of microaggregates is usually associated with the production of adhesive metabolites by bacteria and the fungal polysaccharides are more stable to degradation than those derived from bacteria.

The fungal-based food webs often support large populations of microarthropods (especially mites and Collembola) and earthworms. In some soils, especially those of temperature regions, the feeding activity of litter- dwelling microarthopods may be responsible for significant accumulations of fecal pellets in the surface soil.

Earthworms also influence the stabilization of aggregates both directly, through the rearrangement of particles and the deposition of mucus, and indirectly through the stimulation of microbial, especially fungal activity. With proper conditioning the fecal pellets and casts of these organisms form stable aggregates that contain their excretory byproducts, as well as fragmented plant and microbial debris. Once stabilized, these fecal aggregates can function as anaerobic microsites where conditions are suitable for high levels of denitrification. In bacterial-based food webs, the feeding of protozoa on bacteria in the micropores of aggregates may contribute to increased organic matter turnover and reduce aggregate stability.

The changes in soil structure that result from intensive cultivation are often the associated with significant losses of SOM. The magnitude of these effects depends greatly on the intensity of cultivation and the quantity and quality of fertilizers and organic residues returned to the soil.

Minimum and no-tillage soils often support higher standing stocks of SOM and greater soil aggregation than conventionally tilled soils. The formation and stabilization of macroaggregates in no-tillage soils represent important mechanisms for the protection and maintenance of SOM that may otherwise be lost under conventionally tilled practices. This mechanism may be explained in nearly 18 per cent higher standing stock of soil organic matter in no-tillage (30.7 Mg C ha^{-1}) as compared conventionally tilled (26.1 Mg C ha^{-1}) soils at this site.

Various effects of conventionally tilled and no-tillage management on C and N dynamics may be attributed to differences in the composition of their decomposer communities.

The mycelial fungi appear to play a primary role in regulating the decomposition of surface-applied crop residues in no-tillage, while bacteria are more important to the decomposition of incorporated residues in conventionally tilled. Recent studies indicate that mycelial fungi also contribute more significantly to the formation and stabilization of soil aggregates in no-tillage than in conventionally tilled soils.

Table 8.2: Effects of Fungicide on Water-Stable Aggregate Distributions in Surface Soils (0–5 cm) of Conventional and No-Tillage Plots at HSB

Aggregate	*% Water-Stable Aggregates[a,b]*				
	No-tillage		*Conventional Tillage*		*Tillage*
Size-class	Control	Fungicide	Control	Fungicide	(P < 0.05)
> 2,000 μm	65.5	42.3*	43.2*	38.7	**
250–2,000 μm	17.2	31.2*	20.1	23.8	
106–250 μm	5.9	10.0*	9.9	10.4	**
53–106 μm	2.5	5.8*	3.5	2.9	
< 53 μm	8.9	10.6	23.4	24.2	**

a: Values are normalized to a sand-free basis.

b: Asterisks indicate significant (ANOVA/NLSD; $P < 0.05$) effects of the fungicide within tillage and aggregate size-class (*) and the main effects within size-class (**).

The above results are based on studies using the fungicide Captan to manipulate fungi in fields. In the later experiment, a 42 per cent reduction in fungal hyphae in the fungicide-treated soils of no-tillage resulted in a 40 per cent decline in the largest macroaggregates and redistribution of particles into smaller size classes. Despite similar reductions in hyphal densities, there were no significant effects of the fungicide on Water-Stable Aggregate (WSA) distributions in conventionally tilled.

Food Web in Soil

Understanding how agricultural management practices contribute to the sustained fertility and productivity of arable soils requires knowledge of both the structure and function of below ground food webs with particular attention to their spatial and temporal variances. When soils are disturbed by ploughing and the crop residues incorporated, there is much less spatial and temporal differentiation of food web structure and function. Under conventionally tilled practices, bacterial-based food webs play a relatively greater role in regulating SOM dynamics and nutrient cycling. Consisted with their higher rates of decay, incorporated residues retain large populations of both primary and secondary decomposers than do surface applied residues, yet the population dynamics of microbivorous fauna in conventionally tilled appear to be much less tightly coupled to that of the microbial population. In conventionally tilled, the soil microbial biomass also makes up a much higher proportion of the whole soil C than in no-tillage. Furthermore, a greater proportion of this biomass is bacterial in origin and concentrated in the plough layer where it is susceptible to the flushes of mineralization imposed by tillage events and wet/dry cycle. Overall, these properties of cultivated soils appear to contribute to greater organic matter losses and lower nutrient retention.

When cultivation is minimized and crop residues are retained on the soil surface, there is a much greater spatial and temporal differentiation of below ground food webs and the processes that they mediate as compared with conventionally cultivated soils. In no-tillage systems, for example fungi-based food webs often develop on surface-applied crop residues. In this environment, mycelial fungi play a key role in the immobilization of mineral N; however, it is their interactions with fungal-feeding fauna that determine the amount and timing of N releases from decaying residues. The vertical stratification of below ground food webs is also much more pronounced under no-tillage, shifting from a fungal-based food web near the soil surface to a more bacterial-based food web deeper in the profile. Where fungi are concentrated near the mineral soil surface, they play an important role in the stabilization of aggregates formed by the casting activities of meso- and macrofauna (*e.g.*, microarthropods, earthworms). This and other mechanisms of aggregate stabilization can significantly enhance the physical protection of organic matter, yielding greater SOM storage in no-tillage soils.

Determining the extent to which differences in the structure and function of soil food webs can be used to predict changes in the sustainability of agricultural management practices remains an important area for future research.

Formation Mechanisms of Complex Organic Structure in Soil Habitats

Soil contains natural reservoir of biotic debris. It consists of plant remains, dead animals, and microorganisms that are subject to continuous turnover. With time, dead biomass is either mineralized or transformed to yield diverse complex organic substances. Transformations proceeds in two stages:

1. The *First Stage* involves degradative processes that lead to the formation of substances. The more recalcitrant lignin components are selectively preserved.
2. In the *Second Stage*, the substrates and the preserved products are further transformed by synthetic processes to result in the formation of humus.

Formation of humus is a biochemical process. The degradation of biotic debris is catalysed mainly by microbial enzymes. The subsequent synthesis of the humic material can be mediated by both enzymes and abiotic catalysts that are present in the soil. Soil enzymes originate mostly from microorganisms and to a lesser extent from plants, and clay minerals or metal oxides constitute abiotic catalysts. Polycondensation reactions between reducing sugars and amino acids play a role in humus formation. One of the most important catalytic reactions in soil is oxidative coupling that links phenolic products of the degradative processes into humic polymers through enzymatic reaction and mineral catalysis. Oxidative coupling can also result in the binding of xenobiotic to humus. This reaction is usually accompanied by a detoxification effect and may be catalyzed by either enzymes or minerals. Organic chemicals can be bound to soil through various mechanisms involving either physical sorption or covalent binding. By definition, the adsorption process is reversible; in practice, however, xenobiotics are increasingly difficult to desorb as they become gradually sequestered in inaccessible microsites within the soil matrix. The compounds that are bound through stable covalent linkages are believed to the an integral part of humus, thus representing little or no threat to the environment.

Organisms Involved in Organic Matter Formation

Soil contains various living and non-living components, without this the organic matter would not exhibit such richness. Soil has following components:

Living organisms	Less than 1 per cent of the total volume of the soil
Minerals	About 50 per cent
Organic matter	3–6 per cent
Air the water	Up to 50 per cent

Trophic Groups of Living Organisms in Soil

1. Photoautotrophic higher plants, algae and photosynthetic bacteria.
2. Photoheterotrophic bacteria
3. Chemoautotrohpic bacteria
4. Chemoheterotrophic organisms

They include all soil animals, both vertebrates and invertebrates, most bacteria, and most fungi. These organisms are responsible for the biological and biochemical processes of humus formation. Chemoheterotrophs are the only organisms that can perform the two critical functions necessary to the humification process:

1. An active role of transforming raw materials into humus.
2. A passive one of serving as a source of biomass after death.

Chemoheterotrophs (microorganisms) play the most important active role in humification. Plants, however, are most important in passive processes because they serve as the major source of biomass. The tree other groups of organisms play only a passive role in the humification process.

Chemoheterotrophic Animals

These include beetle larvae, centipedes, millipedes, and termites in tropical soils. They are primary decomposers and break up plant litter into smaller pieces.

Saprophytic Bacteria, Fungi and Protozoa

They are also important in the primary decomposition process because these organisms produce extracellular enzymes that catalyze the dissolution of outer protective tissues. Secondary decomposition is performed exclusively by microorganisms, which include most of the soil saprophytic bacteria and fungi. These microorganisms specialize in selectively degrading various large organic molecules such as lignin and cellulose.

The products of microbial degradation can also be linked into humic polymers by the action of the same microorganisms. The microorganisms involved in humification are those that synthesize the oxidoreductive enzymes required to mediate the processes of synthesis.

Oxidoreductive Enzymes

They are designated as either peroxidases or mono-phenol monooxygenases. Peroxidases are found in most of the higher plants and numberous bacteria and fungi include two distinct classes:

Classes of Mono-phenol Monooxygenases	
Laccases	Laccase activity was first detected in the Japanese lacquer tree *Rhus vernicifera*. Microbial laccases were subsequently found in many fungi (*e.g.*, Agaricales, *Aspergillus funigatus*, *Rhizoctonia pratiocola*, *Trametes versicolor*, and others) and in actinomycetes like *Actinomyces albocrustosurs*, *A. galbus*, and *Streptomyces* spp.
Tyrosinases	Tyrosinase activity has been detected in fungi such as *Agaricales* (*Aspergillus nidulans*), brown-rot fungi, *Psalliota arvense*, and *Russula nigricans* and in various actinomycetes.

Both peroxidase and laccase activity have also been detected in soil extracts, apparently as a result of the production of the extracellular oxidoreductases.

Synthetic Reactions in Soil

The microorganisms responsible for the humification process are ubiquitous, therefore, the synthetic reactions are common in soil. The extent of these reactions is limited only by proliferation of the oxidoreductase producers, and the ability of these microorganisms to proliferate appears to be practically limitless. Various studies have shown that essentially all microbial species can be found in most soils, differing only in numbers and proportions. Even in deep layers of soils (up to 500 m) where only anaerobic conditions are expected to prevail, the composition of the microflora is essentially the same as that observed on the surface. Also, the transformation of organic compounds in both types of environments differs little except in the extent of the reactions.

Degradative Processes

Humus formation is a slow decomposition process. It undergoes a constant slow at a rate of about 2–5 per cent a year. Though this process is slow, even than the content of humus in soil remains fairly constant over considerable periods of time due to its systematic replenishment. Plant materials decompose fairly and rapidly, with half-lives varying from days to months. Only a small portion of the total plant input, however, is converted to humic substances. During the first year, about 70 per cent of the plant material is degraded to CO_2, and about 5–10 per cent forms new biomass. Thus, 75–80 per cent of plant substances undergo an initial rapid phase of plant litter decomposition. The material remaining is decomposed more slowly, and only the products of this second phase are incorporated into soil humus.

Types of Plant Material for Decomposition

Alexander divided these substances into six broad categories (Table 8.3).

Table 8.3: Plant Materials, which Decompose in Soil. The type of material with their availability

Plant Material	*Availability in %*
Cellulose	15–60% of the dry weight
Hemicelluloses	10–30%
Lignins	5–30%
Water-soluble sugars, amino acids and aliphatic acids	5–30%
Ether and methanol-soluble fats, oils, waxes, and pigments	
Nitrogen- and sulfur-containing proteins	

The decomposition in the soil may be the *preferential decomposition* (decomposition of easily mineralizable materials) or the *selective preservation* (decomposition of refractory plant components).

Plant components (cellulose, hemicelluloses and proteins) are easily mineralised which lignins and lipids are refractory to decomposition. During the period of rapid decomposition, the former are quickly mineralised to CO, by microorganisms as a source of carbon and energy. Nitrogen contained in the proteins is mineralised to ammonium and nitrates.

The Pathway of Lignin Decomposition

Lignin degradation, however, is critical for humification to occur because the process generates phenols and phenolic acids, which are fundamental constituents of the core humus polymers. The pathway of lignin decomposition to soil is not yet fully understood. Lignin has a complicated structure of phenyl propanoid units connected irregularly by C–O–C and C–C side chain and direct ring-to-ring linkages. The CO_2 released during lignin degradation is the product is the product of cometabolism rather than the result of direct microbial mineralization. Compared to cellulose, very little (less than 1 per cent) lignin carbon is incorporated into new biomass.

Microorganisms for Lignin Biodegradation

White-rot fungi are well known of the lignin biodegraders, because they contain peroxidases that remove electrons from lignin aromatic rings with, completely etherified hydroxyl groups. This reaction yields cationic radicals that are then stabilized by the cleavage of C–C bonds in the lignin side chains. The radicals then react with O_2 or water and become stabilized by the formation of hydroxy or keto derivatives.

Because lignin is highly resistant to biodegradation, it may stabilize reactive components like cellulose, hemicellulose, or proteins by forming chemical or physical linkages to these molecules. A similar protective effect is produced by soil minerals and humus polymers that anchor unstable plant constituents by various adsorption forces or chemical binding.

Highly degradable proteins, for example, may be protected against rapid biodegradation by their nucleophilic addition to aromatic polymers through free NH_2 or SH groups and/or by adsorption on humus or clay.

Acquired Stability

The plant components may not be completely mineralized during the initial rapid decomposition phase. Therefore, many relatively unstable plant constituents, such as polysaccharides, proteins, soluble sugars, ad amino acids, can survive in the soil for a sufficient length of time to participate directly in humus formation.

Synthetic Processes for Humus Formation

The Role of Oxidative Coupling in Humus Formation

Humus (humic acid, fulvic acid, and humin) is a complicated structure. All main fractions of humus have fundamentally similar structures. They have a large aromatic polymer that in nature occurs as micelles. Incorporated into such polymers are various phenols, quinines, nitrogen heterocycles, and benzoic acids having reactive carboxyl, hydroxyl, carbonyl and thiol groups that can, act as binding sites for carbohydrates and amino acids. Humus also contains aliphatic moieties, often hydrophobic in nature, and abounds in stable free radicals that give oxidative properties to humus and enable the attraction of ionized or protonated compounds.

Theories of Humus Formation

1. One of the oldest theories proposes that humic polymers are lignins or other resistant plant materials that have undergone minor modifications under soil conditions.
2. Plant materials serve exclusively as a source of energy suggests that the role of microorganisms is limited to the production of various amino compounds and phenols that, once discharged into soil, polymerize form humus. It seems possible that each of these processes can contribute in the humus formation.

Modern Theories of Humification

This theory is based on the fact that a central role is played by the process of polymerization. The phenolic compounds derived from lignins (ferulic acid, pyrogallol, orcinol, syringic acid, and catechol) are considered to be the most important substrates for these polymerization reactions. This hypothesis was confirmed by many researchers. They have successfully synthesized humic-like substances by polymerizing one or more phenol-derived compounds. The synthetic polymers produced in this way are quite similar to natural humus, particularly with respect to elemental analysis, exchange capacity, total acidity, resistance to microbial degradation.

Phenolic compounds may be polymerized during the process of humus formation through oxidative coupling reactions that, as with most biological processes, require catalysts. This function is fulfilled by a series of biotic catalysts (the oxidoreductive enzymes) produced by soil microorganisms and abiotic catalysts (clay minerals and metal oxides) present in soils. Humus itself also exhibits catalytic properties due to the presence of free radical components.

The mechanism of coupling requires the oxidation of phenolic compounds involving the loss of a proton and an electron from the hydroxyl group.

This loss results in the production of free radicals or reactive quinines that covalently couple to form chains of humic polymers. Such coupling can be shown) in a simple model involving one of the components of humus, namely syringic acid, to demonstrate the formation of various oligomers, ranging from dimmers to hexamers, by a laccase isolated from the fungus *Rhizoctonia praticola*. Mass spectroscopy and nuclear magnetic resonance (NMR) techniques were used to identify the products of an enzymatic reaction in which horseradish peroxidase was used to catalyze the oligomerization of guaiacol. Such studies have contributed to an understanding of mechanisms involved in humus formation.

The syringic acid and guaiacol (polymeric forms) are similar to the structure particularly with regard to the oxygen bridges and the direct carbon-carbon linkages between aromatic rings. It is possible to incorporate a per tide unit similar to that depicted in Stevenson's structure into a simplified synthetic model of humic acid. This incorporation could also be mediated by oxidoreductive enzymes, such as the laccase isolated from *Rhizoctonia proticola*.

Nutrient Dynamics

Biological management of soil fertility depends on the manipulation of organic inputs to the soil more than on any other practice. The organic matter supplies energy and nutrients to the soil biota to stimulate their activities and therefore promote soil fertility.

Organic inputs are processed by a complex web of soil organisms, but biological management is based on the assumption that the outcomes are relatively consistent. These outcomes are multiple and include the generation of inorganic nutrients such as NH_4, NO_3, PO_4, and SO_4 from organic compounds of these elements; the synthesis of soil organic matter (SOM) from precursor compounds in the organic inputs; and modification of the physical structure of the soil as a result of stimulation of biological activities. Quantitative variation in these outcomes occurs as result of following variety of potential aspects:

1. Organic inputs
2. The diversity of organisms involved in the conversion processes.
3. The complexity of the interactions between them.
4. The biological activities are regulated by the quantity and quality of the organic manure added.
5. The environmental conditions under which the processing takes place (including climate and soil type).
6. The status of the soil community.

Increasing insight into the ways in which these factors regulate soil biological activity constitutes the main progress in achieving a predictive capacity for biological management of soil fertility. Most attention has been given to the use of organic mulches to supply nutrients such as nitrogen and to the possibilities of utilizing *resource quality control* as a tool for managing nutrient availability.

Table 8.4: Chemical Composition, Decomposition Rate, and N-Release Rate for Various Agricultural Residues

Material	*Lignin (%)*	*Polyphenol (%)*	*C/N*	*kDM*[a]	*kN*[a]
Gliricidia	12	1.62	13	0.127	0.152
Rice straw	5	0.55	42	0.106	0.052
Maize stover	7	0.56	43	0.085	0.077
Lecuaena	13	5.02	13	0.062	0.055
Acioa	48	4.09	28	0.010	0.001

[a] Decomposition constant ($week^{-1}$) for dry matter (DM) and nitrogen (N), respectively.

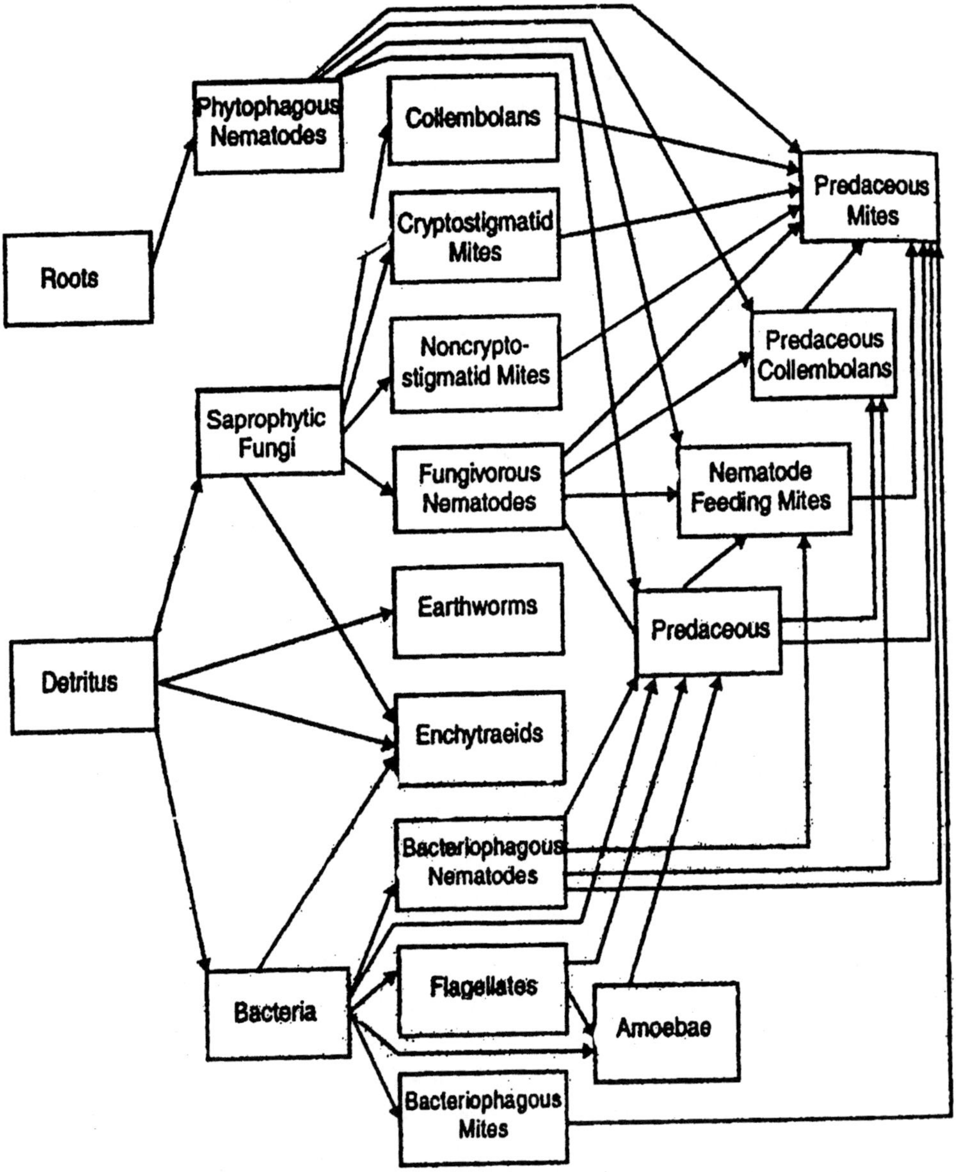

Figure 8.1: A Crop Residue (Detritus)-based Soil Food Web from a Calcaresous Soil Under Arable Cropping in the Netherlands. Under conventional tillage the bacteria-based food web is dominant; under reduced tillage the fungal and earthworms path are more prominent

Resource quality refers to the regulatory effect of the chemical composition of an organic resource on its rate and pattern of decomposition and nutrient release.

Early studies of decomposition processes demonstrated the importance of N concentration and C:N ration as determinants of the N supplying capability of plant residues.

Studies in natural ecosystems subsequently demonstrated that other indices such as the lignin concentration of the lignin: N ratio provided a better predictor of N release patterns. N is not released from forest litters until decomposition of lignin commences, and others have also stressed the importance of the more recalcitrant components as regulators of decomposition rate and nutrient release.

Polyphenols have also been implicated as regulators of N release in materials that, despite being nutrient-rich and low in lignin, are slow to release N. It has been proposed that the polyphenols interact with N to form stable polymers that are resistant to breakdown and therefore delay N release.

The above theories have been used to evaluate organic residues as sources and suppliers of nutrients within cropping systems. Bulky Organic Manures/Animal Manures/Crop Residues/ Composts.

Environmental Regulation of Organic Matter Management

Organic matter chemistry–resource quality–is not the only factor needed to make reliable predictions of the effects of organic matter inputs. The physicochemical environment has control over biological processes from a higher level in the hierarchy than resource quality. Much of this regulation such as that deriving from the climate lies outside the control of the farmer. There are practices, however, that can be manipulated so as to *environmentally tune* the influence, of organic inputs on soil biological processes. These include the location and timing of the application of organic matter. It is well established, for instance, that incorporation of residues in the soil, as opposed to surface mulching, can accelerate decomposition processes significantly and the thus alter the dynamics of nutrient availability. The breaking up of the residues, during tillage may be a factor in this effect, as well as the transfer of the organic matter into the more favorable environment below the soil surface.

It is well established that cultivation changes the composition, diversity, abundance, activity of the soil community and a variety of farming practices involve the management of organic inputs. For instance, there is now a considerable body of data relating to differences in soil biological populations and processes in minimum-tillage systems as compared with conventional tillage. These effects have been shown to be dramatic in relation to many groups of soil fauna, including such keystone groups as earthworms, with subsequent benefits gained in low-till systems to the physical structure and water regime of the soil. The potential exists to utilize these practices to manipulate the biological community and thence to determine the outcome of decomposition processes. A major challenge is to determine pragmatically the extent to which these *fine-tuning* factors of microenvironment and community structure need to be taken into account when developing management systems.

Sustainability of soil fertility and a steady increase in crop productivity would have to be the priority in future for avoiding food scarcity. Exhausted soils with poor or low amounts of nutrient content, impeded drainage causing salinity and alkalinity, erosion due to both water and wind which takes away the nutrients of the soil and increases soil acidity in highly leached condition and continuous fragmentation of holdings making mechanical operation difficult make yield increase difficult.

Soil fertility and productivity have to be restored and maintained at a high level to ensure continuous growth in agriculture. Poor soils are becoming a limiting factor more particularly due to the deficiencies of micronutrients.

Soil amendment/reclamation measures and use of micronutrients have to be simultaneously popularized. A check on further deterioration of soil fertility is critical and improvement of physical and biological properties of the soil coupled with a balance in the nutrient content in the soil alone can ensure sustained crop production. Integrated Nutrient Supply System (IPNS) and Integrated Pest Management (IPM) system have become an accepted strategy to bring about improvement in soil fertility and protecting environment.

A high proportion of organic sources of nutrients, as a renewable source replacement to fertilizers, is considered environment friendly in terms of protecting the quality of underground waters, soil property and the environment in general.

Methods of Organic Matter Recycling

Organic matter is the life of soil and recycling of organic matter is essential for maintaining long term soil fertility.

Organic manures, including animal manures, crop residues, green manures, and composts were traditionally and preferentially used as a major source of organic matter in developing countries until the 1960's when inorganic chemical fertilisers began to gain popularity.

The shift away from organic recycling practices also served to reemphasize the value of, and need for, organic amendments for short-and long-term improvement of cultivated soils and maintenance of soil productivity. Without regular additions of adequate amounts of organic materials to soils, there is often a concommitant increase in nutrient runoff, erosion, and gradual deterioration of soil physical properties.

There are three ways for maintaining and recycling the organic matter in the soil in other words these are the methods of applying manure in the soils.

In-situ Manuring

In this method the manure is provided by animal or plant sources at site. Thus it is the direct recycling of the organic matter available in the soil and useful for making nutrient in the available form.

Ex-situ Manuring

The organic matter and plant nutrients taken out in terms of biological yield need to be returned to the soil for sustaining the soil fertility. Incorporation of decomposed plant and animal wastes are the major source for this type of manuring.

Biological Manuring

Besides the above two methods there are several types of fauna and flora which either help in mobilising available organic matter and nutrient or act as nutrient provider form other sources *e.g.*, Nitrogen fixing microbes.

Chapter 9

Pest Management in Organic Rice Fish System

Pests have plagued agriculture ever since people began domesticating plants and animals. Over the centuries, farmers have developed a wide range of methods to combat these pests, but with varying degrees of success. In the 20th century, however, the introduction of commercial pesticides revolutionised pest control. These modem pesticides have helped to control and reduce crop losses to a remarkable degree. The use of these pesticides has, created some of today's major environmental and health problems. It reduced the abundance and diversity of wildlife, resulted in human health hazards, associated with acute or chronic exposure to dangerous products in the workplace and contaminated air, food and water. Most of the social costs are unevenly distributed within and between countries. About half of all pesticide poisoning of people and 80 per cent of the pesticide related deaths are occur in developing countries where only 15–20 per cent of pesticides are consumed.

The self-defeating nature of the chemical control strategy that dominates today's crop protection efforts has also become more apparent in recent years. Repeated application of synthetic pesticides have selected pesticide resistant pests world wide and there are now at least 450 species of insects and mites, 100 species of plant pathogens, 48 species of weeds resistant to one or more products. In addition, the death of natural enemies has allowed previously harmless organisms to reach pest status. The impression is that more and more pesticides have to be used to achieve less and less.

For these reasons, crop protection specialists are increasingly being asked to develop pest control methods that are more compatible with the goals of a sustainable, productive, stable and equitable agriculture. To meet these aims, research must seek to integrate a range of complementary pest control methods in a mutually enhancing fashion, namely, Integrated Pest Management. In integrated pest management insecticides are used as the last resort. In order to achieve cent per cent toxicant free food, organic farming is encouraged. Highly toxic organic products should not be used for pest management in rice fish system as these compounds may be toxic to fish. Much research work has not been carried out on the impact of organic pesticides on fish. Thus pest management in rice fish system is achieved by optimizing pest control, create and maintain situation in which insects are prevented from causing significant damage to crops by cultivating tolerant varieties, adopting mechanical, cultural and natural control methods, by using microbial pesticides, pheromone traps and augmentation of parasitoids. This strategy is best suited for the management of pests in organic rice fish system

Major Pests of Rice and their Management in Organic Rice Fish System

Large number of insects are attacking rice plant in the field. Among them only six are considered as major pests in wetland ecosystem. These pests can cause considerable damage to rice crops in certain areas at times under favorable climate conditions. The intensity of damage by these pests to the crop varies with varieties. The loss assessment and adoption of remedial measures have to be done after understanding the activity of the pests, their nature of damage, their specificity of damage to stage of the crop, agro-ecosystem etc. The major pests found in wetland ecosystem are:

Rice Stem Borer *Scirpophaga incertulas* (Walker), Pyraustidae

This pest is widespread in occurrence. Adults are quiet during daytime, hiding among rice plants or weeds near the field. When disturbed they fly only a few meters. They are active at night and lay their eggs on the upper side of the leaf. They are strong fliers and can fly 2 km at a stretch. Each female lays 200–300 eggs during a lifetime of 4 days. Eggs are laid in masses on the upper side of the leaf near leaf tips and are covered with hairs. The emerged larvae bore through the base of stem and only one larva occurs in a stem. This pest is most abundant in aquatic habitats, where flooding occurs and multiple rice crops are grown annually. Larvae seal entrance holes with silks to make them watertight. The larvae feed only on rice and related wild rice.

Management

Cultural Control

Ploughing the field and destruction of rice stubbles immediately after harvest reduces stem borer attack considerably as the pest has no other alternate host.

Cultivation of Tolerant Varieties

Kanchana, Aruna, Makom, Bhagya, Swarnaprabha, Dhanya, Nila, Kanakom, Lakshmi etc. are the tolerant varieties released from Kerala Agricultural University, cultivation of which reduces the yield reduction due to pest infestation.

Mechanical Control

Collection and destruction of adult moths by using light traps or by sweep nets before laying eggs in the field is an effective method for the suppression of the pest. Collection of egg masses which are normally infested with parasitoids and keeping them away from the field, could reduce larval population and increase the activity of parasitoids.

Biological Control

There are different types of parasites and predators present in the rice ecosystem. The stem borer eggs are parasitised by *Trichogramma japonicum*, *Telenomus* sp., *Tetrastichus* sp., *Telenomus* sp. parasitises stem borer eggs while the moth is in the act of oviposition before the eggs are covered with hair. The parasitic wasp attaches itself to the tuft of anal hair near the ovipositor of the moth and waits for the moth to lay eggs. Egg masses are the food of tettigonid and gryllid predators also. The larval and pupal stages are attacked by large number of parasites but parasitisation rates are often very low as they are inside the plant (Braconids and Ichneumonids). Carabid beetle (*Ophionea nigrofasciata*) preys on the larvae and spiders on the adult moths.

If natural enemies are less in the ecosystem, augment with egg parasitoids @ 50,000/ha. Placing of pheromone traps @ 20 traps/ha is also recommended for the effective management of rice stem borer.

Brown Plant Hopper [*Nilaparvata lugens* (Stal), Delphacidae]

It is a serious pest of rice in Asia. Its feeding causes plants to wilt and the symptom is called hopper burn. It also transmits grassy stunt and ragged stunt virus diseases. The pest is common in wetlands and rare in upland rice. Eggs are inserted in to the midrib or leaf sheath. Newly emerged nymphs are very small and whitish and seen near the base of the tillers. Nymphs and adults suck plant sap from phloem vessels. It sucks more plant sap than it requires. The excess plant sap is expelled as honeydew. During the act of feeding it produces feeding sheaths in the plant tissues. The feeding sheath blocks the flow of plant sap. This results in the drying up of the plants. Brown plant hopper out breaks are associated with cultivation of susceptible varieties and indiscriminate use of insecticides that kill all the natural enemies.

Management

Cultural Control

The alternate hosts like wild rice, *Leersia hexandra* should be destroyed to reduce the infestation of the pest.

Cultivation of Tolerant Varieties

Tolerant varieties like Pavithra, Panchami, Triveni, Matta Triveni, Jyothi, Aruna, Makom, Bharathi, Jayathi, Athira, Aiswarya, Arathi, Bhadra, Nila, Asha, Remya and Lekshmi should be cultivated to reduce brown plant hopper infestation and to increase yield:

Biological Control

Many parasites and predators attack all stages of brown plant hopper and effectively control this pest under most of the situations. Eggs are parasitized by *Anagrus* sp., *Paracentrobis* sp. and *Tetrastichus* sp. and predated by mirid bug (*Cyrtorhinus lividipennis*) and phytoseiid mite (*Amblyseius* sp.). Nymphs and adults are parasitized by *Pseudogonotopus* sp. and *Haplogonotopus* sp. (dryinids) and *Elenchus* sp. (strepsipteran), predated by water bugs, nymphs of dragon flies and damsel flies in water, beetles (*Phaederus* sp., *Micraspis* sp. and *Opheonia* sp.) and spiders (*Lycosa* sp., *Araneus* sp. and *Atypena* sp.), present among rice plants.

Cultivation of tolerant variety and conservation of existing natural enemies are found to be the best methods for the management of *N. lugens*.

Rice Gall Midge [*Orseolia oryzae* (Wood-Mason), Cecidomyiidae]

The maggot like larva feeds inside developing tillers causing the silver shoot formation. Gall midge becomes highly abundant during the rainy season in wetland rice and not a pest in upland rice areas. Gall midge abundance is favoured by cloudy or rainy weather. Gall midge adults emerge at the start of the monsoon rains from alternate hosts (wild rices, *Echinochloa* sp., *Leersia* sp., *Ischaenum* sp., *Panicum* sp., *Brachiaria* sp. and *Paspalum* sp.). Eggs are laid on the under surface near the base of leaf blades. Eggs require high RH (80–90 per cent) for development and hatching. The maggot like larva takes about six hours to move down the leaf blade on a film of dew. Larvae die if humidity is low for more than 24 hours. They move between the leaf sheaths until they reach the growing-point. The larvae feed inside the developing buds, it is converted into tubular galls, enlarges, elongates and emerges as tubes. Pupation is inside the gall. The pupa has abdominal spine which it uses to wriggle to the top of the gall in preparation for emergence as adult. By the time galls are observed, the adults have emerged. Galls appear with in a week after the larvae enter the growing point. The gall midge damages rice from the seed bed to the end of the tillering stage.

Management

Cultural Control

Destruction of alternate hosts will be useful in reducing the infestation as the insect lives in the plant when rice is not available in the field.

Avoid Staggered Planting

This pest attacks rice in the vegetative phase. If staggered planting is practiced it can grow and multiply several times and attack become more serious.

Practice Late Planting

Gall fly causes severe damage at the time of tillering. This pest requires cloudy and rainy weather and high humidity for its faster multiplication. If these weather conditions could be avoided during tillering stage, the pest infestation could be reduced to minimum. This could be achieved by late planting.

Cultivation of Tolerant Varieties

Cultivation of tolerant varieties like Kairali, Kanchana, Aruna, Makom, Dhanya, Nila, Remya or Reshmi to suppress gall fly infestation in endemic areas.

Biological Control

Phytoseiid mite attacks gall fly eggs, Platygaster oryzae parasitises the larvae and pupae and spiders *Tetragriatha* sp. and *Argiope* sp. feed on adult midge in the field.

Rice Leaf Folder [*Cnaphalocrocis medinalis* Guen., *Marasmia patnalis* (Bradley), Pyraustidae]

Rice leaf folder occur in all environment and are more abundant in rainy season. Moths may migrate to several kilometers. Adults hide on rice and grassy weeds during the day to escape predation. Eggs are laid at night singly or in small groups. Larvae on hatching crawl to the base of the youngest unopened leaves and begin to feed. They migrate to older leaves from the second instar onwards. Usually one larva in one fold (*G. medinalis*) and more than one in *M. patnalis*. The larvae feed on wild rice and graminaceous weeds and crops. Infestation is severe in shady areas.

Management

Cultural Control

If possible destroy alternate hosts when the crop is not in the field, which will reduce severity of attack in the coming crop.

Cultivation of Tolerant Varieties

Tolerant varieties released from KAU which showed tolerance to rice leaf folders are Kairali, Makom, Bhagya, Jayathi, Neeraja, Lakshmy and Reshmi. Cultivate anyone of the variety in leaf folder endemic areas to reduce the severity of pests.

Mechanical Control

The field should be monitored frequently and the leaf folds should be opened with stick in order to activate predation.

Biological Control

High natural mortality occurs at every growth stage of leaf folder. *Trichogramma* sp. develop one each inside an egg killing the leaf folder larvae before it hatches. Many species of wasps (*Apantiles* sp., *Temclucha* sp., *Trichomma* sp., *Brachymeria* sp. and *Elasmus* sp.) parasitise the larval and pupal stages. Crickets prey on eggs, damsel flies, ants and beetles prey on larvae and spiders capture adult moths. Augmentation of egg parasitoid by placing Tricho cards @ 50,000 nos. parasitised eggs) as soon as adult population is seen in the field.

Rice Case Worm [*Nymphula depunctalis* Guen., Pyraustidae]

Case worm is an aquatic larva. The damaging stage is the larvae that live in sections of leaves cut from young rice plants and rolled into tubes called cases. In a field, damaged plants occur in patches. They normally recover from attack. Case worm occurs only in rice fields with standing water and is more prevalent in rainy season. Case worm moth hides in rice field during the day and lays eggs at night. Eggs are pale yellow, disc like and irregular in shape. Eggs are laid on the underside of leaves floating on water. Newly hatched larvae scrape the surface of young leaves. Grown up larvae have branched thread like gills along the sides of their bodies and can take in oxygen from water. Larvae make their cases from leaves. First of all, the larvae cut the tip of the leaves and discarded. Next portion of the leaf is cut and roll the leaf section around their bodies and secure them with silk. Water is trapped inside the leaf cases which are open only at the head end by falling on the standing water for getting oxygen for respiration through gills. At night the larvae with the cases crawl up the rice plant to feed. Cases are replaced with each moult. When the larva is ready to pupate, it crawls on a plant and it attacks its case on tiller above the water level. Damage can begin in flooded seed bed but does not occur after maximum tillering. Case worm attack can be distinguished from that of other pests in three ways *viz.*, ladder like appearance of the chlorophyll removed leaf tissues, leaves cut at right angles as with a pair of scissors, one inch long leaf tips found floating on water. Maximum damage will be noticed in low lying fields. Damage can recover easily but maturation may be delayed by 10 days. Rice is the main host of this pest but it survive on several grassy weed species like *Echinochloa* sp., *Paspalam* sp., *Cynadon* sp., *Leersia* sp. and *Leptochloa* sp.

Management

Cultivation of Tolerant Variety

Only one variety (Bhagya) is released from KAU which showed tolerance to rice case worm. In endemic areas cultivation of this variety safeguard yield.

Cultural Control

Destruction of alternate hosts will reduce the severity of infestation. Dewatering the field for 48 hours kills case worm larvae as non-availability of water dissolved oxygen for their respiration.

Biological Control

Snails feed on the eggs of case worm moth from the leaves, water beetles feed on the larvae and spiders prey on adult moths.

Rice Bug [*Leptocorisa acuta* (Thunb), Alydidae]

It is a pest in wet land and upland rice. Adults are active in late afternoon and early morning. They rest in grassy areas during bright sunshine periods. During the dry season adults move to garden lands and rest on trees, remain dormant. Disc shaped eggs are laid in straight rows along the midrib on the upper surface of a leaf. Nymphs and adults feed on liquid endosperm of rice grains and also suck plant sap. While feeding the secreting liquid form stylet sheath that hardens around the point of feeding. Rice bug do not bore hole in the rice hull and the mouth parts enter the rice grain through the space between lemma and palea. Rice bug do not directly cause unfilled grains because they cannot remove all the liquid endosperm from a developing grain. They produce only half filled grains.

Management

Cultural Control

Many plants act as alternate hosts but wild rice and *Echinochloa* sp. are the most important hosts and they should be destroyed.

Avoid Staggered Cultivation

As far as possible cultivate whole rice field together as staggered planting increases the bug attack.

Biological Control

Grass hoppers prey on the eggs, spiders are found feeding on young nymphs. The fungus *Beauveria bassiana* is effective in suppressing both nymphs and adults of rice bug.

Thrips [*Stenchaetothrips* (*Baliothrips*) *biformis* (Bagnall), Thripidae]

It is a small insect barely visible by naked eyes. Feeding by nymphs and adults causes leaves to roll along the longitudinal axis to form a protected chamber. Thrips outbreaks are normally small scale and plants can recover from much of the damage. Thrips are present in all rice environment but are more abundant during periods of dry weather. Heavy rainfall washes off them from the plants. Adults are day flying insects. Eggs are partially inserted into the leaf tissues. Larvae and adults have rasping mouth parts and they suck leave sap. Damaged leaves curl upwards longitudinally and the leaf tips then dry up particularly when the crop is under drought stress.

Management

Cultural Control

Flooding the field to submerge the plants for two days effectively control thrips.

Cultivation of Tolerant Variety

The rice variety Nila released from KAU showed resistance to thrips.

The overall management of the pests in organic rice fish system is to be carried out by understanding the major pests of a particular area and cultivate varieties tolerant for these pests. The field should be monitored at least at weekly intervals so as to find out the incidence of the pest in the initial stage of attack and adopt mechanical, cultural and biological methods to suppress the population effectively. Various scientists from South East Asia supported the view and the found that cultivation of tolerant varieties and conservation of natural enemies in wet land rice ecosystem were the 'best mix' for the management of rice pests and protecting the yield.

Chapter 10

Changing Concepts in the Residual Toxicity of Pesticides

Introduction

Pesticide is an integral component of the scientific agricultural production programme. However, synthetic organic chemical pesticides are powerful biological poisons and can cause:

1. Direct toxicity to the applicator or consumer, especially to pregnant women and children.
2. Development of strains of pests that are resistant to a pesticide.
3. Destruction of non-target organisms such as parasites and predators of pests; honey bees and other pollinators; fish, birds, and other wildlife.
4. Outbreak of other pests that are no longer controlled by their natural enemies.
5. Accumulation of harmful residues in crops.
6. Harmful impact on domestic animals, wildlife, and the environment.
7. Potential to contaminate surface and underground water.
8. Multiple destructive impacts on human health, especially the behaviour and development of children and future generations.

The widespread use of agrochemicals as crop protection agents, wood preservatives, vector control agents or industrial/medicinal disinfectants, may result in exposure of man, animals and the environment to these chemicals. Humans may be exposed during production or application and to a lesser extent as consumers of food products, which, may contain trace levels of residues. Pesticides have neurotoxic, immunotoxic, carcinogenic, hormone disruptive and other health damaging impact.

Classical Concepts in Toxicology of Pesticides

Toxicological Evaluation of Pesticides and Safe Residue Limits

The World Health Organization (WHO) and the Food and Agricultural Organization (FAO) have conducted programmes designed to protect the health of the consumer for more than 35 years. Since

1961 experts from governments and academia have evaluated approximately 230 pesticides. Monographs are annually issued containing detailed reviews of studies relevant to the establishment of maximum permitted levels of pesticide residues in foods. It should be noted that foods containing residues at the maximum permitted levels are considered safe for consumption. However, the permitted levels are not to be considered as safety limits.

All substances exert some degree of toxicity to various forms of life, depending on the exposure level of the substance. Toxicity studies are focused on the characterization of the precise nature and extent of toxic effects and on the determination of dose levels, which do not exert adverse effects. Data are generally obtained from experiments with laboratory animals and cultured cells.

Acceptable Daily Intake (ADI)

Occasionally data from occupational or accidental exposure of animals and man to agrochemicals may be used to supplement laboratory animal data. Based on these data and using extrapolation or uncertainty factors, an Acceptable Daily Intake for humans may be calculated, *i.e.* the amount of a substance which may be consumed by humans on a daily basis during the entire life span, without appreciable risk for the occurrence of adverse effects. Comparison of these values with potential exposure of humans to a compound in the environment, including dietary exposure or intake from all foods, will indicate whether the use of an agrochemical may be permitted.

Animal Studies

The purpose of toxicity studies primarily carried out in laboratory animal species is to characterize the toxicological profile of a test compound. Usually rodents (rats, mice, guinea-pigs), dogs and non-human primates are used.

Prior to animal experimentation, test compounds must be characterized with respect to chemical identity, purity and stability. Different types of toxicological studies have been designed in order to characterize a compound:

Pharmacokinetic and Metabolism Studies

These studies are performed to determine the 'fate' of a compound upon entering the animal body, *i.e.* absorption, distribution and elimination. In many cases foreign compounds may be converted into substances (metabolites) which are more readily eliminated from the body than the original compound, and therefore information must be obtained concerning the nature of these metabolites. Furthermore, information must be obtained whether a test compound or metabolite accumulates in the body at specific sites and whether the compound interferes with biochemical pathways.

(Sub) Acute Toxicity Studies

These studies are performed to screen for potential toxic effects resulting from (unexpectedly) high exposure to a compound. Animals treated with a single or multiple dose of the test compounds via the oral, dermal or inhalation route are examined for signs of acute toxicity, skin or eye irritation or sensitization. These studies are of particular relevance for humans involved in the production or application of agrochemicals. These data are used for classification and labeling of compounds.

Long-term Toxicity Studies

These studies are designed to investigate the effects induced by a chemical in animals exposed during a substantial part of their normal life span. Extensive analyses are carried out during and upon termination of the experiment. Blood and urine analysed, food and water intake and changes in body and organ weights are recorded, and organ functions evaluated, and histopathological analysis of tissues and organs is performed. Insight in the mechanisms of observed toxicity may be obtained from these studies. These studies when properly conducted should provide information on toxic responses caused by increasing dose levels of the test compound and may demonstrate a dose level, which does not produce observable adverse effects (NOAEL).

Reproduction Studies

Reproduction studies are performed in order to verify that agrochemicals do not affect the reproductive capacity of male and female animals nor affect new borns or young animals. To this end, one or multi-generation studies may be performed with specific attention to potential adverse effects on male and female fertility, incidence of resorptions and abortions, litter size, sex ratio, birth weight and growth of new borns.

Carcinogenicity Studies

Of great concern is the potential of certain chemicals to interfere with replication processes of cells, which may lead to the formation of tumors. Specific animal studies are designed to screen for the potency of a test compound to induce tumors. These studies are usually carried out in two rodent species orally exposed to various dose levels of the test compound during most of the entire lifespan. Besides evaluation of physiological, clinical and histopathological parameters, tumors are identified on the basis of their histogenic origin. Carcinogenicity studies are performed when:

1. Relevant amounts of residues of a pesticide may be expected in foods.
2. Structural similarity of the test compound or its metabolites with other known carcinogens has been noted.
3. Experimental evidence suggests that the test compound may induce early signs of carcinogenicity.
4. The test compound has been shown to interact with cellular DNA or chromosomes (mutagenic effects) or
5. The test compound exhibits new structural characteristics.

Mutagenicity Studies

Specific tests have been designed to screen for the genotoxic potency of compounds, *i.e.* the capacity of a substance to react with the genetic material of living cells (DNA, chromosomes), which may lead to tumor formation or heritable defects.

Specialized Toxicity Studies

Special studies have been designed to assess the potential adverse effects of a chemical on the immune and neural system, and to test for skeletal malformations and other alterations induced by chemicals in foetuses during the period of organogenesis (teratogenic effects). These studies are carried out when the structural properties of a compound or data from toxicity studies indicate potential adverse effects, or in case of compounds with completely new structures.

In-vitro Studies

The use of isolated cells, subcellular fractions, or perfused organs and tissues derived from different animal species has become common practice when the toxic properties of agrochemicals are studied. In particular the use of isolated liver epithelial cells (hepatocytes) has been successful in identifying the biotransformation profiles of compounds, and for rapid screening of their toxic potency. Proper use of these systems may complement and reduce whole animal experimentation.

Evaluation of Data

The available data provide the basis for the safety assessment of compounds and should include NOAELs for the most sensitive animal species. Knowledge on the comparative metabolism of compounds in animals and man, and on differences in responses to toxic effects of substances, is taken into account when the NOAEL is established. From the NOAEL, normally expressed as mg/kg body weight per day, the Acceptable Daily Intake (ADI) for humans may be calculated, i.e. the amount of a pesticide residue in food and drinking water, which can be ingested daily over a lifetime by humans without appreciable health risk:

$$\text{NOAEL animal studies (mg/kg body weight)} \times \text{SF (safety factor)}$$
$$= \text{ADI for humans (mg/kg body weight/day)}$$

In order to extrapolate results from animal studies to humans, safety factors (SF) are applied to allow for potential differences in toxic responses between animal species and man, and to account for possible differences in sensitivity within the human population. Normally a safety factor of 100 is used: 10 (extrapolation between animal species), 10 (differences in sensitivity within the human population). However, depending on the available data and their quality, lower or higher factors may be chosen.

Maximum Residue Limit (MRL)

Maximum concentration of a residue that is legally permitted or recognized as acceptable in, or on, a food, agricultural commodity or animal feedstuff as set by Codex or a national regulatory authority.

The scheme for calculating MRL for a pesticide in a food commodity from the ADI value by considering the body weight of an average citizen and the average intake of that particular food commodity by a person per day is given below.

Decisions on the levels of pesticide residues in foods and the amount of residues likely to be consumed are part of the pesticide evaluation and approval process. The results of two important sets of studies, toxicology studies and studies on likely residues in foods arising from the use of the pesticide, must be reconciled before pesticide uses are approved. Reconciling dietary intakes of residues likely to occur in practice and acceptable intakes derived from toxicology studies is known as the risk assessment process.

Calculation of MRL

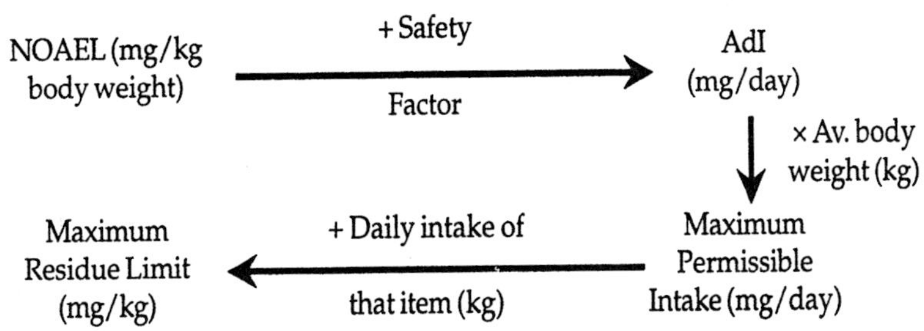

Supervised residue trials are designed to produce reliable data on residues occurring in food and feed commodities under normal commercial practice. The trial conditions are chosen from the efficacy studies, which demonstrate the use pattern (application rate, method, timing, etc) necessary for pest control, but using no more pesticide than necessary. Because the supervised **trials** are intended to generate residue data to support the establishment of MRLs (maximum residue limits) we choose the conditions from the efficacy studies which will be the maximum allowable conditions on the registered label. The supervised trials also provide the basic residue data for dietary intake estimation.

MRLs are prescribed by each country for each pesticide in each food commodity. MRLs fixed by the Ministry of Health and Family Welfare, Government of India of important pesticides in principal food commodities is given in the Table 10.1.

Table 10.1: Maximum Residue Limits (ppm) of Important Pesticides in Principal Food Commodities (Fixed by India)

	Carbaryl	*Carbo-furan*	*Chlor-pyriphos*	*DDT*	*Dime-thoate*	*Dithiocarbo-mates*	*Endo-sulfan*	*Fenvale-rate*	*Lindane*	*Mala-thion*
Food Grains	0.50	0.10	0.05			0.20			2.25	4.00
Vegetables	10.0	0.10	0.20	3.50	2.00	3.00	2.00	2.00	3.00	3.00
Fruits		0.10	0.50	3.50	2.00	3.00	2.00	2.00	3.00	4.00
Eggs				0.50					0.10	
Milk and Products		0.05	0.01	1.25				0.01	0.20	

In conclusion, the FAO and WHO are taking all efforts to assess the safety of every pesticide chemical. The suitability of continued use of the chemical by member countries is reassessed by reviewing the data generated on the environmental behaviour of such chemicals from time to time. In supervised field trials, the member countries make their own risk assessment in different situations as they exist in their own countries. In reconciliation of these two sets of data, MRLs are prescribed by each country for each pesticide in every food commodity. It is considered that commodities with a residue load below the MRL are safe for consumption from the point of view of that chemical alone. However in assessing the safety of a commodity, the combined residue load of all the toxicants will have to be considered together.

Recent Concepts in Toxicology of Pesticides

The classical concepts in the toxicology of pesticides is poised to take a sharp turn in the light of certain research findings about the astonishing effect of even miniscule quantities of pesticides on the health and well being of man and wildlife. These ultra low levels of residues which were hitherto considered as "safe" to man and the environment have been shown to be "large" enough to cause "insult" on certain vulnerable tissues like the reproductive tissue, developing embryo etc which proved to be detrimental. The result is congenital malformations, infertility, nervous disorders, proneness to cancer and a multitude of other ill effects. It may be true that the low level exposure is safe to the exposed population. But it can cause deleterious effects on the next generation of the exposed population.

A summary of such observations is given below.

Embryonic Exposure on the Physical, Emotional and Mental Development of the Offspring

1. The developing foetus is especially vulnerable to pesticides:
   (*a*) Almost all pesticides and other pollutants cross the placenta, the natural protective shield for the foetus.
   (*b*) Pesticides and other pollutants have been detected in amniotic fluid and body tissues of human foetuses even during the early stages of prenatal life.
   (*c*) Exposure to pesticides causes many health problems including the risk of miscarriage.
   (*d*) The foetus is particularly susceptible to impacts of pesticides and other pollutants because its cells divide rapidly, its blood brain barrier is more permeable and its detoxification enzymes are not yet developed.
   (*e*) Prenatal exposure to pesticides acting as hormone disruptors can cause defective sexual development and other abnormalities; including birth defects.
   (*f*) Intrauterine exposure to pesticides may cause cancer anytime in the first few decades of life.
   (*g*) Such exposure may also induce cancer and other health problems in the offspring and subsequent generations of those individuals.

Pesticides Play a Role in Increasing the Incidence of Cancers (Testicular, Prostate and Breast), Brain Tumors and other Malignancies

Cancer is characterized by the partial or complete loss of control over multiplication of cells, and the development of a local tumor mass, which often spreads, into distant organs through a process called metastasis. Cancer kills mainly through local spread destroying an essential body function, or through the effects of metastases on other vital organs, such as the brain, liver or lungs. Cancer is the most important cause of premature mortality in both sexes. It is the second most common cause of death in Ontario after all forms of cardiovascular disease. The growth of cancer begins when the ontogenesis (genes controlling cell growth and multiplication) in a cell or cells are transformed by agents known as carcinogens. Once a healthy cell is transformed into a tumor-forming type (malignant transformation), the change in its genes is passed on to all offspring cells. A small group of abnormal cells is thus established, and these abnormal cells divide more rapidly than the normal surrounding cells. Usually the abnormal cells show a lack of differentiation–that is, they no longer perform the specialized task of the cells of their host tissue and may escape normal body control. The effects of carcinogens on tissue appear irreversible. Exposure of small doses of a carcinogens over period of time results in a summation or ostentation of the carcinogenic effect. It is important to realize that the incidence of cancer today reflects exposure that occurred fifteen, twenty or even more years ago. An increase of carcinogenic contaminants in the environment today will lead to carcinogenic effects some fifteen to twenty years later for this reason it is urgent that every effort to detect and control sources of carcinogenic contamination be made well before damaging effects become apparent.

Since the widespread introduction of chemical pesticides, including an expected latency period (a time interval between exposure to a carcinogen and the clinical appearance of cancer), there has been a steady increase of incidence of certain cancers in adults as well as in children:

1. Farmers, manufacturers and applicators of pesticides have an increased risk of certain types of malignancies, especially cancer of the lip, prostate cancer, testicular cancer, lymphoma, leukemia, brain tumors, pancreatic cancer, sarcoma and multiple myeloma.
2. Apart from cancer of the lip, the same type of malignancies appear to be rising also -although less sharply -in the global population, especially in immunodeficient patients.
3. These epidemiological findings suggest that the globally rising trend of prostate cancer, testicular cancer, lymphoma, leukemia, brain tumors, pancreatic cancer, sarcoma and multiple myeloma is related to pesticide exposure.
4. In 1998, according to projected data, a typical North American male had an estimated 50 per cent risk of developing invasive cancer.
5. It is generally accepted that 70 to 90 per cent of cancer is due to external factors and is therefore preventable.

In adults, the majority of cancers are carcinomas. They arise in the membranes of those organs which are in contact with the outer environment, *i.e.*, in the bronchi due to exposure to tobacco smoke and polluted air, in the colon due to nitrates, pesticide residues in foods, and other chemicals such as food additives and food colors; hormone dependent cancers occur in excretion glands–the breast and the prostate. In adults–in contrast to childhood malignancies–the majority of cancers are the result of damage by external factors. The malignancies starting in deep tissues, bones, muscles and connective tissues (sarcomas) are rare.

Exposure to pesticides increases the risk of many types of cancer, especially hormone dependent cancers like testicular cancer, prostate cancer and breast cancer. The incidences of testicular and prostate cancer are significantly elevated in pesticide applicators as well as in young farmers and positively correlated with the amount of pesticides they were handling. Breast cancer has become one of the major causes of death in women in North America, and many other countries. In the past 40 years, the incidence of breast cancer has been steadily and dramatically increasing all over the world. Known factors account for approximately one third of the breast cancer cases; two thirds of breast cancers are due to unknown causes.

Together with its dramatic steadily increasing incidence of breast cancer, the breast cancer is also becoming more malignant because it is affecting women at a younger and younger age. It is well documented that the younger the woman at the onset of the breast cancer, the greater and more rapid the tendency of the breast cancer to spread to other parts of the body and to kill the patient.

Childhood types of cancer differ from those in later life. In children, greater proportions of cancer are found in the deep tissues of the body in the brain-brain tumors; in the bone marrow-leukemia; in lymphatic glands-lymphoma; and in the bones-sarcoma.

The location of childhood cancers in the deep tissues of the body apparently reflects an exposure to pesticides and other pollutants, during the prenatal development of these tissues well before birth:

1. Exposure to carcinogens such as pesticides, prenatally or at an early age, may cause cancer also in offspring and subsequent generations.
2. As mentioned previously, it is generally accepted that 70 to 90 per cent of all cancers are due to external causes. They are therefore preventable by avoiding carcinogens, especially tobacco, pesticides, and other environment carcinogens.

3. A National Cancer Institute study reported that children are as much as six times more likely to get childhood leukaemia when pesticides are used in the home and garden. This study can also explain why the incidence of leukaemia is about six to seven times more frequent in children from higher social classes–their exposure to pesticides is higher because of "manicured" lawns and gardens surrounding them.
4. With the growing number of childhood and early adulthood cancers, it can be expected that the ratio between cancer due to external and hereditary causes will be changed, because there will be a growing number of individuals who will transfer the cancer tendency to their children, and subsequent generations.
5. Prior to the introduction of pesticides about half a century ago, cancer in children was practically nonexistent. This was especially the case in Europe, where in the past the population was less exposed to various chemicals, because industrialization, including the use of pesticides, started later than in North America. The rarity of childhood cancer at that time can be documented by cancer registry statistics and the results of autopsies, which were mandatory in some European countries. Presently, with the use of pesticides, and rising pollution, the incidence of childhood and adult cancer has been steadily increasing, especially in Eastern Europe. In Canada today almost all hospitals for children have a special cancer wing. The steady rise of various types of cancers is also observed in wildlife and domestic animals.

Neurotoxic Effects of Pesticides on the Human Brain

1. The majority of the most widely used pesticides are designed to disrupt the normal functioning of the nervous system of the target organism.
2. Because our cells are basically similar to the cells of plants, insects and animals, it is impossible to achieve target selectivity of pesticides; this basic similarity is also valid at the sub cellular level of the human brain, which is therefore also vulnerable to pesticides.
3. The brain contains large amounts of fatty substances, and therefore fat-soluble pesticides and other fat-soluble pollutants have a tendency to accumulate in the brain in increased quantities.
4. When residues of fat-soluble pesticides and other pollutants are detected in blood, there is a significantly greater concentration of the same compound in other body tissues as a result of complex equilibrium mechanisms. In many cases, the concentrations in tissues with high fat content may be 100–300 times greater than in the blood. Because the brain is an organ with a high fat content, it is not surprising that behavioural changes, such as emotional imbalance, depression, concentration and memory problems are the most sensitive indicators of exposure to fat soluble chemicals such as certain pesticides.
5. The brain is especially vulnerable to pesticides and other pollutants during the early period of development (conception to puberty).
6. Exposure to pesticides can impair concentration, learning abilities and memory, and can trigger aggression, mood changes and other behavioural disturbances.
7. Research carried out at Harvard Medical School shows that human exposure to minute amounts of some pesticides can alter brain activity and cause prolonged periods of irritability, insomnia, loss of libido and a reduced ability to concentrate.

8. The first subtle signs of pesticide exposure are emotional imbalance, and difficulties recalling names and words.
9. There is a steadily growing number of Canadian children, as well as children worldwide, suffering from hyperactivity and learning disabilities, labeled today as attention deficit disorder (ADD) and attention deficit disorder with hyperactivity (ADDH).
10. Individuals occupationally exposed to pesticides (farmers, manufacturers and pesticide applicators, etc.) have an increased risk of neurotoxic deficits, including a significantly higher incidence of Parkinson's disease;
11. The brain damage may manifest itself only in later life, for example, as either memory impairment or dementia, that frequently is labeled as Alzheimer's disease.

Pesticides Disrupt the Action of Hormones in the Body

Hormones are chemical messengers essential for proper prenatal and postnatal development, especially the development of the reproductive system, brain and immunity. Hormones bind with receptors, specifically tailored cell proteins. The receptors and its hormones activate the genes. Hormone disrupting chemicals interfere with the way and timing of this procedure. If the hormone signals come too early, too late or not at all, the organism may be unable to start, or develop or function normally. For example:

1. Female offspring of women who took diethystilbestrol (DES) (a synthetic estrogen administered to prevent miscarriages) encounter more difficulty in conceiving, and have more miscarriages, spontaneous abortions, and tubal pregnancies. Male offspring have a higher incidence of reproductive tract anomalies such as abnormally small penises, undescended testicles at birth, and low sperm counts.
2. Prenatal exposure to pesticides acting as hormone disruptors can cause defective sexual development and other birth defects.
3. Traditional testing of pesticides on fully developed adult animals does not detect the devastating antiandrogen effects that contribute to de-masculinization of males.
4. There are no standardized or validated screens for the impact of pesticides on the embryo and foetus.
5. Insecticide used widely in household and pet care products, and agriculture formulations, when fed to pregnant animals acted as both an estrogen and an anti-androgen, thus feminizing and demasculinizing the male pups.
6. The incidences of "young males' cancer"–testicular cancer, as well as other hormone dependent cancers such as of the prostate and breast cancers are rising.
7. Average human male sperm counts have dropped by almost fifty per cent between 1938 and 1990, and they continue to decline.

Pesticide Levels in Blood have been found to be Directly Correlated with the Incidence of Diseases in Man

1. A correlation has been observed between levels of 5–10 ppb of pesticides in blood serum and various biochemical abnormalities and illnesses.
2. Elevated levels of pesticides are present in people with all types of cancer or in the terminal stage of various diseases, in comparison with healthy people.

Astonishing information about the hidden dangers of pesticide residues below the socalled "safe" limits is flowing in. There is an urgent need to reverse the present day trend of employing chemicals wherever possible for the sake of speed, convenience and profits. Two agencies that oppose the moratorium on pesticides are (1) the pesticide manufacturer MNCs–for obvious reasons and (2) the agricultural scientists and policy makers who argue that a high-tech agricultural production programme with pesticide as one of the important inputs is absolutely essential for sustaining food production in the war against hunger and production of raw materials for many of the industries. The wholesome truth that is pending admittance is that if the present status of use of pesticides in food production is continued, it will not be too far that there will not remain enough mouths for consuming the food that you so produce.

It is in this context that an agricultural production procedure with least employment of chemicals and null involvement of pesticide is evolved. Let all the energies be directed towards decontamination of our already contaminated environment and working out a sound food production method *sans* pesticides.

Chapter 11
Rice Diseases Under Organic Cultivation

Background

The early 1960s witnessed a dramatic increase in the global population and a mood of despair regarding the world's ability to cope with the food–population balance. Concerned at this impending crisis, several organizations sought to promote rice cultivation in the developing, rice growing regions of the world. High yielding, fertilizer responsive cultivars like IR-8, which could be grown throughout the year, were therefore introduced for cultivation in several parts of Asia, where 92 per cent of the worlds rice is grown. Today, 23 countries contribute more than one million tons of rice in the global scenario.

This acceleration of growth in rice production has been possible largely due to the replacement of traditional agricultural practices with modern ones. While the deployment of high yielding cultivars contributed to a direct enhancement of grain yield, the development of effective strategies for pest and disease management minimized crop losses and consequently increased the net.

Annual losses of up to 40 per cent are said to be incurred due to biotic stresses like insects pests, pathogens and weeds. Among several diseases caused by bacterial, fungal and viral pathogens that devastate rice yields all over the world, *Magnaporthe grisea* (rice blast), *Rhizoctonia solani* (Sheath blight), *Sarocladium oryzae* (Sheath rot), *Xanthon!onas oryzae* pv. *oryzae* (bacterial leaf blight) and the rice tungro virus (Tungro disease) are considered serious constraints.

Disease management strategies aimed at reducing losses and averting outbreak of epidemics have been developed in the past and have been in practice ever since. The use of chemicals, specially compounds of mercury and copper and diverse antibiotics were widely resorted to, in order to achieve high levels of disease suppression. However, the persistent, injudicious use of chemicals has been discouraged owing to their toxic effects on non-target organisms and due to the undesirable changes they inflict upon the environment. Many of these chemicals are also too expensive for the resource-poor farmers of India and Asia. Though the exploitation of host resistance and introgression of R-genes into local high yielding cultivars appear promising, the large-scale and long term use of resistant cultivars is bound to result in significant shifts in the virulence characteristics of pathogens, culminating in resistance breakdown.

Biological control therefore assumes special significance in being an eco-friendly and cost-effective strategy for disease management, which can also be used in integration with other strategies to afford greater levels of protection and sustain rice yields under organic cultivation. Though fungi, viruses, insects or any organism (other than the damaged host or the pathogen causing the disease) can be used as agents mediating biological control, bacterial antagonists to various plant pathogens have received enormous attention and are considered ideal candidates for biological control, due to obvious reasons like rapid growth, easy handling and aggressive colonization of the rhizosphere. These bacteria may mediate biocontrol by the production of antibiotics, iron-scavenging siderophores, lytic enzymes and microbial cyanides or by initiating a cascade of events resulting in the induction of systemic resistance in the host. This chapter aims at highlighting the usefulness of biological control agents, in particular *Pseudomonas fluorescens / P. putida* in suppressing some major diseases of rice.

Pseudomonas Among Other Candidate Bacterial Strains for Biological Control of Rice Pathogens

Diverse groups of microbes exist in nature. Obviously, biological control agents are not limited to any specific group–however, only very few groups of microbes have received attention and have been widely acclaimed as ideal candidates for biocontrol. Bacterial antagonists in general, *Pseudomonas* and *Bacillus* in particular are thought to be the most appealing candidates for biological control. Bacilli are Gram positive endospore producing bacteria that are tolerant to heat and desiccation–a feature that makes them very attractive for effective deployment. The pseudomonads are Gram negative rods and have simple nutritional requirements. They are known to be excellent colonizers and are widely prevalent in the rice rhizosphere. This chapter describes for a larger part, the use *Pseudomonas* as bacterial biocontrol agents for rice disease management.

Pseudomonas fluorescens was introduced as research tool for the management of rice diseases in India back in 1982 in our laboratory at the University of Madras. From then onwards a handful laboratories have intensely research in them. Both fluorescent and non-fluorescent bacteria have been implicated in the suppression of rice diseases (Table 11.1). A number of antagonistic bacteria identified from the rice rhizosphere soils of upland and lowland fields, diseased and healthy plants and from rice field flood waters have been broadly categorized as fluorescent or non- fluorescent strains. Among them, 91 per cent of the former and 33 per cent of the latter inhibited mycelial growth of *R. solani in vitro.* When used for seed bacterization, these strains reduced rice sheath blight (ShB) severity in greenhouse and field tests.

In the Philippines, Rosales have identified different groups of bacterial antagonists to seed-borne, foliar and sclerotium-forming rice pathogens. These antagonists belonged to the genera *Bacillus, Pseudomonas, Serratia* and *Erwinia.* All of 23 antagonists screened, inhibited mycelial growth of *R. solani,* while few of them could inhibit growth of other fungal pathogens like *Sclerotium oryzae, Helminthosporium oryzae, Pyricularia oryzae, Sarocladium oryzae* and *Fusarium moniliforme.*

Table 11.1: The Major Rice Diseases, Their Causal Organisms, Symptoms and Biocontrol Agents Identified

Disease	*Causal Organism*	*Symptoms*	*Biocontrol Agents*
Blast	*Pyricularia oryzae* Teliomorph: *Magnaporthe grisea*	Small, water soaked, whitish-grey spots that enlarge forming elliptical lesion with pointed ends and brown margins. In the panicle, the neck blackens and rots causing it to fall as the grain sets.	*Pseudomonas fluorescens*
Brown spot	*Drechslera oryzae*/ *Helminthosporium oryzae.* Teliomorph: *Cochliobolus miyabeanus*	Initiated as a brown, oval or round spot on leaf or sheath. Spots coalesce in severe cases. Black or brown spots seen on glumes as well.	*Pseudomonas* *P. aeruginosa* *Bacillus* sp. *B. subtilis*
Bacterial blight	*Xanthomonas oryzae* pv. *oryzae*	Dull green water soaked lesions on leaf tip or margin that enlarges forming characteristic yellow lesions with wavy margins that extend up to the sheath. Leaves later turn straw-like and blight.	*Bacillus* spp, (*B. lentus, B. substilis, B. coagulans* and *Bacillus* sp.) and DAPG–producing *P. fluorescens*
Stem rot	*Sclerotium oryzae*/ *Helminthosporium sigmoideum* Teliomorph: *Nakatara sigmoidea*/ *Magnaporthe salvinii*	Small black irregular lesions on the outer leaf sheath. Lesions advance to the inner leaf sheath and leaf rots. Inter nodes rot causing stem to lodge	*P. fluorescens* *P. aeruginosa* *B. subtilis* *B. pumilus*
Sheath blight	*Rhizoctonia solani* Teliomorph: *Thanatephorus sasaleii*	Ellipsoid to ovoid lesions on leaf sheath with a greyish white center and brown margin. Lesions enlarge and extend to the leaves causing severe blight symptoms. Sclerotia formed during advanced stages.	*P. fluorescens* *P. putida,* *P. fluorescens,* *Bacillus sp.* *B. subtilis,* *B. laterosporus,* *B. polymyxa,* *B. pumilus,* *Serratia marcescens,* *Pseudomonas,* *P. aeruginosa*
Sheath rot	*Sarocladium oryzae*	Rotting of the upper most leaf sheath enclosing young panicles. Lesions begin as oblong, irregular spots with brown margin and grey centre, which enlarge and coalesce later. Young panicles remain within sheath or only partially emerge	*P. fluorescnes* *B. subtilis* *P. aeruginosa* *Pseudomonas*
Tungro Disease	*RTV Vector: Nephotettix virescens*	Stunting, discolouration of leaves to various shades of yellow to orange. Young leaves mottled, old leaves have rusty specks of various sizes.	*P. fluorescens* For green leaf hopper vector

Also, studies from our laboratory have revealed that a large number of bacterial strains possess the ability to protect rice plants from diseases such as blast, sheath blight, sheath rot and stem rot. About 40 bacterial isolates antagonistic to the rice bacterial blight pathogen were identified through dual plate assays. Treatment of susceptible rice plants with these antagonists appears to effect statistically significant reductions in bacterial blight lesion lengths (Table 11.2) as evident from net-house and field assays. Also, there seems to exist a direct correlation between the endophytic survival of a biocontrol agent applied to rice plants through various treatments and the extent of bacterial blight suppression by *Pseudomonas putida* strain V14i. The exploitation of endophytes for biocontrol is therefore an exciting possibility, especially for the control of vascular pathogens.

Table 11.2: Evaluation of 2,4-Diacetylphloroglucinol (DAPG)–Producing *Pseudomonas fluorescens* Strains for Biological Suppression of BB in Rice cv IR24 in a Field Experiment, RARS, Pattambi, Kerala

Sl.No.	Name of Bacterial Treatment	Mean of BB Lesion Length (cm)	Per cent BB Disease Suppression	LSD Value	
				5%	1%
1.	KAD7	10.15	53.93**	3.4	4.5
2.	IMV14	9.53	56.74**	3.4	4.5
3.	IMV2	14.72	33.18**	3.4	4.5
4.	BGR19	12.62	42.71**	3.4	4.5
5.	PTB9	7.83	64.46**	3.4	4.5
6.	MON1	13.85	37.13**	3.4	4.5
7.	TVM8	16.85	23.71**	3.4	4.5
8.	VEL17	11.07	50.75**	3.4	4.5
9.	VEL10	18.17	17.52*	3.4	4.5
10.	GDY4	12.90	41.44**	3.4	4.5
11.	GDY7	12.05	45.30**	3.4	4.5
12.	TRP5	14.42	34.54**	3.4	4.5
13.	TRP18	11.38	48.34**	3.4	4.5
14.	MDR9	16.90	23.29**	3.4	4.5
15.	MDR7	10.04	54.43**	3.4	4.5
16.	STR7	14.09	36.04**	3.4	4.5
17.	VGP13	11.19	51.21**	3.4	4.5
18.	MDR16	11.87	46.12**	3.4	4.5
19.	PDY7	9.50	51.88**	3.4	4.5
20.	VLB7	20.03	9.08	3.4	4.5
21.	KVR5	133.42	39.08**	3.4	4.5
22.	TN113	15.82	28.19**	3.4	4.5
23.	KOV8	19.89	9.71	3.4	4.5
24.	RJP31	21.92	0.50	3.4	4.5
25.	KOV8	19.89	9.71	3.4	4.5
26.	PDU1	19.43	11.80	3.4	4.5
27.	PDU9	21.58	2.04	3.4	4.5
28.	Control	22.03	0.00	0.0	0.0

Each figure is a mean of 40 observations.

**Reduction in lesion length significant at 1 per cent by statistical analysis by LSD.

*Reduction in lesion length significant at 5 per cent by statistical analysis by LSD.

Table 11.3: Suppression of Some Major Rice Diseases by Biological Agents and Modes of Disease Reduction

Rice Diseases Studied	*Bacteria Used*	*Mechanism (s) Identified*
Blast	*Pseudomonas fluorescens Pf7*-14	Antifungal antibiotics (phenazines, antifungal proteins), induced systemic resistance (ISR)
Sheath blight	*Pf7*-14, *Bacillus* spp.	Chitinase, antibiotics, ISR
Bacterial blight	*P. fluorescens, Bacillus* spp. (*B. lentus, B. circulans, B. coagulans, Bacillus* sp.)	Antibacterial antibiotics (kanosomine), 2,4-diacetylphloroglucinol
Rice tungro virus	*P. fluorescens*	Antibiotics that kill the insect vector

Methods of Bacteria Application

The assessment of the potential of any organism to bring about disease suppression demands that it is applied onto the plant system in suitable form. This may be achieved either by direct inoculation (dipping seeds in a bacterial culture; aerial spraying or spreading the bacteria in sowing furrows by drip systems) or by the use of various solid-phase inoculants.

Also, recent work from our laboratory has demonstrated the insecticidal activity of *P. fluorescens* strains to the rice green leafhopper (GLH), *Nephotettix virescens*. The GLH is the insect vector of the rice tungro virus (RTV) causing the Tungro disease. This is one of the most devastating rice diseases, whose successful management has been a challenge to rice growers all over the world. Bacterial strains showed maximum toxicity to the insect vector, bringing about death of 90 per cent of the insects that were fed on bacteria treated rice leaves for 7 days.

Conclusions and Future Directions

Inconsistent performance of biocontrol agents in the field developed thus far has plagued researchers and their efforts to exploit them for commercial application. Therefore, there is a compelling need to identify efficient and dependable biocontrol agents to be used singly or as mixtures, so as to ensure consistent performance in the farmer's field. For all these, *Bacillus* strains are attractive candidate agents, as they withstand desiccation and storage conditions better than fluorescent pseudomonads or other bacteria.

Chapter 12
Mycorrhizae in Transplanted Crops

Vesicular arbuscular mycorrhizal (VAM) fungi are known to improve growth and yield of many crop plants mainly through phosphorus nutrition. However application of this technology in commercial crop production is minimal as VAM fungi are obligate symbionts and are not cultured under laboratory conditions. The obligate symbiotic nature of VAM fungi presently dictates that all VAM inoculum be grown on roots of an appropriate host plant. The best way to utilize VAM fungi for crop production may be to concentrate on transplanted crops like chilli and tomato which are normally raised in nursery beds. They can be easily inoculated to nursery beds at the time of sowing and precolonized seedlings can be transplanted to main field to harness the benefits of mycorrhization. Some researchers have observed wide variations among and with in different species of VAM fungi in their ability to promote plant growth. Recently some scientists have noticed host preference for VAM endophytes. Looking into these findings it seems advantageous to select VAM fungus suitable for a particular host-soil-climate combination. Hence the present investigations were carried out to select an efficient VAM fungus each for chilli and tomato, to determine their optimum dares of inoculation and to asses their performance on plant growth, and yield at different P levels in versitol.

Materials and Methods

Glomus fasciculatum, *G. macrocarpum*, *Gigaspora margarita*, *Acaulospora laevis*, and *Sclerocystis dussii* were tried for the selection of efficient VAM fungi for chilli var. Byadagi and tomato var L-15 based on their inoculum potential. Earthen pots of 30 cm diameter were filled with unsterile vertisol which was P-deficient (28kg P_2O_5 per hectare). VAM inocula were inoculated to pots with uniform number of infective propagules. Comparable uninoculated control pots were maintained. Fertilizer NPK were given at the recommended dose (Chilli 150 : 75 : 75 and Tomato 115 : 100 : 60 kg NPK/ha). Bold healthy seeds of chilli var. Byadagi or tomato var L-15 were sown to pots. All agronomic practices were followed. After crop harvest, shoot dry weight and weight of fruits were recorded. Per cent mycorrhizal root colonization and mycorrhizal spore counts were estimated. Shoot P concentration was determined by vanadomolybdate phosphoric yellow method.

After analysing the results of this pot trial. nursery beds of chilli and tomato were raised. The efficient VAM inoculum for each crop was tried at different levels (0, 0.5. 1.0, 1.5, 2.0 and 2.5 kg per bed) to know the optimum dose of inoculum. The size of each nursery bed was one square meter. Comparable uninoculated control beds were maintained, precolonized seedlings were transplanted to microplots of size 5.4 × 5.4 square meters for chilli and 4.5 × 3.75 square meters for tomato. Fertilizer NPK were given at recommended level for each crop. The plant growth, yield and mycorrhizal parameters were estimated by following standard procedures.

After determining the optimum level of efficient inoculum for each crop, another field trial with varied levels of P (0, 20, 40, 60, 80 and 100 per cent of recommended dose for each crop) was conducted to workout the P-savings with inoculation of efficient VAM fungi. All agronomic practices and parameters determined were similar as in the previous trial.

Results and Discussion

Chilli and tomato responded well to the inoculation of VAM fungi. Among different VAM fungi tried, *Glomus macrocarpum* and *G. fasciculatum* caused significantly maximum root colonization, spore number, shoot dry weight, shoot P concentration and fruit yield in chilli and tomato respectively (Tables 12.1 and 12.2) as compared to other VAM fungi. Inoculation of respective efficient VAM fungi to the nursery beds resulted in precolonized seedlings. Per cent root colonization and spore numbers increased with increase in inoculum levels in both the crops. Both these parameters did not differ significantly between 2.0 and 2.5 kg (Tables 12.3 and 12.4). A matching trend was observed in plant dry weight fruit yield and plant P concentration also (Tables 12.3 and 12.4) in both the crops. Thus inoculation of 2.0 kg of respective efficient VAM inoculum was found be optimum.

Table 12.1: Effect of Different VAM Fungi on Mycorrhizal and Plant Parameters in Chilli var. Byadagi

VAM Fungi	*Per Cent Root Colonization*	*Spore Count per 50 g Soil*	*Shoot P Conc. (%)*	*Shoot Dry Weight (g/plant)*	*Weight of Green Fruits (g/pot)*
Glomus fasciculatum	88	446	0.19	19	106
Gigasora margarita	77	396	0.16	15	89
Acaulospora lacvis	70	361	0.14	11	73
Sclerocystis dussii	71	369	0.14	11	76
Glomus macrocarpum	90	478	0.24	22	135
Uninoculated Control	41	98	0.06	6	48
CD at P: 0.05	4.5	101.8	0.02	1.06	4.89

Table 12.2: Influence of Different VAMP Fungi on Mycorrhizal and Plant Parameters in Tomato var. L-15

VAM Fungi	*Per Cent Root Colonization*	*Spore Count per 50 g Soil*	*Shoot P Conc. (%)*	*Shoot Dry Weight (g/plant)*	*Weight of Fruits (kg/pot)*
Glomus fasciculatum	93	476	0.35	23	1.10
Gigasora margarita	29	455	0.27	20	0.98
Acaulospora lacvis	86	439	0.24	19	0.98
Sclerocystis dussii	82	438	0.22	19	0.85
Glomus macrocarpum	90	470	0.33	22	1.00
Uninoculated Control	38	86	0.17	15	0.60
CD at P: 0.05	4.1	9.8	0.02	3.0	0.26

Table 12.3: Effect of Different Levels of Efficient VAM Fungus, *Glomus macrocarpum* on Mycorrhizal and Plant Parameter Under Field Conditions in Chilli var Byadagi

Inoculum Level (kg/bed)	*Per Cent Root Colonization*	*Spore Count 50 g Soil*	*Plant P Conc. (%)*	*Plant Dry Biomass (g/Plant)*	*Fruit (g/Plant)*	*Yield (Q/ha)*
0	26	70	0.18	49	132	20.2
0.5	47	103	0.23	63	137	20.5
1.0	61	141	0.30	72	151	21.5
1.5	71	183	0.40	80	166	24.0
2.0	79	202	0.46	87	182	26.2
2.5	79	202	0.47	89	185	26.6
SEm (±)	1.13	2.19	0.01	0.97	2.03	0.43
CD at P: 0.05	3.42	6.61	0.02	2.94	6.10	1.31

Note: Size of nursery bed 1 m x 1 m.

Table 12.4: Influence of Different Levels of Efficient VAM Fungus, *G. fascilulatum* on Mycorrhizal and Plant Parameters in Tomato var. L-15 Under Field Conditions

Inoculum Level (kg/bed)	*Per Cent Root Colonization*	*Spore Count 50 g Soil*	*Plant P Conc. (%)*	*Plant Dry Biomass (g/Plant)*	*Fruit (g/Plant)*	*Yield (Q/ha)*
0	27	72	0.18	46	1950	40.1
0.5	49	125	.0.26	54	1980	41.0
1.0	65	164	0.32	67	2096	43.4
1.5	73	195	0.38	75	2232	46.2
2.0	79	224	0.48	83	2318	48.06
25	79	225	0.48	83	2328	48.27
SEm (±)	1.36	2.09	0.01	0.96	13.6	0.74
CD at P: 0.05	4.10	6.30	0.012	2.90	48.9	2.22

In the last trial, per cent root colonization and spore counts were found to increase with increase in P-level up to 80 per cent of recommended dose beyond which these parameters decreased in both the crops (Tables 12.5 and 12.6). The plant dry weight, plant P concentration and fruit yield increased with increase in P level. However these parameters did not differ significantly between 80 and 100 per cent of recommended P in both the crops (Tables 12.5 and 12.6).

Glomus macrocarpum and *G. fasciculatum* exhibited host preference in chilli and tomato respectively perhaps because of optimum in-flow rates of P. Two kilogrammes of respective efficient VAM inoculum seems to contain optimum number of infective propagules to cause maximum plant growth and yield in both chilli and tomato. Thus it was found to be an economical and optimum dose. Inhibition of per cent root colonization and spore counts at higher P level is well documented. This may be ascribed to decrease in membrane mediated root exudated. The benefits of mycorrhization in terms of plant growth and yield can be obtained at 80 per cent of recommended P in both the crops suggesting the possibility of savings in P to an extent of 20 per cent of recommended dose.

Table 12.5: Performance of Efficient VAM Fungus, *G. marcrocalpum* at Different P Levels in Chilli var. Byadagi Under Field Conditions

P-Level (% Rec. Dose)	*Per Cent Root Colonization*	*Spore Count 50 g Soil*	*Plant P Conc. (%)*	*Plant Dry Biomass (g/Plant)*	*Fruit (g/Plant)*	*Yield (Q/ha)*
Inoculated						
0	49	70	0.09	32	95	15.5
20	51	71	0.14	37	103	18.3
40	65	92	0.22	62	151	26.5
60	73	139	0.31	81	196	34.5
80	82	189	0.40	93	254	44.2
100	69	144	0.41	96	258	45.0
Uninoculated						
100	44	54	0.36	91	230	39.0
SEm±	1.71	5.9	0.01	1.5	6.2	1.3
CD at P: 0.05	5.30	18.3	0.04	4.6	19.0	3.9

Table 12.6: Performance of Efficient VAM Fungus, *G. fasciculatum* at Different P Levels in Tomato var. L-15 Under Field Conditions

P-Level (% Rec. Dose)	*Per Cent Root Colonization*	*Spore Count 50 g Soil*	*Plant P Conc. (%)*	*Plant Dry Biomass (g/Plant)*	*Fruit (g/Plant)*	*Yield (Q/ha)*
Inoculated						
0	53	79	0.08	26.4	403	15.1
20	59	112	0.15	38.5	619	24.6
40	70	152	0.27	54.7	1091	34.3
60	84	196	0.34	65.7	1790	49.9
80	69	135	0.42	71.4	2395	56.7
100	68	114	0.45	78.0	2513	58.2
Uninoculated						
100	44	35	0.39	69.8	2334	53.7
SEm±	1.8	6.4	0.01	2.4	55.2	0.7
CD at P: 0.05	5.6	19.8	0.04	7.5	170.3	2.3

Thus these trials clearly revealed *Glomus macrocarpum* and *G. fasciculatum* to be efficient VAM fungi for chilli and tomato, two kg of respective efficient VAM inoculum to be optimum dose to harness the benefits of inoculation in terms of P savings to an extent of 20 per cent in both the crops.

Summary

The screening and selection of efficient vesicular arbuscular mycorrhizal (VAM) fungi was carried out for chilli and tomato in vertisol. Among various VAM fungi tried, *Glomus macrocarpum* and *G. fasciculatum* were found to be efficient for chilli and tomato respectively which promoted uptake of P, plant growth and yield. The optimum dose of VAM inoculum was found to be two kg per nursery bed of one square meter size. Upon inoculation of optimum dose of respective VAM inoculum to nursery beds and transplantation of precolonized seedlings to main field, a net P saving to an extent of 20 per cent of recommended doses was achieved.

Chapter 13

Biochemical and Genetic Characterisation of Mineral Phosphate

Phosphate solubilizing microorganisms (PSMs) are considered to play a significant role in making available soil phosphates for the growth of plants. PSMs solubilize mineral phosphates by bringing about a drop in the pH of the surrounding either by proton extrusion or by the secretion of mono, di and tri-carboxylic acids.

We have earlier reported that two PSMs, *Citrobacter koseri* and *Bacillus coagulans*, could not solubilize mineral phosphates present in alkaline vertisol soil when carbon and nitrogen sources were provided. They also did not solubilize rock phosphate when the medium was buffered. These PSMs secreted low levels of organic acids whereas 20–50 fold higher concentration of organic acids are necessary for solubilizing soil phosphates. This factor may therefore be responsible for the variations in the efficacy of PSMs in plant-PSMs inoculation experiments.

Since alkaline soils have a high buffering capacity, we screened rhizobacteria which could grow and solubilize rock phosphate under buffered media conditions. We could isolate three isolates from pigeon pea (*Cajanus cajan*) rhizosphere which secrete approximately 50mM gluconic acid. These bacteria have been identified as *Enterobacter asburiae* and could solubilize soil phosphates when they were grown in the presence of glucose as C source and ammonia or nitrate as the N source.

It is known that pyrroloquinoline quinone (PQQ)-dependent glucose dehydrogenase (GDH) is responsible for gluconic acid secretion in *Erwinia herbicola, Pseudomonas cepacia* and *Acinetobacter calcoaceticus, A. calcoaceticus* has two GDH isozymes. GDH-A is a monomer of 82Kd and is localized in the plasma membrane whereas GDH-B is a homodimer of 50 Kd and is present on the outer side of periplasmic membrane. While both the enzymes can use glucose as the substrate, GDH-B can also act on disaccharides like maltose and lactose whereas deoxyglucose is a specific substrate of GDH-A. The GDH from *E. coli* and *Pseudomonas* is like GDH-A. GDH from *P. aeruginosa* is 20 per cent active with galactose as substrate as compared to GDH from *P. cepacia* which is as active with galactose as with glucose.

In this chapter, we show that the activity of GDH of P-solubilizing *E. asburiae* is increased by about 5-fold under P starvation conditions and protein synthesis is required for this increase. The GDH is similar to the B type enzyme found in *A. calcoaceticus.*

Material and Methods

Culturing Procedures

The *E. asburiae* strains were cultured on media containing 100mM glucose, 25mM $MgSO_4$, 10mM NH_4Cl and the following micronutrients (mg/L) $FeSO_4.7H_2O$ (3.5), Zn SO_4 $7H_2O$ (0.16), $CuSO_4.5H_2O$ (0.08), H_3BO_3 (0.5), $CaCl_2$ $2H_2O$ (0.03) and $MnSO_4$ $4H_2O$ (0.4). Phosphate source was 0.1mM KH_2PO_4. The medium was buffered with 100mM Tris-Cl pH 8.0. These cells were use-d to inoculate to fresh medium as described below.

Introduction of *GDH* and Alkaline Phophatase (*AP*) of Soil Isolates

The isolates were inoculated in minimal medium containing 100m-M glucose, 10mM NH_4Cl, 10mM K_2HPO_4 and the micronutrients. The medium was buffered with 100mM Tris-HCl pH 8.0. The cells were harvested after growth to 0.4 O.D. and washed twice with normal saline. The cells were then resuspended in same medium with 100µ M P. To monitor the effect of protein synthesis inhibitor on the induction of GDH, chloramphenicol was added at 170 µg/ml. The cells were harvested at different time intervals, washed thrice with saline and resuspended in 50mM Tris-HCl pH 8.0 and were used for GDH and alkaline phosphatase assay. To study the levels of P concentration that was sufficient for GDH repression, different concentrations of K_2HPO_4 were added. The cells were harvested after 6 hours washed and used for GDH assay. Glucose was replaced with other carbon sources at 100mM concentration for determining their effect on GDH. The cells were harvested by centrifugation in a table top centrifuge, washed twice with 0.9 per cent saline and resuspended in the above described media either with 10mM Pas P sufficient or with 10mM Pas P deficient conditions.

Enzyme Assays

The assay mixture for GDH consisted of 1 ml of 50mM Tris-HCl buffer pH 8.75, 0.1 ml of 6.7 mM 2.6 Dichlorophenolindophenol (DCIP), 0.1 ml 20mM phenazine methosulphate (PMS) and 0.1 ml of cells in a final volume of 3 ml. GDH activity was determined by measuring the decrease in absorbence at 600 nm of DCIP mediated with PMS using glucose to begin the reaction. For alkaline phosphatase assay 0.1 ml of the cells were added in the assay system consisting of 1 mg/ml P-Nitrophenylphosphate (pNPP) in 30mM Tris-HCl pH 9.0. The activity was determined by estimating P-Nitrophenol liberated from pNPP and comparing it to absorbence of standard P-nitrophenol at 420 nm. Protein estimations were done by the dye binding method.

Results

Induction of *GDH* and *AP*

The specific activity of GDH was found to increase upon P starvation. There was basal level of GDH activity which was increased to 5 fold in 6 hours. By P starvation in all the isolates. The increase in GDH activity correlated with the decrease of media pH. GDH of these isolates was induced only upon P starvation and was independent of the carbon source used for the growth of the cells (Table 13.1). The increase in activity upon P starvation required *de novo* protein synthesis as addition of chloramphenicol abolished the induction of GDH (Table 13.2).

Substrate Specificity of *GDH* of Soil Isolates

In order to determine whether the GDH of these isolates was of GDH-A or GDH-B type the GDH activity was assayed with different substrates. The whole cells of all the three soil isolates could show activity with maltose and galactose as substrates but could not use deoxyglucose as a substrate. The activity with these substrates also increased upon P starvation similar to the increase in GDH (Table 13.3).

Discussion

The P-solubilizing *E. asburiae* isolates under study, secrete high concentration of gluconic acid and showed the induction of glucose dehydrogenase activity upon P starvation. The induction required *de novo* protein synthesis and different concentration of free P was required to repress the GDH activity with 5mM P repressing the activity to about 50 per cent for isolates 1 and 2 and more than 75 per cent for isolate 3. This indicates that the GDH of these isolates is regulated by the P concentration in the media and could be part of phosphate starvation inducible mechanism of these bacteria. That the soil isolates expressed genes induced for P starvation under conditions allowing maximum GDH activity is seen from the induction of alkaline phosphatase under the same conditions. It is known that various strains of *Rhizobium* are P limited at 100–150 μm concentration of P as determined by the depression of AP activity.

Table 13.1: GDH Activity of Soil Isolates with Various C Sources

C Source	*GDH Activity (Units/mg Total Protein)*	
	+P	*–P*
	Glucose	
Isolate 1	262±4.1	1390±149
Isolate 2	265±7.5	1737±125
Isolate 3	285±8.5	1500±55
	Glycerol	
Isolate 1	242±7.5	1303±100
Isolate 2	274±10	1600±60
Isolate 3	236±5.1	1418±125
	Mannitol	
Isolate 1	225±6.0	1193±130
Isolate 2	250±6.5	1395±75
Isolate 3	240±5.0	1158±100
	Gluconate	
Isolate 1	200±10.0	1000±100
Isolate 2	220±5.5	900±65
Isolate 3	250±5.0	1200±120

Values are expressed as mean ± S.D. of three independent experiments.

In *E. herbicola* the P solubilizing property was shown to be inducible by P starvation. 20mM P was required to completely abolish HAP solubilization by *E. herbicola*. The mineral phosphate solubilizing gene involved in PQQ biosynthesis cloned from *E. herbicola* in *E. coli* also showed P induction/ repression phenotype indicating it to be part *of* phosphate starvation system. Similarly the DCP solubilization by *E. coli* is repressed by P concentration. However, the Ed pathway enzymes are not induced by P starvation in *E. coli.* In Pseudomonas GDH is induced by gluconate or 2-ketogluconate. The GDH of the soil isolates reported here was induced only by P starvation as the activity was similar with all the C sources used for the growth of the bacterial strains.

Table 13.2: Effect of Chloramphenicol on Induction of GDH

	GDH Activity (Units/mg Total Protein)	
	Chl+	*Chl–*
Isolate 1	300±55	1000±100
Isolate 2	250±40	1200±125
Isolate 3	200±50	900±75

Values are expressed as mean ± S.D. of three independent experiments.

Table 13.3: Substrate Specificity of GDH of High-Potency PS Bacteria

Substrate	*GDH Activity (Units/mg Total Protein)*					
	Isolate 1		*Isolate 2*		*Isolate 3*	
	+P	*–P*	*+P*	*–P*	*+P*	*–P*
Maltose	100	600	150	700	150	750
Galactose	50	200	50	150	60	250
Lactose	UD	UD	UD	UD	UD	UD

Values are expressed as mean of three independent experiments.

The GDH of these isolates was active with maltose as substrate and could not use deoxyglucose indicating it to be like GDH-B of *A. calcoaceticus* which is a periplasmic protein. The whole cells also showed activity with galactose although it was substantially less as compared to glucose.

It would be of interest to characterize further the gluconic acid secretion by these *E. asburiae* strains and also clone the gene (s) involved in the overproduction of gluconic acid from these soil isolates. The cloned gene (s) could then be transferred to PGPRs like *Rhizobium* and *Pseudomonas* allowing them to provide P to the plants they colonize.

Summary

We have isolated *Enterobacter asburiae* strains capable of releasing Pi from alkaline vertisol and which can secrete 50mM gluconic acid. We show here that the phosphate solubilization is P-starvation inducible and correlates with an increase in the glucose dehydrogenase (GDH) activity of these strains. The GDH enzyme is similar to the B-type GDH of *Acinetobacter claocoaceticus.* GDH activity reaches a maximum value in 6 hours at 0.1mM P concentration. This is five-fold higher than that in the presence of 10 mM Pi and requires *de nova* protein synthesis.

Chapter 14

Effect of Phosphobacterium

Swordbean, otherwise called Jackbean is a dual crop and contains a protein content of around 24.5 per cent. It can play a vital role in filling up the protein shortage. It is nontraditional pulse and relatively an unexplored legume with higher productivity and wider adaptability. The tender pods are delicious and become fibrous on maturity. Seeds are large and contain an alkaloid canavalin. This crop can also be grown as an ornamental hedge or border plant or as a forage legume.

Material and Methods

The present study on swordbean, variety SBS-1 (Swordbean Selection-1), was designed to study the effect of phosphobacterium (*Bacillus megaterium*) under graded levels of phosphorus at the farm of Gandhigram Rural Institute (Deemed University), Tamil Nadu, during the year 1995–96. The experiment was carried out in randomised block design with three replications. There were three methods of phosphobacterium application *viz.*, seed treatment (500 g of inoculum for 120 kg of seed per hectare), soil application (2000g of inoculum per hectare mixed with dried cowdung manure) both basal along with an untreated control. Three levels of phosphorus were tried *viz.*, 25, 50 75 kg P_2O_5 per hectare with a common dose of 25 kg N as basal. Sowing was done in plots of size 4.5 × 3.0 m at a spacing of 45 × 30 cm under ridges and furrow system. Observations on 25 randomly selected plants per plot were recorded for plant height (cm), number of flowers and fruits per plant, fruit setting percentage, number of seeds per pod and seed yield (kg per hectare). One hundred seed weight (g) and germination percentage were assessed from each replication. Seeds were sown in germination trays in incubator at 25°C and 95 per cent RH. Seedling count was taken after seven days and germination was expressed in percentage.

Results and Discussion

Results obtained indicated that there was a significant effect of seed inoculation of phosphobacterium on swordbean in increasing the fruit setting percentage (49.27 per cent), number of seeds per pod (10.52) and seed yield (1410 kg per ha) as compared to soil application and untreated control (Table 14.1). This may be attributed to the presence of seed borne inoculum within rhizosphere where root hairs absorb the nutrients. Different levels of phosphorus application showed no significant difference except in influencing the number of fruits and fruit setting percentage. They were found to be more with either 50 or 75 kg P_2O_5 application as compared to 25 kg P_2O_5 per hectare. One hundred seed weight and germination percentage of seeds were not influenced by different treatments.

Table 14.1: Effect of Phosphobacterium on Growth and Seed Yield of Swordbean

Treatment	*Plant Height at Harvest (cm)*	*No. of Flowers per Plant (75th Day)*	*No. of Fruits per Plant (75th Day)*	*Fruit Setting Percentage*	*No. of Seeds per Fruit*	*One Hundred Seeds Weight (g)*	*Seed Yield (kg/ha)*	*Germination (%)*
			Phosphobacterium					
P_0–Untreated control	111.2	36.18	14.73	40.28	8.67	132.60	1050	98.7
P_1–Seed inoculation	113.1	39.44	18.33	49.27	10.52	136.70	1410	99.1
P_2–Seed application	112.1	36.98	15.20	40.64	9.83	136.20	1250	99.6
CD (P = 0.05)	NS	0.58	0.89	1.64	0.04	0.51	140	NS
			Phosphorus Levels					
F_1–25 kg/ha	111.7	37.69	15.47	41.91	9.63	132.60	1200	99.1
F_2–50 kg/ha	111.2	37.20	16.33	43.51	9.70	133.80	1250	99.6
F_3–75 kg/ha	111.6	37.71	16.47	44.74	9.66	133.80	1260	99.7
CD (P = 0.05)	NS	NS	0.86	1.64	NS	0.51	NS	NS

NS: Not Significant

Application of phosphobacterium seemed to be environment friendly and the study revealed the efficiency of seed inoculation of phosphobacterium in reducing the phosphorus loss due to insolubility thereby increasing the seed yield of swordbean.

Summary

The study on Swordbean variety SBS-1 was designed to study the effect of phosphobacterium (*Bacillus megaterium*) under graded levels of phosphorus. There were three methods of phosphobacterium application *viz.*, seed treatment (500g of inoculum per hectare), soil application (2000g per hectare) and an untreated control. The levels of phosphorus tried were 25, 50 and 75 kg P_2O_5 per hectare. The results revealed that the seed inoculation of phosphobacterium increase fruit setting percentage (49.27 per cent), number of seeds per pod (10.52) and seed yield of swordbean (1410 kg) as compared to soil application and untreated control. Different levels of phosphorus application showed no significant difference except in influencing the number of fruits and fruit setting percentage. They were found to be more with either 50 or 75 kg P_2O_5 application as compared to 25kg P_2O_5 per hectare. One hundred seed weight and germination percentage of seeds were not influenced by different treatments.

Chapter 15
Inoculation of Phosphomicrobes

Groundnut requires large amount of phosphorus for growth and high phosphorus availability to the crops in the form of different phosphatic fertilizer is of great economic importance, especially in country like ours, where the soils are poor in available phosphorus content and super phosphate in expensive. Insoluble inorganic compounds of phosphorus are largely unavailable to the plants. Phosphomicrobes have a capacity in bringing the insoluble phosphate into solution. Apart from increasing the availability in soil, they are known to produce growth promoting substances which in turn enhance growth and yield. The fertility status of the soils is likely to go down unless adequate quantities of plant nutrients are added to the soil. The information on the use of phosphomicrobes on the growth and yield of crops in vertisols of Karnataka is scanty and inconsistent. Therefore long term field trial investigation was undertaken to know the performance of phosphomicrobes on yield and nutrient uptake in groundnut.

Material and Methods

A field experiment conducted at Main Research Station, University of Agricultural Sciences, Dharwad for three consecutive years of Kharif 1992, 1993 and 1994 on medium black soils. The soil having a pH of 8 organic carbon 0.54 per cent, available N, P_2O_5 and KO were 254, 35 and 335 kg/ha respectively. The treatments consisted of phosphorus application (50 kg P_2O_5/ha) through two sources *viz.*, Mussoorie rock phosphate (MRP) and single super phosphate (SSP) with 0, 25, 50, 75 and 100 per cent of the recommended dose. The phosphomicrobes includes *Aspergillus awamori* and *Pseudomonas striata*. Total of 18 treatments were replicated thrice in a Randomised Block Design. The fertilizers N and K_2O were given as per the recommenced dose (each 25 kg/ha). After harvest of the crop yield and yield components were recorded. The nodule number and weight was taken at 45 DAS. The nutrient uptake studies were made at harvest. The nitrogen in the plant sample was estimated by modified Kjeldahl's method and the phosphorus by vanadomolybdo phosphoric yellow colour method using spectrophotometer at 470 nm as described by Jackson.

Results and Discussion

Yield and Yield Components

The pod yield recorded with inoculation of *A. awamori* (3399 kg/ha) of *P. striata* (3416 kg/ha) alone were 12 and 13 per cent higher respectively when compared to control (3026 kg/ha). The beneficial effect of P-solubilizers on grain yield of rice have been reported by Gaur. Inoculation of *P. striata* along with the application of 75 per cent MRP + 26 per cent SSP recorded (4648 kg/ha) and differed significantly over uninoculated treatments (Table 15.1). Similar studies were corroborated by Manjaiah. The number of pods per plant and pod weight per plant followed the similar trends as that of pod yield per hectare.

Table 15.1: Number of Pods Per Plants, Pod Weight (g) Per Plant, Pod Yield (kg/ha) and Haulm Yield of Groundnut as Influenced by Phosphomicrobes (Average of Three Years)

Sl.No.	Treatments	Pod Numbers per Plant	Pod Weight (g) per Plant	Pod Yield (kg/ha)	Haulm Yield (kg/ha)
1.	Control	12.86	14.23	3026	3137
2.	MRPO: SSP100	17.93	19.16	3543	3743
3.	MRP25: SSP75	17.20	20.10	3492	3627
4.	MRP50: SSP50	18.10	20.10	3413	3637
5.	MRP75: SSP25	16.60	20.63	3368	3605
6.	MRP100: SSPO	16.63	17.90	3336	3445
7.	*Aspegillus awamori* (A)	17.70	19.80	3399	3360
8.	MRPO: SSP100 + A	18.63	22.03	3613	3871
9.	MRP25: SSP75 + A	17.96	21.76	4100	3504
10.	MRP50: SSP50 + A	17.73	20.70	3600	3761
11.	MRP75: SSP25 + A	20.70	20.36	3869	3173
12.	MRP100: SSPO + A	17.36	20.43	3518	3187
13.	*Pseudomonas striata* (P)	18.43	21.23	3416	3623
14.	MRPO: SSP100 + P	20.40	22.53	4445	3853
15.	MRP25: SSP75 + P	18.66	21.70	3842	3731
16.	MRP50: SSP50 + P	19.30	21.43	3865	3531
17.	MRP75: SSP25 + P	22.40	23.90	4648	3900
18.	MRP100: SSPO + P	10.06	20.50	3571	3464
	S.Em±	1.40	0.47	139	300
	C.D. at 5 per cent	4.10	1.35	401	NS

MRP: Mussoorie rock phosphate; SSP: Single sugar phosphate; NS: Not significant.

The haulm yield of groundnut was unaffected due to various treatments. But inoculation of phosphomicrobes significantly influenced the nodule number and nodule dry weight at 45 DAS. Inoculation of *P. striata* with the application of 75 per cent MRP + 25 per cent SSP recorded significantly higher nodule number per plant (110.20) when compared to uninoculated treatments. Nodule dry weight was highest in *Pseudomonas striata* with 100 per cent SSP (119 mg/plant). Similarly Manjaiah reported that phosphorus application increased the nodule number and nodule dry weight and SSP was more effective than MRP.

Uptake of Nitrogen and Phosphorus

Inoculation of *P. striata* with 75 per cent MRP + 26 per cent SSP recorded significantly highest nitrogen uptake (83.29 kg/ha) over control (46.80 kg/ha) indicating 1.7 fold increase in N uptake by groundnut (Table 15.2). The increased in N-uptake by groundnut crop was ascribed due to the application of phosphorus and their synergistic interaction.

Table 15.2: Nodule Number, Nodule Dry Weight (mg/plant), Nitrogen and Phosphorus Uptake as Influenced by Phosphomicrobes (Average of Three Years)

Sl.No.	*Treatments*	*Nodule Number/ Plant at 45 DAS*	*Nodule Dry Weight (mg/pl at 45 DAS)*	*N-uptake (kg/ha)*	*P-uptake (kg/ha)*
1.	Control	49.73	46	46.80	5.85
2.	MRPO: SSP100	78.73	81	66.20	9.00
3.	MRP25: SSP75	64.66	75	63.00	8.50
4.	MRP50: SSP50	80.83	65	60.25	8.22
5.	MRP75: SSP25	90.73	50	55.40	6.90
6.	MRP100: SSPO	75.50	83	59.50	7.26
7.	*Aspergillus awamori* (A)	66.00	75	61.25	7.00
8.	MRPO: SSP100 + A	98.80	104	71.00	10.50
9.	MRP25: SSP75 + A	77.20	68	68.25	9.25
10.	MRP50: SSP50 + A	90.86	70	60.00	9.90
11.	MRP75: SSP + A	93.93	99	63.28	10.25
12.	MRP100: SSP + A	81.83	76	57.94	10.00
13.	*Pseudomonas striata* (P)	73.06	84	64.00	8.30
14.	MRPO: SSP100 + P	115.46	119	80.25	11.31
15.	MRP25: SSP75 + P	90.60	69	70.20	10.20
16.	MRP50: SSP50 + P	102.60	80	69.60	10.70
17.	MRP75: SSP25 + P	110.20	89	83.29	13.00
18.	MRP100: SSPO + P	88.06	93	69.40	9.65
	S.Em. ±	4.07	2.50	0.84	0.40
	CD at 5%	11.72	7.45	2.50	1.20

MRP: Mussoorie rock phosphate; SSP: Single sugar phosphate; NS: Not significant.

Inoculation of *A. awamori* or *P. striata* alone recorded 20 (7 kg/ha) and 30 (8.3 kg/ha) per cent higher total P-uptake by groundnut plant than control (5.85 kg/ha). The higher P-uptake may be due to the increased availability of phosphorus in the soil due to solubilization of native phosphorus by phosphomicrobes. Ahmed reported increased P-uptake by wheat and gram with the inoculation of phosphobacterin. Among all the treatments, inoculation of *P. striata* with 75 per cent MRP + 25 per cent SSP recorded highest P-uptake (13.00 kg/ha).

The results clearly indicates the use of phosphomicrobes to augment the yield and nutrient uptake in groundnut crop. The Indian farmers succumbs to a greater loss due to the high cost incurred by application of phosphorus through SSP. Hence application of 75 per cent of the recommended dose of P through MRP and 25 per cent through SSP along with the inoculation of *P. striata* is more lucrative than applying 100 per cent of phosphorus through SSP alone.

Summary

A field experiment was conducted during Kharif seasons of 1992, 1993 and 1994 in vertisols of the Main Research Station, University of Agricultural Sciences, Dharwad to know the effect of phosphomicrobes on yield and nutrient uptake in groundnut. The results vividly indicates that, inoculation of *Pseudomonas striata* along with the application of the recommended phosphorus (50 kg P_2O_5/ha) through 75 per cent Mussoorie rock phosphate (MRP) and 25 per cent single super phosphate (SSP) recorded significant higher pod number, pod weight and yield compared to control and all other uninoculated treatments. The former recorded more number of nodules and nodule dry weight. The uptake of nutrients were significantly superior in the inoculated treatments than the uninoculated treatments.

Chapter 16
Phosphocompost from Neem

Decomposition of organic matter is a natural process responsible for the breakdown of complex organic compounds. During this process, inorganic minerals locked up in organic residue released are made available to plants. The activity of microorganisms and various physico-chemical agencies bring about change in the structure and chemical composition of organic matter which in turn determines species composition of late colonizing microorganisms. Microbial succession continues until all the organic substances are mineralized.

The composting of neem along with low grade rock phosphate and 'P' solubilizers is known to increase the solubility of insoluble phosphates. Although the extent of solubilization can vary with the kind of wastes. Neem compost enriched with rock phosphate and *Aspergillus awamori* produced a good quality of phosphocompost.

Material and Methods

Neem leaves were collected from the university campus and placed in specialized pots for composting.

Phosphate Solubilizing Microorganism

Aspergillus awamori was selected as it is an efficient solubilizer of insoluble phosphates.

Phosphatic Fertilizer

Mussoorie rock phosphate served as the phosphate source and was added at 100 µg/g concentration to the compost on the 17th day.

Mycorrhizal Fungi

Glomus fasciculatum was used to treat the plants along with the compost. *Abelmoschus esculentus* (Okra, bhendi) was used as a test plant.

The treatments given were as follows:

1. C (Control)

2. NC + A.a + M (Neem compost inoculated with *Aspergillus awamori* and *Glomus fasciculatum*)
3. NC + RP (Neem compost with rock phosphate)
4. NC + RP + M (Neem compost with rock phosphate and *Glomus fasciculatum*)
5. M (Mycorrhiza alone-*Glomus fasciculatum*)
6. PSM + M (*Aspergillus awamori* + *Glomus fasciculatum*)
7. PSM (*Aspergillus awamori* alone)
8. NC (Neem compost alone)
9. NC + A.a (Neem compost with *Aspergillus awamori*)

Experiments were carried out in replicated randomized block design with 3 replications for each treatment. Plant growth, 'P' content and yield were recorded at the maturity of the crop (70 days). The yield obtained were compared with respective treatment control.

Results and Discussion

The data obtained showed that when bhendi plants inoculated with neem compost along with *Aspergillus awamori* and VA Mycorrhiza. as given in the histogram showed highest fruit yield.

Neem compost along with VAM showed significant increase in 'P' content of soil and shoot 'P' than did alone.

The data indirectly show that the insoluble rock phosphate incorporated to neem waste is solubilized and is available to bhendi plants due to biochemical and microbial activity during decomposition of organic wastes. Since the neem compost also contain efficient 'P' solubilizers it improves the phosphorus used efficiently and VAM activity for maintenance of soil productivity which is very much essential for ecofriendly and sustainable agriculture.

Chapter 17

Cyanobacteria on Soil Characteristics

Soil salinity has become an alarming problem in Indian agriculture. Sufficient technology is developed for reclamation of such soils in Indo-Gangetic plains. The salinity problem is however of different nature in peninsular region needing some different integrated approach. Apparently salinity research has acquired many dimensions. The response of cyanobacteria (BGA) as biological ameliorant is being increasingly understood. In Maharashtra the salt affected area is 30,000 ha. In Vidarbha Region, it is located in Puna River basin mostly under rainfed cropping, having Black Cotton soil with an average annual rainfall 827mm. The conventional recuperative measures are long known, and are found to suffered from limited adoptability due to economic reasons.

The object of present studies is endeavoured to assess the feasibility of usingcyanobacterial inoculant (Blue Green Algae) as a bioameliorant in problem soil under protective irrigation conditions as a possibility of harmless, low cost, renewable, easily adoptable measure to augment conventional methods.

Material and Methods

A field experiment was conducted in problem saline soil at Washim Road Block of P.K.V. Akola in Rabi 1994, using randomized block design, with five replications and eight treatments combinations. Gram crop was sown in October 1994 and harvested in February 1995. Treatments consisted four levels of Blue Green Algae (0, 25, 50 and 75 kg/ha) incorporation with three levels of gypsum (0, 1 and 2t/ha) addition to soil at sowing.

Soil based inoculant culture of BGA was obtained from Agricultural Microbiologist, Mahatma Phule Krishi Vidyapeeth, Pune which consisted mixture of *Tolypothrix* sp. (G4), *Calothrix* sp. (C3) and *Nostoc calcicola* (B5).

Soil samples were collected from surface layer up to 9 cm depth from each of the plot. Samples from five replications were mixed thoroughly to get a composite sample for each of the treatment. Soil parameters *viz.*, pH, EC, Organic Carbon, ESP, available N and available P were quantified using standard procedures. Pretreatment soil sample was used as a check. Grain yield of Gram Cv. CHAFFA from each of the net plot was recorded on sun drying for 4 days after harvest (20.2.95). Weight of 1000 grains from each of the plot was measured to assess effect on quality/boldness.

Results and Discussion

Yield Responses

Increase in grain and biomass yield of Gram (*Cicer arietinum*) crop due to single application of BGA and/or gypsum as soil ameliorant are recorded with a synergistic response (Table 17.1). The rise in grain yield being 3.94 to 5.85 q/ha due to BGA and 3.72 q/ha due to gypsum with an additive effect of 0.232 q/ha under protective irrigation cultivation on a problem patch of soil. Increase in yield of rice and sugarcane after first year of reclamation of Usar soils with native algal flora are reported by Singh (1961); and in paddy in saline soils by Kaushik and Subhashini. Kaushik have reported a significant yield increase in paddy in second and third year of algalization to sodic soils of Delhi. Similar promising yield response to paddy crop in salt affected soils at Govt. Experimental farm, Gohad in M.P in 1985 and 1986 due to algalization is reported by Sharma.

Table 17.1: Effect of BGA and Gypsum Incorporation to Saline and Soil on Yield of Gram Crop (Rabi 1994) and Soil Parameters

Treatments	*Grain Yield kg/ha*	*Weight of Grains g/1000*	*pH*	*EC dSm^{-1}*	*Org. Carbon*	*ESP %*	*Available N kg/ha*	*Available P kg/ha*
BOGO	1312	152.2	9.70	0.66	0.42	25.24	191.2	8.9
B25GO	1706	152.8	9.48	0.59	0.46	23.33	206.9	10.6
B50GO	1833	153.2	9.36	0.55	0.52	22.50	213.2	10.6
B75GO	1897	153.2	9.30	0.50	0.58	20.98	225.7	12.4
BOG1	1684	153.7	9.35	0.49	0.46	21.80	197.5	11.5
BOG2	1731	154.0	9.25	0.44	0.48	19.83	202.1	16.9
B25G1	1729	154.0	9.34	0.47	0.48	21.80	206.9	12.1
B75G2	1845	154.0	9.24	0.40	0.50	19.93	222.6	13.5
Pretreatment count			9.72	0.68	0.40	25.48	190.4	8.8
SE(m)±	11.2	0.2						
CD at 5%	32.3	0.577						

Grain quality is noticed to have improved in present studies as evidenced by statistically, significant increase in weight of grains. It was 154.02g and 153.97g when B 25 G1 and B 75 G2 is applied respectively as compared to 152.26g in control (Table 17.1).

Soil Properties

Promising improvement in soil properties due to application of BGA and/or gypsum are observed in gram crop in these studies. pH is reduced to 9.30 due to BGA 75, and 9.24 due to gypsum application at the rate of 2 tonnes as against 9.70 in control. EC is also reduced from 0.66 dSm^{-1} to 0.50 dSm^{-1} when BGA was applied and 0.44dSm^{-1} when gypsum was applied. The ESP is also markedly reduced in present trial from 25.24 per cent in control to 19.93 per cent when B75 G2 is applied and 19.83 per cent with BO G2 application. Reduction in soil pH due to algalization in salt affected soils is reported by Hemantkumar.

Summary

Feasibility of using cyanobacterial inoculant (BGA) as bio-ameliorant to a problem saline soil was tried under rainfed (arid) field conditions of PKV Akola (Maharashtra) during 1994–95. Four levels of BGA inoculant with and without gypsum were assessed using Randomised Block Design.

Increase in grain and biomass yield of gram (*Cicer arietinum*) due to single application of BGA was evidenced. The rise in grain yield being 3.94 q/ha due to BGA (soil application at the rate of 25 kg/ha) 3.72 q/ha due to gypsum (at the rate of 1 t/ha). The yield in control being 13.13 q/ha. The weight of grains was 154.02 g/1000 and 153.97 g/1000 when B25 G1 and B75 G2 was applied respectively as compared to 152.26 g in control. Soil pH was reduced to 9.30 and 9.24 when BGA at the rate of 75 kg/ha and gypsum at the rate of 2t1ha respectively was applied as against the initial pH 9.70. Similarly EC was also reduced from 0.66 dSm^{-1} to 0.50 dSm^{-1} when BGA was applied and 0.44 dSm^{-1} when gypsum was applied. Improvements in available C, N, and P were also observed.

Chapter 18

Crop Response to Algalization in Rice Variety

Incorporation of Blue Green Algae in standing water in paddy crop have been considered as a sure remedy for scarcity of chemical nitrogen. The performances of BGA application in rice fields have aptly been reviewed and further emphasized the need for more thrust through state extension agencies. The technology has already been introduced in many states since 1975. The impact could not reach to many rice growing areas in Tamil Nadu due to absence of emphatic extension propaganda.

Material and Methods

Blue Green Algae was applied to paddy crop one week after transplanting by broadcasting in July 1991 in a replicated field trial at Tharsa, Distt. Nagpur, BGA inoculant was obtained from the National Facility Centre for BGA; Indian Agricultural Research Institute, New Delhi, which consisted *Anabaena* sp., *Nostoc* sp., *Tolypothrix* sp., and *Cylindrospermum* sp. Grain and straw yield from each net plot was recorded on harvesting in November 91. Weight of 1000 grains, and protein content were also recorded by routine methods.

Results and Discussion

Results (Table 18.1) indicated that grain yield was increased from 11.84 per cent to 41.58 per cent with different treatments and the straw yield increased from 6.05 per cent to 63.96 per cent over control. Incorporation of BGA evoked a statistically significant response in plant height; number of tillers per hill; grain weight and chlorophyll content; both with and without chemical nitrogen. The substitution effect was however; not noticed perhaps due to slow availability of biologically fixed N in the cropping system. Highest residual nitrogen *i.e.* 82 kg^{-1} and 41 kg^{-1} was recorded in two graded doses of BGA respectively as compared with -01.00 in chemical N and 08.00 in absolute control.

Table 18.1: Response of Kharif Rice to BGA with and without Chemical Nitrogen

Treatment (Quantities/ ha)	*No. of Tillers/ Plant*	*Plant Height (cm)*	*Chl. (mg. g^{-1})*	*Grain Yield (q.ha^{-1})*	*Straw Yield (q.ha^{-1})*	*Grain Weight (g/1000)*	*Protein (g%)*	*Residual N in Soil Over Initial (kg.ha^{-1})*
Control (B_0N_0)	15.7	81.2	1.35	45.0	51.1	16.10	8.51	–08.0
BGA 10 kg (B_1N_0)	17.4	86.8	1.56	54.1	54.4	16.31	9.21	11.00
BGA 20 kg (B_2N_0)	17.9	89.9	1.83	53.7	56.3	16.35	9.47	23.00
BGA 10 kg+ N-50 (B_1N_1)	19.2	93.6	2.06	54.4	67.9	17.33	10.33	41.00
BGA 20 kg+ N-50 (B_2N_1)	19.4	98.4	2.20	63.8	77.0	17.40	10.75	82.00
N-50 alone (B_0N_1)	18.6	19.4	2.00	50.4	58.6	16.64	9.50	–02.00
N 100 alone (B_0N_2)	20.6	99.5	2.22	61.9	83.7	17.69	9.82	–01.00
SE	0.08	0.10	0.07	1.42	1.17	0.06	0.18	
CD at 5%	0.25	0.32	0.23	4.50	2.50	0.20	0.56	

B_0: No BGA
N_0: No chemical nitrogen through urea

B_1: BGA @ 10kg
N_1: 50 kg N ha^{-1} through urea

B_2: BGA @ 20kg
N2: 100 kg N ha^{-1}
Chl = chlorophyll

Dar reported 20 per cent increase in grain yield; 12.4 per cent in straw yield; besides appreciable improvement in grain weight; plant height and tillering. Singh put review on crop response to rice. Ahluwalia indicated that BGA playa very important role in supplementing part of chemical N. Mandhare observed in a field trial for three years that the highest yield of paddy was obtained due to application of 20 kg BGA + 100 per cent chemical N. Application of BGA could save 25 per cent of chemical N required for paddy. However Roy through similar experiments for three years reported a saving of 50 per cent chemical N. Similar trends are reported in saving of chemical N; plant height; tillering; chlorophyll content over two years at Pusa in Bihar by Jha.

Chapter 19

Biofertilizers in Banana Fields

Studies on soil algal have received much attention in these days because of its eco-friendly behaviour with other organisms. Investigations on soil algae have done by Stokes. Algae are the important constituent of the soil microflora of temperate and tropical regions. Native soil-algal floristics and their distribution in relation to soil fertility have been studied in detail across the globe. From the point view of morphology, physiology and nutritional behaviour they are heterogeneous and create variation in the elemental requirements. The native algal flora of crop fields, other than paddy fields is not well documented, except few references.

The present exploratory survey was undertaken to fill the lacuna of native algal composition in the Banana plantations.

Material and Methods

The Bangalore urban district includes eleven talukas, the present study area is located in Anekal Taluk with 533.9 sq km area and situated south-eastern part of the district. It receives moderate rainfall (80 cm/yr) and the mean annual temperate is 23.6°C. The overall soil characteristic is red-"ragi soil", black soils are also found in some pockets. The climate is mostly semi-arid type. The common crops are ragi, paddy, beans, horse gram, tomato, sunflower, groundnut and maize. The commercial crops like Papaya and Banana are also of common occurrence in this region.

The surface soil samples were collected from wetland as close to the vicinity and also between rows of plants from different points all over the field and a composite sample was made. The composite samples were studied in the laboratory for their physico-chemical properties using standard methods. Whereas, for assessment of the native algal flora, soil-water biphasic medium and moist cultures were maintained in triplicate using De's modified Beneck medium. Cultures were maintained in natural day-light at north-east window of the laboratory for a period of three months. The periodic harvesting of algae started on 15th day of incubation and was continued up to ninety days. The voucher specimens are preserved in Lugol solution. Algae were identified with the help of camera lucida drawings and photographs by making use of standard reference books.

Results and Discussion

Physico-chemical parameters of the soils are presented (Table 19.1). It is clear that the pH ranges from 6.3 to 7.8, indicating moderately acidic to alkaline. The other nutrients like Nitrogen, phosphate, Potassium and sodium are in low quantities, whereas carbonates, chlorides and organic carbon occur moderately.

Table 19.1: Physico-chemical Composition of Banan Fields Soils from Anekal (Chemical composition in terms of % of dry soil except pH)

Sample	*pH*	*CO_3*	*PO_4*	*Cl*	*Org. C*	*K*	*Na*	*N*	*Ca*	*C/N*	*Algal Spp. per Sample*		
											Cy	*Ch*	*Ba*
ABS_1	6.2	0.08	0.002	0.40	0.54	0.0062	0.0110	0.080	0.009	6.78	7	1	0
ABS_2	7.3	0.40	0.004	0.34	0.34	0.0050	0.0090	0.128	0.006	2.34	21	2	0
ABS_3	6.7	0.32	0.006	0.036	0.90	0.0572	0.0501	0.064	0.018	9.47	9	1	0
ABS_4	7.8	0.62	0.005	0.044	1.20	0.0357	0.0550	0.095	0.025	18.75	23	3	1

Cy: Cyanophyceae; Ch: Chlorophyceae; Ba: Bacillariophyceae.

Table 19.2: Distribution Pattern of Algal Forms from Anekal Soil

Sl.No.	*Species*	*Soil Collection Number*
1.	*Chroococcus turgidus* (Kutz.) Nag	2, 4
2.	*C. indicus* Zeller	1
3.	*Gloeocapsa decorticans* (A.Br.) Richter	1, 2, 4
4.	*G. quatemata* (Breb.) Kutz.	3
5.	*Gloeothece membranancea* (Rabenh.) Bomet	2
6.	*Aphanocapsa biformis* A.Br.	4
7.	*Aphanothece pallida* (Kutz) Rabenh	1, 2
8.	*Spirulia subsalsa Oerst* ex. Gomont	2
9.	*Oscillatoria curviceps* Ag. ex Gomont	2, 4
10.	*O. cholrina Kutz.* ex Gomont	1, 2, 4
11.	*O. proboscidea* Gomont	1, 3, 4
12.	*O. limosa* Ag.ex Gomont	
13.	*Phormidium foveolareum* (Mont) Gomont	2, 4
14.	*P. relzii* (Ag.) Gomont	2, 4
15.	*P. favosum* (Bory) Gomont	2, 4
16.	*P. tenue* (Menegh) Gomont	1, 3
17.	*Lyngbya kuetzingiana* Kirdner	2, 4
18.	*L. trunicicola* Ghose	3, 4
19.	*L. corticola* Bruhlet Biswas	2, 3, 4
20.	*L. limnetica* Lemmermann	3, 4

Contd...

Table 19.2–Contd...

Sl.No.	Species	Soil Collection Number
21.	*L. rubida* Fremy	2
22.	*Microcoleus paludosus* (Kutz) Naegeli	2, 4
23.	*Cylindrospermum musicola* Kutzing ex Bom. et Flah	4
24.	*Nostoc muscorum* Ag. ex Bomet et Flah	2, 3, 4
25.	*N. linckia* (Roth) Bomet ex Bomet Falt	2, 4
26.	*N. piscinale* Kutzing ex Born et Flah	2
27.	*N. entophytum* Born et Flah	4
28.	*Anabaena anomala* Fritsch	4, 2
29.	*Plectonema puteale* (Kirchner) Hansgirg	
30.	*Scytonema dilatatum* Bhardwaja	3, 4
31.	*S. schmidtii* Gom.	2
32.	*S. hofmanni* Ag. ex bom. et Flah	4
33.	*Tolypothrix distorta* Kutz. ex Bom et. Flah	2
34.	*Calothrix africa* Fremy	2, 3
35.	*Cholorococcum humicola* (Naegeli) Rabenhorst	1, 3, 4
36.	*Chlorella vulgaris* Biejerinck	4
37.	*Protococcus viridis* Aqandh	2, 4
38.	*Oedogonium* sps.	2
39.	*Nitzschia diserta* Hustedt	4

The distribution pattern of algae in the study area is given (Table 19.2). In to the number of algal species is 39, out of these, 34 belong to Cyanophyceae, 4 to Ch1orophyceae and 1 to Bacillariophyceae. Among these, *Oscillatoria* and *Nostoc* had appeared in all types of soils and represented with four species each, whereas *Chroococcus indicus* present only in ABS_1 soil. On the other *Chlorella vulgaris* and *Nitzschia diserta* from ABS_4 soil only.

It was observed that during the growth of algae, the species cyanophyceae exhibited maximum growth and predominated over other groups in all the soils. An increase in the number of cyanophyceae may be attributed to the presence of elevated levels of pH and carbonates (Table 19.1). Whereas the sample ABS_2 and ABS_4 showed maximum amount of nitrogen and had harboured 21 and 23 species respectively. Cameron opined that the abundance of algal species may have direct bearing on soil moisture and the extent of time it is available. Banana plantations needs no waterlogging unlike paddy soils.

In the present investigation, the predominance of blue green algae was observed and it is in agreement with paddy maize, jowar, wheat and sugarcane soils. This may be due to neutral alkaline nature of these soils. During culture, it was observed that Cyanophyceae always dominated over Chlorophyceae and Bacillariophyceae.

It is too early to conclude that, the banana crop may lodge more number of cyanophyceae members in the absence of literature on algae of banana fields.

Summary

Soil samples from Anekal banana plantations indicated the pH range from 6.2 to 7.8. Total 4 samples from Anekal were subjected to physico-chemical and biological characteristics. Totally 39 species were enumerated out which, 34 species were assigned to Cyanophyceae, four to Chlorophyceae and one to Bacillariophyceae. The abundance of native algal flora were correlated to nutritional requirements.

Chapter 20

Isolation of Pesticides from Wheat Crop

Use of industrial effluent and sewage sludge in agricultural fields has become a common practice. The effluents contain a variety of organic and inorganic chemical like pesticides and heavy metals which affect microbial population in soil as well as the processes mediated by these organisms. The biological effects vary depending on the types of these pollutants, their concentration and the duration of treatment.

The common pollutants of soil *viz.*, pesticides and heavy metals adversely affect the growth and survival of the plant as well as its rhizosphere microflora. One of the most important nitrogen fixing microorganisms, the Azotobacter has been shown to be sensitive to these toxicants.

In the present chapter an attempt has been made to overcome the problem of toxicity of pesticides and heavy metals by isolating the tolerant strains of *Azotobacter chroococcum*. These studies can serve as an appropriate biofertilizer in the polluted soil.

Material and Methods

Experimental Background

The soil in which the study was conducted is a sandy clay loam having a pH of 8.0. The physico-chemical properties of soil are given in Table 20.1. The soil has received no exogenous input of metals. The soil was properly mixed, sieved and heavy metals like Zn^{2+}, Pb^{2+}, Cu^{2+}, Ni^{2+}, Cd^{2+}, Cr^{3+} were added in solution to the soil before sowing. Sufficient water was added to bring the soil to 50 per cent water holding capacity, pesticides like Mancozeb, 2,4-D and Malathion were also added in solution after ten days of sowing of wheat seeds.

Pot Experiment

Three and a half kilogram of soil was taken in each pot for different treatments and three replicates were taken for each treatment. The soil was fertilized with 120 : 60 : 50 kg/ha of nitrogen, phosphorus and potassium (NPK). The carrier based strain inoculant of *Azotobacter chroococcum* obtained from IARI, New Delhi was used to treat the seeds of wheat (*Triticum aestivum* L. var. PDW 154). Wheat seeds were surface sterilized in 0.1 per cent mercuric chloride solution and washed with six changes of sterile distilled water. Four seeds of wheat were sown in each pot. Seedlings were thinned to two plants per pot. Plants were grown for three months. Pots were watered as an when required.

Table 20.1: Properties of the Experimental Soil

Property	*Nature/Value*
Type	Alluvial
Soil class	Sandy clay loam
Water holding capacity	29 per cent
pH	8.0
Organic carbon	0.62 per cent
Cation exchange capacity	
CEC/cmol kg^{-1}	11.7

Isolation of *Azotobacter chroococcum* from Rhizosphere Region of Wheat

A total of fifteen Azotobacter strains were isolated from the rhizospheric region of wheat plant in the flowering stage by soil dilution and plate count method on Jensen' s agar from the soil containing different amounts of heavy metals and pesticides. The isolates were characterized up to the species level by different biochemical tests and pigment production.

Determination of Minimum Inhibitory Concentration of *Azotobacter chroococcum* Isolates

MIC of all the isolates was determined on Jensen's agar. The medium was solidified with 2 per cent agar and the pH was adjusted to 7 with NaOH. Filter sterilized solution of heavy metals and pesticides were added to molten Jensen's medium and the plates were allowed to solidify. The individual isolates (10^4 cells) was spot inoculated onto the metal amended plates and kept for five days at 28±2°C. The minimum concentration of each metal and pesticides that inhibited the growth of Azotobacter was defined as MIC.

Results and Discussion

MIC values of various strains of *Azotobacter chroococcum* to heavy metals and pesticides from metal and pesticide contaminated soil is summarized in Tables 20.2 and 20.3. It is evident from the Tables 20.2 and 20.3 that the isolates exhibited high MIC and some were found to be multiple tolerant to high concentration of heavy metals and pesticides. Thus, the order of decreasing toxicity of the metals was as follows:

$$Cd^{2+} > Ni^{2+} > Zn^{2+} > Cu^{2+} > Pb^{2+} > Cr^{3+}$$

The pesticides toxicity was in the following order Mancozeb Malathion 2,4-D. Growth of all kinds of organisms including both gram positive and gram negative bacteria, actinomycetes and fungi was affected by Cd(II). The Cd(II) toxicity is believed to be the consequence of inactivation and inhibition of certain vital enzymes of living cells. It seems that the multiple metal resistance possessed by these isolates might be due to the increased production of extracellular polysaccharide. The production of polysaccharide is thought to be responsible for binding the Cr(III) thus making it non toxic. Moreover, it was suggested to be necessary adjunct to the survival of bacterial cells in polluted environment. Heavy metals were found to be more toxic to these microorganisms which are involved in the biogeochemical cycles *e.g.* nitrogen fixation, nitrification etc.

The data presented in Tables 20.2 and 20.3 show that 2,4-D is the least toxic pesticide. Earlier reports also demonstrated the degradation of endosulphan and 2,4-D by *Azotobacter chroococcum* thus making them least toxic.

Table 20.2: MIC Values of *Azotobacter chroococcum* Strains for Heavy Metals

Heavy Metals	*MIC (ppm)*
Ni^{2+}	150.8
Cu^{2+}	182.5
Zn^{2+}	158.6
Cd^{2+}	75.0
Pb^{2+}	1500
Cr^{3+}	1750

Table 20.3: MIC Values of *Azotobacter chroococcum* Strains for Pesticides

Pesticides	*MIC (ppm)*
Mancozeb	225
2,4-D	125
Malathion	200.7

The application of industrial effluents and indiscriminate use of pesticides to agricultural land which are used for cultivation of cereals require caution. However, these isolates can serve as suitable biofertilizers in highly polluted agricultural field for seed inoculation.

Summary

Azotobacter is a free living heterotropic bacterium which helps in nitrogen fixation and thus enhances the crop production. However, very little information is available regarding the effect of agro-chemicals *viz.*, pesticides and heavy metals on the free living microbes. Therefore, a study was conducted to examine the toxic effect of these agrochemicals on the population of *Azotobacter* in the rhizosphere region of wheat crop and to isolate the tolerant strains of *Azotobacter chroococcum*. Pesticides and heavy metals in the soil significantly affected the *Azotobacter* population and the inhibition varied with the nature of toxicant. BHC and Cd were found to be the most toxic to the *Azotobacter* population. Minimum inhibitory concentration (MIC) of the tolerant isolates were found to be much higher than the amount of pesticides and heavy metals added to the soil. The results show that these isolates were multiply tolerant to different pesticides and can be used as biofertilizers in the pesticides treated fields and/or where the fields are irrigated with sewage sludge or industrial effluent. Molecular biological studies of the plasmids present in the tolerant isolates are in progress.

Chapter 21

Azotobacter Isolated from Mulberry

Free living nitrogen fixation is closely associated with the growth of the nitrogen fixing microorganisms. Among free living nitrogen fixers importance of *Azotobacter chroococcum* has been stressed since the time it was discovered by Beijerinck. Members of the Azotobacteraceae are primarily characterized as non-symbiotic, aerobic, heterotrophic bacteria whose main property is the ability to fix molecular nitrogen in nitrogen free or poor medium with an organic carbon as energy source.

In the present investigation efforts have been made to isolate, characterize and identify the *Azotobacter* strain isolated from mulberry rhizosphere soil and to screen efficient nitrogen fixing *Azotobacter* strain.

Material and Methods

Rhizosphere soil samples of mulberry was collected at six locations two each from Kolar, Mysore and Ramanagar as per the procedure outlined by Dobereiner. The collected soil samples were brought to the laboratory, air dried, powdered passed through 0.27 mm sieve and used for isolation. Seven *Azotobacter* strains from Kolar, five each from Mysore and Ramanagar were isolated by employing dilution plate technique and maintained on Walksman No. 77 medium.

Characters like colony growth, intensity and pigmentation were studied by streaking the purified cultures on Walksman No. 77 medium. The isolation were further tested for gram reaction (Hucker's modified method). Formation of microcysts studied in three weeks old cultures stained with acridine orange and observed under oil immersion. Utilisation of various carbohydrates–Mannitol, Glucose, Fructose, Sucrose, Rhamnose, Malate and Citrate as carbon sources were studied separately by adding to Walksman No.77 medium at one per cent w/v. The total nitrogen fixed per gram of carbon Source utilised was estimated by micro Kjeldahl method.

The authentic *Azotobacter* culture obtained from collection of the Department of Agricultural Microbiology, University of Agricultural Sciences, G.K.V.K. Bangalore was used for comparison and identification of isolated cultures. This standard isolate was gram negative, microcyst former, produced brown black pigment and had fixed 6.6 mg of Nitrogen per gram of Mannitol used.

Results and Discussion

Cultured cells were morphologically found to be ovoid, motile, gram negative and all were microcyst formers these observations are comparable to that of Parker. All the strains produced brown black pigment which was water insoluble. The details are given in Tables 21.1–21.3. Among the seven carbohydrates tested as a carbon source Mannitol and Glucose were the most preferred carbohydrates, where as Fructose and Sucrose were moderately preferred carbon sources and none of the strains were able to utilise Rhamnose, Malate and Citrate as carbon source Table 12.4, which is in accordance with the results described by Tchan.

Table 21.1: Growth Characters for Authentication and Nitrogen Fixing Ability of *Azotobacter* Strains Isolated from Kolar Soil

Azotobacter Strains	*Grain Strain*	*Formation of Microsysts*	*Pigmentation*	*Cultural Characters*	*mg N-fixed/g of C Source (Mannitol)*
K1	Gram-ve	+	On fourth day light brwn	Good growth flat, entire, wrinkled at the edge of the colony	6.1
K2	Gram-ve	+	On third day dark brown	Good growth, flat entire, wrinkled at the edge of the colony	6.9
K3	Gram-ve	+	On fifth day pale brown	Good growth, flat entire, wrinkled at the edge of the colony	6.3
K4	Gram-ve	+	On eighth day dark brown	Moderate growth, flat entire	5.8
K5	Gram-ve	+	On sixth day light brown	Moderate growth, flat entire	5.6
K6	Gram-ve	+	On seventh day dark brown	Good growth, flat entire, wrinkled at the edge of the colony	5.8
K7	Gram-ve	+	On fourth day pale brown	Moderate growth, flat entire	5.2
SEM ±					0.1046
CD at 5%				0.3078	

All values are means of four replications. Gram-ve: Gram negative; +: Microcyst formed.

Based on these studies the isolates were identified as *Azotobacter chroococcum* and labelled as K1, K2, K3, K4, K5, K7, MI, M2, M3, M4, R1, R2, R3, R4, and R5.

The nitrogen fixed per gram of carbon source varied among the strains. It ranged from 5.3 mg to 6.9 mg in isolates of Kolar, 5.6 to 6.8 mg in Mysore and 5.3 to 6.8 mg in Ramnagar soils.

The nitrogen fixing ability of seventeen *Azotobacter* strains was low to moderate. *Azotobacter* strain CK_2 fixed maximum amount of nitrogen followed by CM_3 and CR_2 strains. However, this was not up to the maximum level of fixation.

Table 21.2: Growth Characters for Authentication and Nitrogen Fixing Ability of *Azotobacter* Strains Isolated from Mysore Soil

Azotobacter Strains	*Grain Strain*	*Formation of Microsysts*	*Pigmentation*	*Cultural Characters*	*mg N-fixed/g of C Source (Mannitol)*
M1	Gram-ve	+	On fifth day pale brown	Good growth flat, entire, wrinkled at the edge of the colony	6.3
M2	Gram-ve	+	On sixth day dark brown	Good growth, flat entire, wrinkled at the edge of the colony	6.0
M3	Gram-ve	+	On fourth day dark brown	Good growth, flat entire, wrinkled at the edge of the colony	6.8
M4	Gram-ve	+	On third day light brown	Moderate growth, flat entire, wrinkled at the edge of the colony	6.3
M5	Gram-ve	+	On seventh day light brown	Moderate growth, flat entire	5.6
SEM ±					0.0658
CD at 5%					0.1984

All values are means of four replications. Gram-ve: Gram negative; +: Microcyst formed.

Table 21.3: Growth Characters for Authentication and Nitrogen Fixing Ability of *Azotobacter* Strains Isolated from Ramnagar Soil

Azotobacter Strains	*Grain Strain*	*Formation of Microsysts*	*Pigmentation*	*Cultural Characters*	*mg N-fixed/g of C Source (Mannitol)*
R1	Gram-ve	+	On fifth day dark brown	Moderate growth flat, entire, wrinkled at the edge of the colony	6.2
R2	Gram-ve	+	On sixth day dark brown	Good growth, flat entire, wrinkled at the edge of the colony	6.8
R3	Gram-ve	+	On fourth day dark brown	Good growth, flat entire, wrinkled at the edge of the colony	6.1
R4	Gram-ve	+	On third day light brown	Moderate growth, flat entire	5.4
R5	Gram-ve	+	On seventh day light brown	Good growth, flat entire, wrinkled at the edge of the colony	5.3
SEM±					0.0707
CD at 5%					0.2131

All values are means of four replications. Gram-ve: Gram negative; +: Microcyst formed.

Table 21.4: Utilisation of Different Carbon Sources and Authentication of *Azotobacter* Strain Isolated from Soils of Kolar, Mysore and Ramanagar

Azotobacter	Carbon Source						
	Mannitol	Glucose	Fructose	Sucrose	Rhamnose	Malate	Citrate
K1	4+	3+	2+	2+	–	–	–
K2	4+	4+	3+	3+	–	–	–
K3	4+	4+	2+	2+	–	–	–
K4	3+	3+	2+	2+	–	–	–
K5	3+	3+	+	+	–	–	–
K6	3+	2+	2+	2+	–	–	–
K7	3+	3+	+	+	–	–	–
M1	4+	4+	2+	2+	–	–	–
M2	3+	2+	2+	–	–	–	–
M3	4+	2+	3+	–	–	–	–
M4	3+	+	+	–	–	–	–
M5	3+	+	+	–	–	–	–
R1	4+	4+	2+	2+	–	–	–
R2	4+	4+	2+	3+	–	–	–
R3	3+	3+	+	+	–	–	–
R4	3+	3+	2+	2+	–	–	–
R5	3+	3+	+	+	–	–	–

4+: Maximum; 3+: Good growth; 2+: Moderate growth; +: Less growth; –: No growth.

Summary

An investigation was conducted to isolate, characterize and identify *Azotobacter* spp. from mulberry rhizosphere soil. The seventeen *Azotobacter* strains isolated were gram negative, motile, avoid rods, produced brown black pigment. Formed microcyst and indicated marked variations in growth between different carbohydrates. They were identified as *Azotobacter chroococcum*. The maximum nitrogen fixed per gram of carbon source (Mannitol) was 6.9 mg by *Azotobacter* strain CK_2 followed by CM_3 and CR_2 with 6.8 mg of nitrogen each.

Chapter 22

Sulphatic Biofertilizer on Pigeonpea

The sulphur requirements of legume plants are comparable to that of phosphorus and therefore, it is recognised the fourth major nutrient in Indian Agriculture. In India 25 million hectares crop land have been reported as sulphur deficient. Plants utilise the sulphur in the form of sulphate (SO_4^{-2}) and microorganisms are mainly responsible for oxidizing unavailable, elemental or reduced sulphur to available sulphate. There is a possibility of using heterotrophic sulphur oxidizing microorganisms as inoculant to increase the rate of oxidation of sulphur fertilizer in soil which are slow to oxidize. Sulphur oxidizing fungi appear particularly useful for this purpose since their spores or mycelium can be produced economically in large quantities and have good survival characteristics, both in the inoculant and soil. However, no information is available regarding the presence of heterotrophic oxidizing micro-organisms in soil, their sulphur oxidizing potentiality and effect on plant growth. Therefore, present study was undertaken to assess the effect of artificial inoculation of different sulphur oxidizing microorganisms on growth parameters, yield and quality of Pigeonpea.

Material and Methods

A pot culture experiment was conducted in the glass house during 1992. The design of experiment was a factorial and completely randomised (FCRD) with five replications. The treatments comprised, seven cultures as main factors and two sulphur levels (Table 22.1) as sub factors. Ten seeds of *Cajanus cajan* CV.BDN-1 were sown per pot and after germination only five plants were finally kept to record the growth parameters. The recommended dose of NPK (25 : 50 : 0 kg/ha) and 100 kg. elemental sulphur/ha was distributed evenly in the soil before sowing. Five most efficient microorganisms which were isolated from 150 soil samples of different types and locations were multiplied and inoculated in the soil at the rate of 250 ml/pot having 10^7 cells or spores/ml in liquid medium culture.

The soil had an electrical conductivity of 0.5 dSm^{-1} exchangeable sodium percentage of 8.54, pH 7.7 and sulphate content of 9.0 ppm. The soil and plant samples were analysed for different chemical properties by standard methods. The microbial population in soil *viz.*, bacteria and nitrogen fixers were recorded by Most Probable Number (MPN) method and fungi and actinomycetes were recorded by dilution plating technique.

Table 22.1: Effects of Sulphur Oxidizing Microorganisms with and without Sulphur on Growth of Pigeonpea

Treatments	No. of Nodules/Plant			Dry Matter Weight of Nodules (mg/plant)			Dry Matter Weight (g/plant)			Grain Weight (g/plant)		
	With S	Without S	Mean	With S	Without S	Mean	With S	Without S	Mean	With S	Without S	Mean
Scolecobasidium constrictus	24.5	20.0	22.25	250	175	212	5.00	3.90	4.45	8.32	7.20	7.77
Myrothecium cinctum	25.0	20.0	22.50	260	180	220	5.15	4.05	4.60	8.50	7.25	7.87
Aspergillus terreus	26.0	21.0	23.50	280	185	232	5.40	4.25	4.82	8.65	7.45	8.05
Thiobacillus thioparus	25.0	20.5	22.75	265	180	222	5.20	4.10	4.65	8.50	7.30	7.90
Thiobacillus thiooxidans	27.0	21.5	24.25	285	190	237	5.50	4.30	4.90	8.70	7.50	8.10
Composite culture	29.5	23.0	26.25	300	210	255	5.70	4.45	5.07	8.90	7.65	8.27
Control	24.0	19.0	21.50	220	170	195	4.60	3.80	4.20	7.80	7.00	7.40
Mean	25.8	20.7		265.7	184		5.22	4.12		8.48	7.33	
	S.E±	CD 5%		S.E±	CD 5%		S.E±	CD 5%		S.E±	CD 5%	
Culture	0.31	0.96		3.53	10.60		0.07	0.21		0.08	0.25	
Sulphur	0.52	1.67		1.92	5.77		0.31	0.90		0.26	0.86	
Culture × Sulphur	0.44	1.35		1.10	3.31		0.09	0.30		0.15	0.48	

Results and Discussion

The effects of different sulphur oxidizing culture with and without sulphur on the growth and yield of pigeonpea (Table 22.1) revealed that all the six cultures with and without sulphur addition increased significantly the total number of nodules, dry matter and grain yield/plant over control. However, inoculation of soil with culture along with sulphur reported more than the culture without sulphur. The maximum response was observed in composite culture with sulphur, followed by *T. thiooxidans, A. terreus, T. thioparus, M. cinctum, S. constrictum,* sulphur alone and minimum with the uninoculated control.

The total number of nodules and dry matter of nodules/plant were increased from 19.0 to 29.5 and 170 to 300 mg/plant, respectively, due to inoculation of soil with the various cultures with sulphur over uninoculated control.

The sulphur had significant role in increasing nodulation and nitrogen fixation in grain legumes.

The dry matter weight and grain yield of pigeonpea were found to be increased, significantly over control due to inoculation of culture plus sulphur. The dry matter weight ranged from 3.80 to 5.70 g/plant whereas grain weight ranged from 7.00 to 8.90 g/plant. These results are in agreement with the results reported by Badiger. They reported that application of sulphur resulted in increasing the dry matter and grain yield of different pulses, including pigeonpea from 20 to 50 per cent.

Aulakh reviewed that, elemental sulphur and other forms of sulphur which could improve the straw and grain yield as well as quality of grain of pigeonpea.

The results regarding sulphur uptake by pigeonpea and protein per cent in grains (Table 22.2) indicated that sulphur application and culture inoculation both together could increase the sulphur uptake ranging from 260 to 390 mg/100 g dry weight. Whereas protein percentage ranged from 22.60 to 23.85. Sulphur had a vital and important role in grain legumes for their protein synthesis.

Table 22.2: Effects of Sulphur Oxidizing Microorganisms with and without Sulphur on Sulphur Uptake and Protein Content of Pigeonpea

Treatments	*Sulphur Uptake (mg/100g)*			*Protein(%)*		
	With Sulphur	*Without Sulphur*	*Mean*	*With Sulphur*	*Without Sulphur*	*Mean*
Scolecobasidium constrictum	330	265	297	23.35	22.70	23.02
Myrothecium cinctum	340	270	305	23.40	22.72	23.06
Aspergillus terreus	360	285	322	23.60	22.85	23.22
Thiobacillus thioparus	345	270	307	23.50	22.75	23.12
Thibacillus thiooxidans	375	290	332	23.70	22.90	23.30
Composite culture	390	300	345	23.85	23.00	23.42
Control	310	260	285	23.15	22.60	22.87
Mean	350	277		23.50	22.78	
	S.E. ±	C.D. 5%		S.E. ±	C.D. 5%	
Culture	3.85	11.48		0.38	N.S.	
Sulphur	2.20	6.56		0.46	N.S.	
Culture × Sulphur interaction	3.21	9.71		0.68	N.S.	

The effects of cultures with and without sulphur on microbial population (Table 22.3) revealed that sulphur oxidizing bacteria, fungi, actinomycetes as well as nitrogen fixers were increased due to inoculation of soil with cultures and application of sulphur. However, culture in conjunction with sulphur was more effective that without sulphur. Further, it was noticed that total microbial population was more at 60 days than at harvesting state of the crop. The composite culture of bacteria and fungi with sulphur treatment showed maximum microbial population amongst all the cultures under study.

Table 22.3: Effects of Inoculation on Soil Microbial Population in *Pigeonpea* Crop

Treatments	*At 60 days*				*After Harvesting*			
	Bacteria ($\times 10^5$)	*Fungi* ($\times 10^3$)	*Actino-mycetes* ($\times 10^5$)	*Nitrogen Fixers* ($\times 10^4$)	*Bacteria* ($\times 10^5$)	*Fungi* ($\times 10^3$)	*Actino-mycetes* ($\times 10^5$)	*Nitrogen Fixers* ($\times 10^4$)
C_1 *Scolecobasidium constrictus*	13.16	16.50	5.16	7.83	10.33	12.83	4.83	6.50
C_2 *Myrothecium cinctum*	14.50	18.00	5.33	8.16	11.50	15.00	5.16	7.50
C_3 *Aspergillus terreus*	14.83	21.83	6.16	8.66	12.00	18.83	6.00	7.83
C_4 *Thiobacillus thioparus*	18.33	10.50	6.16	8.33	16.00	9.66	5.83	8.00
C_5 *Thiobacillus thiooxidans*	20.16	12.66	6.33	8.83	17.16	10.33	6.00	8.00
C_6 Cosposite culture	21.66	22.16	7.66	9.16	19.33	20.00	6.83	8.50
C_1 + Sulphur	17.66	21.83	6.83	8.16	15.16	17.50	6.50	8.00
C_2 + Sulphur	18.50	22.16	7.16	9.50	16.50	19.33	6.83	8.33
C_3 + Sulphur	20.66	25.50	8.16	9.66	17.16	22.00	7.50	8.83
C_4 + Sulphur	23.66	13.00	8.33	9.50	21.00	11.50	7.50	8.83
C_5 + Sulphur	26.33	14.66	8.50	9.83	22.66	13.50	8.00	9.16
C_6 + Sulphur	28.16	26.33	8.83	10.16	25.50	2.66	8.16	9.83
C_0 + Sulphur	10.50	8.33	4.16	6.00	9.16	8.00	3.83	5.50
$C_0 + S_0$ (control)	7.50	6.50	3.50	4.16	7.16	6.00	3.16	4.00

Summary

A pot culture study on sulphur oxidizing microorganisms in pigeonpea crop revealed that all the growth parameters *viz.*, total number of nodules on roots, dry matter weight of nodules and plants and grain yield were increased, significantly due to inoculation of sulphur oxidizing microorganisms with or without sulphur. The composite culture of two bacteria *viz., Thiobacillus, Thioparus, Thiooxidans* and three fungi *viz., Scolecobasidium constrictum, Myrothecium cinctum* and *Aspergillus terreus* in conjunction with elemental sulphur found to be significantly, superior over individual culture and sulphur treatment. The effects of bacterial culture (*Thiobacillus thiooxidans*) were on par with fungal culture (*Aspergillus terreus*). The increases in yield up to 27.14 per cent sulphur uptake up to 50 per cent and protein content ranging from 22.60 per cent to 23.85 per cent due to inoculation of composite culture with sulphur fertilizer over uninocul ated control.

Chapter 23
Biodynamic Agriculture

Biodynamic agriculture is an advanced organic farming system that is gaining increased attention for its emphasis on food quality and soil health. Biodynamic agriculture developed out of eight lectures on agriculture given in 1924 by Rudolf Steiner (1861–1925), an Austrian scientist and philosopher, to a group of farmers near Breslau (which was then in the eastern part of Germany and is now Wroclaw in Poland). These lectures, as well as four supplemental lessons, are published in a book titled Spiritual Foundations for the Renewal of Agriculture, originally published in English as an Agricultural Course.

The Agriculture Course lectures were taught by Steiner in response to observations from farmers that soils were becoming depleted following the introduction of chemical fertilizers at the turn of the last century. In addition to degraded soil conditions, farmers noticed deterioration in the health and quality of crops and livestock. Thus, biodynamic agriculture was the first ecological farming system to develop as a grassroots alternative to chemical agriculture.

A basic ecological principle of biodynamic is to conceive of the farm as an organism, a self-contained entity. A farm is said to have its own individuality. Emphasis is placed on the integration of crops and livestock, recycling of nutrients, maintenance of soil, and the health and well being of crops and animals; the farmer too is part of the whole. Thinking about the interactions within the farm ecosystem naturally leads to a series of holistic management practices that address the environmental, social, and financial aspects of the farm. A comparison of objectives between biodynamic and conventional agriculture systems summarizes these ideas in table format.

A fundamental tenet of biodynamic agriculture is that food raised biodynamical is nutritionally superior and tastes better than foods produced by conventional methods. This is a common thread in alternative agriculture, because other ecological farming systems make similar claims for their products. Demeter, a certification program for biodynamically grown foods, was established in 1928. As such, Demeter was the first ecological label for organically produced foods.

Today biodynamic agriculture is practiced on farms around the world, on various scales, and in a variety of climates and cultures. However, most biodynamic farms are located in Europe, the United States, Australia, and New Zealand. In India there are over 3500 farms practicing biodynamic agriculture.

While biodynamic parallels organic farming in many ways–especially with regard to cultural and biological farming practices–it is set apart from other organic agriculture systems by its association with the spiritual science of anthroposophy founded by Steiner, and in its emphasis on farming practices intended to achieve balance between the physical and higher, non-physical realms; to acknowledge the influence of cosmic and terrestrial forces; and to enrich the farm, its products, and its inhabitants with life energy is a table that illustrates cosmic and terrestrial influences on yield and quality.

In a nutshell, biodynamics can be understood as a combination of "biological dynamic" agriculture practices,. "Biological" practices include a series of well-known organic farming techniques that improve soil health, "Dynamic" practices are intended to influence biological as well as metaphysical aspects of the farm (such as increasing vital life force), or to adapt the farm to natural rhythms (such as planting seeds during certain lunar phases).

The concept of dynamic practice–those practices associated with non-physical forces in nature like vitality, life force, ki, subtle energy and related concepts–is a commonality that also underlies many systems of alternative and complementary medicine. It is this latter aspect of biodynamics which gives rise to the characterization of biodynamics as a spiritual or mystical approach to alternative agriculture. See the following table for a brief summary of biological and dynamic farming practices.

The higher, non-physical realms include etheric, astral, and ego. It is the complicated terminology and underlying metaphysical concepts of Steiner which makes biodynamics hard to grasp, yet these are inherent in the biodynamic approach and therefore they are listed here for the reader's reference.

Table 23.1: Biodynamic Farming Practice

Biological Practices	*Dynamic Practices*
Green manures	Special compost preparations
Cover cropping	Special foliar sprays
Composting	Planting by calendar
Companion planting	Peppering for pest control
Integration of crops and livestock	Homeopathy
Tillage and cultivation	Radionics

Life energy is a colloquial way of saying etheric life force. Again, Steiner's use of terms like etheric forces and astral forces are part and parcel of biodynamic agriculture. Biodynamic farmers recognize there are forces that influence biological systems other than gravity, chemistry, and physics. Dr. Andrew Lorand provides an insightful glimpse into the conceptual model of biodynamics in his Ph.D. dissertation Biodynamic Agriculture–A Paradigmatic Analysis, published at Pennsylvania State University in 1996. Lorand uses the paradigm model described by Egon Guba in The Alternative Paradigm Dialog to clarify the essential beliefs that underpin the practices of biodynamics. These beliefs fall into three categories: Beliefs about the nature of reality with regard to agriculture (ontological beliefs); Beliefs about the nature of the relationship between the practitioner and agriculture (epistemological beliefs); and Beliefs about how the practitioner should go about working with agriculture (methodological beliefs). Lorand's dissertation contrasts the ontological, epistemological, and methodological beliefs of four agricultural paradigms: Traditional Agriculture, Industrial Agriculture, Organic Agriculture, and Biodynamic Agriculture.

The Biodynamic Preparations

A distinguishing feature of biodynamic farming is the use of nine biodynamic preparations described by Steiner for the purpose of enhancing soil quality and stimulating plant life. They consist of mineral, plant, or animal manure extracts, usually fermented and applied in small proportions to compost, manures, the soil, or directly onto plants, after dilution and stirring procedures called dynamizations.

The original biodynamic (BD) preparations are numbered 500–508. The BD 500 preparation {horn-manure) is made from cow manure (fermented in a cow horn that is buried in the soil for six months through autumn and winter) and is used as a soil spray to stimulate root growth and humus formation. The BD 501 preparation (horn-silica)is made from powdered quartz (packed inside a cow horn and buried in the soil for six months through spring and summer) and applied as a foliar spray to stimulate and regulate growth. The next six preparations, BD 502–507, are used in making compost.

Finally, there is BD preparation 508 which is prepared from the silica-rich horsetail plant (*Equisetum arvense*) and used as a foliar spray to suppress fungal diseases in plants.

The BD compost preparations are listed below:

1. No. 502 Yarrow blossoms (*Achillea millefolium*)
2. No. 503 Chamomile blossoms (*Chamomilla officinalis*)
3. No. 504 Stinging nettle (whole plant in full bloom) (*Urtica dioca*)
4. No. 505 Oak bark (*Quercus robur*)
5. No. 506 Dandelion flowers (*Taraxacum officinale*)
6. No. 507 *Valerian flowers* (*Valeriana officinalis*)

Biodynamic preparations are intended to help moderate and regulate biological processes as well as enhance and strengthen the life (etheric) forces on the farm. The preparations are used in homeopathic quantities, meaning they produce an effect in extremely diluted amounts. As an example, just 1/16th ounce–a level teaspoon–of each compost preparation is added to seven- to ten-ton piles of compost.

Biodynamic Compost

Biodynamic compost is a fundamental component of the biodynamic method; it serves as a way to recycle animal manures and organic wastes, stabilize nitrogen, and build soil humus and enhance soil health. Biodynamic compost is unique because it is made with BD preparations 502-507. Together, the BD preparations and BD compost may be considered the cornerstone of biodynamics. Here again, "biological" and "dynamic" qualities are complementary: biodynamic compost serves as a source of humus in managing soil health and biodynamic compost emanates energetic frequencies to vitalize the farm.

The traditional manner in which the biodynamic compost is made is rather exacting. After the compost windrow is constructed; Preparations 502–506 are strategically placed 5–7 feet apart inside the pile, in holes poked about 20 inches deep. Preparation No. 507, or liquid valerian, is applied to the outside layer of the compost windrow by spraying or hand watering.

Valerian (507) is mixed into a liquid; a portion is pored into one hole, and the rest is sprinkled over the top of the compost pile.

Because compost is often at a premium on farms, European biodynamic researcher Maria Thun developed Barrel Compost. Consisting of fresh cow manure that has been treated with the original preparations as well as egg shells and basalt rock dust–then allowed to ferment in a pit for about 3 months, finished Barrel Compost is diluted in water and applied directly to the fields as a spray. Use of Barrel Compost compensates to some degree for lack of sufficient compost. A variation on Barrel Compost is mixing stinging nettle with fresh cow manure in a 50 : 50 volume to volume ratio.

Some notable concepts and practices relating to soil and compost management from the biodynamic experience:

Microbial Inoculation

Dr. Ehrenfried Pfeiffer's work with composts in the 1940's and 50's led to the development of the BD Compost Starter, one of the earliest compost inoculants in commercial use in the United States.

Soil in Compost

The audition of soil to compost was an early biodynamic practice prescribed by Steiner. Dr. Pfeiffer discussed the reasons and benefits for adding soils to compost in the 1954 edition of Biodynainics Journal. (Vol 12, No. 2) in an article titled "Raw Materials Useful for Composting." He said that soil is an essential ingredient to compost and should be added at 10 per cent–20 per cent of the windrow volume.

Mineralized Compost

The addition of rock powders (greensand, granite dust) to compost piles is a long-time biodynamic practice known as mineralized compost. The dusts add mineral components to the compost and the organic acids released during the decomposition process help solubilize minerals in the rock powders to make nutrients more available to plants.

Phases of Compost

An outgrowth of Dr. Pfeiffer's compost research was a clearer understanding of the Breakdown and Buildup compost phases:

The Breakdown Phase

In the breakdown phase organic residues are decomposed into smaller particles. Proteins are broken down into amino acids, amines, and finally to ammonia, nitrates, nitrites, and free nitrogen. Urea, uric acids, and other non-protein nitrogen-containing compounds are reduced to ammonia, nitrites, nitrates, and free nitrogen. Carbon compounds are oxidized to carbon dioxide (aerobic} or reduced methane (anaerobic). The identification and understanding of breakdown microorganisms led to the development of a microbial inoculant to moderate and speed up the breakdown phase. The BD Compost Starter developed by Dr. Pfeiffer contains a balanced mixture of the most favorable breakdown organisms, ammonifiers, nitrate formers, cellulose, sugar, and starch digesters in order to bring about the desired results. The microbial inoculant also works against organisms that cause putrefaction and odors.

The Buildup Phase

In the build-up phase simple compounds are re-synthesized into complex humic substances. The organisms responsible for transformation to humus are aerobic and facultative aerobic, sporing and non-sporing and nitrogen fixing bacteria of the azotobacter and nitrosomonas group. Actinomycetes and streptomycetes also play an important role. The addition of soil, 10 per cent by volume, favors the development and survival of these latter organisms. The development of humus is evident in color changes in the compost, and through qualitative tests such as the circular chromatography method.

Compost and Soil Evaluation

Biodynamic research into compost preparation and soil humus conditions has led to the development or specialized use of several unique qualitative tests. A notable contribution of biodynamics is the image-forming qualitative methods of analysis; *e.g.*, circular chromatography, sensitive crystallization, capillary dynamolysis, and the drop-picture method. Other methods focus on the biological-chemical condition, *e.g.* The Solvita Compost Test Kit and The Solvita Soil Test Kit, colorimetric humus value, and potential pH.

Cover Crops and Green Manures

Cover crops play a central role in managing cropland soils in biological farming systems. Biodynamic farmers make use of cover crops for dynamic accumulation of soil nutrients, nematode control, soil loosening, and soil building in addition to the commonly recognized benefits of cover crops like soil protection and nitrogen fixation. Biodynamic farmers also make special use of plants like phacelia, rapeseed, mustard, and oilseed radish in addition to common cover crops like rye and vetch. Cover crop strategies include undersowing and catch cropping as well as winter cover crops and summer green manures.

Green manuring is a biological farming practice that receives special attention on the biodynamic farm. Green manuring involves the soil incorporation of any field or forage crop while green, or soon after flowering, for the purpose of soil improvement. The decomposition of green manures in soils parallels the composting process in that distinct phases of organic matter breakdown and humus buildup are moderated by microbes. Many biodynamic farmers, especially those who follow the guidelines established by Dr. Ehrenfried Pfeiffer, spray the green residue with a microbial inoculant (BD Field Spray) prior to plowdown. The inoculant contains a mixed culture of microorganisms that help speed decomposition, thereby reducing the time until planting. In addition, the inoculant enhances formation of the clay-humus crumb which provides numerous exchange sites for nutrients and improves soil structure.

Crop Rotations and Companion Planting

Crop rotation–the sequential planting of crops–is honed to a fine level in biodynamic farming. A fundamental concept of crop rotation is the effect of different crops on the land. Koepf, Pettersson, and Schaumann speak about "humus-depleting" and "humus-restoring" crops; "soil-exhausting" and "soil-restoring" crops; and "organic matter exhausting" and "organic matter restoring" crops in different sections of BioDynamic Agriculture: An Introduction.

Seemingly lost to modem agriculture with its monocrops and short duration corn-soybean rotations, soil building crop rotations were understood more clearly earlier in this century when the USDA published leaflets like Soil-Depleting, Soil-Conserving, and Soil- Building Crops in 1938.

Companion planting, a specialized form of crop rotation commonly used in biodynamic gardening, entails the planned association of two or more plant species in close proximity so that some cultural benefit (pest control, higher yield) is derived. In addition to beneficial associations, companion planting increases biodiversity on the farm which leads to a more stable agroecosystem

Liquid Manures and Herbal Teas

Herbal teas, also called liquid manures or garden teas, are an old practice in organic farming and gardening–especially in biodynamic farming–yet little is published on this topic outside of the practitioner literature. A complementary practice is the use of compost teas.

In reality, herbal teas usually consist of one fermented plant extract, while liquid manures are made by fermenting a mixture of herb plants in combination with fish or seaweed extracts. The purpose of herbal teas and liquid manures are many fold; here again, they perform dual roles by supporting biological as well as dynamic processes on the farm; *i.e.*, source of soluble plant nutrients; stimulation of plant growth; disease–suppression; carrier of cosmic and earthly forces. To reflect their multi-purpose use, they are sometimes referred to as immune-building plant extracts, plant tonics, biotic substances, and biostimulants.

Further insight into foliar-applied plant extracts, liquid manures, and compost teas can be understood by viewing biological farming practices in the way they influence the rhizosphere or phyllosphere. (Those microbially-rich regions surrounding the root and leaf surfaces). Herbal teas and liquid manures aim to influence the phyllosphere; composts, tillage, and green manures influence the rhizosphere. In addition to physical modification of the leaf surface to inhibit pathogen spore germination or the promotion of antagonistic (beneficial) microbes to compete against disease-causing organisms (pathogens), foliar-applied biotic extracts can sometimes initiate a systemic whole plant response known as induced resistance.

Horsetail Tea

Horsetail tea is extracted from the common horsetail (*Equisetum aruense*), a plant especially rich in silica. Horsetail is best seen as a prophylactic (disease-preventing, not disease-curing) spray with a mild fungus-suppressing effect. During the months when green, plants are not readily available, you can prepare an extract by covering dry plants with water and allowing them to ferment in a sunny place for about ten days. Dried equisetum, available through the Josephine Porter Institute for Applied Biodynamics in Wool-wine, Virginia, can also be used to make horsetail tea.

Stinging Nettle Tea

Stinging nettle tea is extracted from whole nettle plants (*Urtica dioica*) at any stage of growth up to seed-set. To make nettle tea, use about three pounds of fresh plants for every gallon of water, allow the mixture to ferment for about ten days, then filter it and spray a diluted tea. Dilution rates of 1:10 to 1:20 are suggested in the biodynamic literature. A biodynamic nettle tea is prepared by adding BD preparations 502, 503, 505, 506, and 507 prior to the soaking period.

Chamomile Tea

is derived from the flowers of true chamomile (*Matricaria chamomilla*) which have been picked and dried in the sun, Fresh flowers may be used too, but they are only available during a short part of the growing season. To prepare the tea, steep about one cup of tightly packed flowers per gallon of hot water. Stir well, and spray the filtered tea when cool. Chamomile is high in calcium, potash, and sulfur; it is good for leafy crops and flowers and promotes health of vegetables in general.

Comfrey Tea

Comfrey tea is another tea commonly used in organic farming and gardening. Comfrey is a rich source of nutrients; it is especially good for fruiting and seed filling crops. It can be made by packing a barrel three-quarters full with fresh cut leaves, followed by topping the barrel full of water. It is allowed to steep for 7–14 days, then filtered and diluted in half with water prior to use.

Compost teas are gaining wider recognition in biodynamic and organic farming for their disease suppressive benefits as well as for their ability to serve as a growth-promoting microbial inoculant.

Planetary Influences

Lunar and astrological cycles play a key role in the timing of biodynamic practices, such as the making of BD preparations and when to plant and cultivate. Recognition of celestial influences on plant growth are part of the biodynamic awareness that subtle energy forces affect biological systems. A selection of resources are listed below. On examination of the variations in agricultural calendars that have sprung from the biodynamic experience, it is apparent that differing viewpoints exist on which lunar, planetary, and stellar influences should be followed.

Stella Natura: The Kimberton Hills Biodynamic Agricultural Calendar

This is available through BDFGA for $11.95, is the biodynamic calendar edited by Sherry Wildfeur and the most prominently known calendar of this type in the United States. It contains informative articles interspersed with daily and monthly astrological details, and lists suggested times for planting root, leaf, flowering, and fruiting crops.

Working with the Stars

A Biodynamic Sowing and Planting Calendar, available through JPI for $12.95, is the biodynamic calendar based on Maria Thun's research and is more prominently used in Europe. Of the three calendars mentioned here, Thun's calendar relies more heavily on planetary and stellar influences. It contains research briefs as well as daily and monthly astrological details, again with suggested planting times.

Astronomical Gardening Guide, available through Agri-Synthesis in Napa, California for a self-addressed stamped envelope, is the biodynamic gardening guide compiled by Greg Willis of Agri-Synthesis. This calender, which is a simple 2-sheet information leaflet, focuses on lunar phases.

Community Supported Agriculture

In its treatment of the farm as a self-contained entity or farm organism, biodynamics completes the circle with appropriate marketing schemes to support the economic viability of farms. The Demeter label for certified biodynamically grown foods is one avenue. A second outgrowth of this view is the Community Supported Agriculture movement.

Community Supported Agriculture, or CSA, is a direct marketing alternative for small-scale growers. In a CSA, the farmer grows food for a group of shareholders (or subscribers) who pledge to buy a portion of the farm's crop that season. This arrangement gives growers up-front cash to finance their operation and higher prices for produce, since the middleman has been eliminated. Besides receiving a weekly box or bag of fresh high-quality produce, shareholders also know that they are directly supporting a local farm.

Food Quality

A host of biodynamic researchers have looked into the quality of biodynamically grown foods. Though nutritional comparisons between foods raised by organic and conventional production methods is controversial -certainly mainstream science adheres to the view that no differences exist–notable contributions from biodynamic researchers include image-forming qualitative methods of analysis (*e.g.*, sensitive crystallization, circular chromatography, capillary dynamolysis, and the drop-picture method) and studies that report on nutritional analysis.

Summary

Biodynamics uses scientifically sound organic farming practices that build and sustain soil productivity as well as plant and animal health. The philosophical tenets of biodynamics-especially those that emphasize energetic forces and astrological influences are harder to grasp, yet they are part and parcel of the biodynamic experience.

That mainstream agriculture does not accept the subtle energy tenets of biodynamic agriculture is a natural result of conflicting paradigms. In mainstream agriculture the focus is on physical-chemical-biological reality. Biodynamic agriculture, on the other hand, recognizes the existence of subtle energy forces in nature and promotes their expression through specialized "dynamic" practices.

A third view, expressed by a local farmer, accepts the premises that subtle energy forces exist and may affect biological systems, but holds there is not enough information to evaluate these influences nor make practical agronomic use of them.

The fact remains that biodynamic farming is practiced on a commercial scale in many countries and is gaining wider recognition for its contributions to organic farming, food quality, community supported agriculture, and qualitative tests for soils and composts. From a practical viewpoint biodynamics is proven to be productive and yield nutritious, high quality foods.

Chapter 24

Indigenous Knowledge in Rice Organic Farming

Each society has its own indigenous knowledge system that deal with its unique natural and physical environments. They represent the cognitive, ideational and social environments that constitute the immediate social setting. Often coded in the local language and passed down from generation to generation verbally, they constitute the human intellectual gene pool, the basic raw material for local adaptations. Though a systematic body of knowledge acquired by the local people through experiences, informal experiments and intimate understanding of the environment, they have remained outside the domain of the scientific knowledge *per* se for long. In fact, it was the result of the growing disillusionment over the conventional approach in agriculture which stressed intensive input use leading to declining productivity trends and environmental degradation that has brought a shift in agricultural development paradigm towards organic farming. The risk reducing and resource conserving aspects of indigenous knowledge provided a productive and sustainable base for organic farming. Therefore, in recent years the systematic documentation of indigenous knowledge systems into an information bank from which scientists, development workers and farmers can draw insights has gained prominence. When we analyze indigenous knowledge in the context of organic farming, both show a high degree of agreement on the following terms:

1. Both are evolved based on the experiences standardized through experimentation and testing by farmers.
2. Both percolate into the farming system from one generation to another orally
3. Both involve simple measures for ensuring minimal livelihood security through risk aversion and resource use optimization.
4. Both follow location specific strategies to problem solving.
5. Both are internalized through experiential learning process and are related to the cultural traditions and beliefs
6. Both include holistic, dynamic and eco-friendly practices which are adapted through repetitions.

When we analyze indigenous knowledge, we come across three related concepts, *viz.*, Indigenous Knowledge (IK), Traditional Knowledge (TK) and indigenous technical knowledge (ITK).

Indigenous Knowledge

Refers to the unique, traditional and local information existing within and developed around the specific societal conditions, indigenous to a particular geographical area, *e.g.* Inclusion of dietary supplements like ragi, pulses etc. as boarder plants in rice.

Traditional Knowledge

Refer to the knowledge, innovations and practices of indigenous and local communities embodying traditional lifestyles, (owned and used since considerable period), *e.g. Mampookanikkuka, Mallanu vithiduka* etc.

Indigenous Technical Knowledge (ITK)

Refer to the innovations and adaptations made by the farmers in the traditional knowledge integrating the technical knowledge also to solve specific field problem. This is often considered to be a refinement of the indigenous knowledge (developed over time and continue to develop within the system at their own level), *e.g.* rat control using perched coconut husks, use of aquatic weeds as organic manure.

Though the three are distinctly different, for the purpose of this presentations, the three concepts are used interchangeably. Major characteristics or attributes of ITK that makes them ideal components in organic farming include:

Dynamic Nature

The indigenous practices show tremendous potential for constant changes through mechanisms of local creativity and innovation. This attribute provided them with ample flexibility to be adaptive to changes within the system. The indigenous practices remained economical and dynamic over the years by accommodating new options for old problems, *e.g.* there was a practice of using the white inner pseudostem of banana to scare rats from crop fields. This practice has been modified over the years by replacing the banana pseudo stem with the more easily available white plastic bags.

Major Thrust on Risk Aversion

Most of the indigenous practices were simple measures that ensured minimal livelihood security for local people through risk aversion and resource use optimization. Profit maximization was not a major concern in these local innovations. Most of these practices were simple solutions to over come uncertainties related to farming like weather, crop failures etc. and were dependent on local resources, *e.g.* predicting rain by observing sun, moon, wind, flies etc., inclusion of dietary supplements like pulses and other cereals as boarder plants in rice fields etc.

Lack of Abstract Nature

Most of the ITKs lacked the precision and abstractness of scientific technologies. They were non-standardized elaborate mechanisms that depended mostly on the local availability of resources and adaptation to local culture and environmental conditions. Most of the indigenous practices of plant protection use the available materials in available quantity rather than fixing standardized prescriptions, *e.g.* use of indigenous bio-pesticides like leaves of *Oduku*, neem, mango etc. lack exactness in terms of quantity, time etc.

Oral Tradition

Except for the recent documentation efforts made by the mainstream agricultural development process, the indigenous knowledge systems were learned and transmitted mostly through oral communication. Since coded in the local language there existed severe limitations for it to get accepted and comprehended outside the social system, *e.g.* many of the indigenous concepts like *Mampookanikkuka, Oocharal, makarappztli* etc lack transalations.

Location Specific Strategies

Each knowledge system addressed a problem that was unique to a particular agro-ecosystem. They were tuned to the needs of the local people and the quality and quantity of the available resources. Moreover, being time tested through use over generations, they do not require any confirmation trials and are more credible and acceptable to farmers, *e.g.* most of the indigenous rat traps reported from Kerala is the results of local innovations from Kuttanad region where rat is a severe problem.

Use of ITKs in Organic Farming

However, to extract and utilize the full potential of ITKs in organic farming, Indigenous Knowledge should be properly documented. As it is generally passed down by word of month from generation to generation most part of it remain unutilized and uncomphrehended. Moreover, for its wider application in areas like organic farming validation of the documented IK forms a prerequisite. Validation is a logical step to qualify and quantify effectiveness of the indigenous practices. Suitable modifications of these practices, through R&D will help to develop appropriate and acceptable technologies that are more suited to organic farming. Safeguarding IPR issues related to ITK in organic farming forms another area of importance. The names and addresses of the disclosures of various ITK practices in relation to of should be cited in the documents.

Salient ITK in Rice Organic Farming

Seasons

Traditionally three seasons

Sl.No.	Most Parts of the State	Kannur	Kasargode	Wayanad
1.	*Virippu*	Virippur	Enilu	Nariya
2.	*Mundakan*	Puncha	Sukhi	Puncha
3.	*Puncha*	Thavunja	Kolakka	–

Virippu (Kannikrishi)

It denoted the first crop, which starts in Medam and are harvested by Chingam or Kanni. Also referred to as 'Kannikoythu'. If long duration seeds are used they should be broadcasted before vishu (in Revathi njilttuvela)

Mundakan (Makarakrishi)

It forms the second crop, which starts by Chingam-Kanni and ends by Dhanu-Makaram. Referred to as "Makarakoythu". More yield than virippu (both grass and straw).

Puncha

The third crop is often decided by the water availability. It starts by Vrischikam, Dhanu or even in Makaram. Short duration (below 90 days) seeds are used for the crop.

Modankrishi (Kara Krishi)

Broadcast of seed is practiced. Bharani njattuvela of medam is most appropriate.

Paddy Lands

Besides double cropped paddy fields, fields with distinct features are seen is Kerala.

Paduvam

Adjacent to coastal areas in Thiruvananthapuram District. Cultivation is similar to Pokkali– "Thulunadan" variety of seed is used, which can withstand water logging.

Puthayal

Eastern parts of Kallam District potential to raise three crops.

Karappuram

Chertala and Ambalapuzha taluks in Alappuzha District; short duration rice variety and 'Mundakan', long duration rice variety are mixed and sown, which is the traditional practice in this area.

Pokkali

Paravur–Vypeen areas of Ernakulam District and Kodungallur area in Thrissur District. Most traditional method of cultivation, rich in organic content.

Paandi

Adjacent, to Kodungallur in Thrissur District below mean sea level and hence there is intrusion of saline water. Clayey soil rich in organic matter content. Oarpaandi and Swarnapaandi are the main rice varieties.

Kole

Spread over 13,000 ha in Thrissur and Malappuram Districts. Wetland ecosystem with rich biodiversity resources.

Kuttanadan

Various locations. Fields in which Kuttaadan variety of rice is cultivated is referred to Kuttadan fields. Though the duration of this rice variety is nearly one year, its ability to withstand flooding growing taller with the raising of water level has made this variety popular.

Karanjaals

Two types, modan and pallyal lands, Pallyal lands are mote prevalent in Valluvanadu. There are single cropped paddy lands where rainfed rice is cultivated.

Kaipad

Resembles Pokkali fields-coastal areas in Kannur District.

Kolakke

Kasargode District–Three crops are raised.

Farming Calendar

Farming traditionally was undertaken observing the 'Njattuvela' calendar. There are 27 njattuvelas, which begin on the first of the month of Medom. The 27 stars in the calendar are connected with the fifteen days taken by the moon to move around the earth. In this process, the sun will be positioned near a particular star for about 15 days. The time (vela) of the sun (Jnayar) near the star is referred to as njattuvela. During this period, the rising and setting of that particular star coincides with the sun. It is calculated that except Thiruvathira, which is 15 days long, other njattuvelas are of 13 112 days duration.

Each Njattuvela has different patterns of rain, wind, sunshine, cloud formation, etc. A month consists of average of two Njattuvela among which Thiruvathira, Punartham, Pooyam and Ayiliam are important. Bharani Njattuvela is considered to be auspicious for sowing. It is said that 'seeds are kept in Bharani'. It is told that seeds sown after Rohini do not give good results. It is believed that seeds sown in Makayiram will not sprout-because of the pattern of the rain. It is said that Thiruvathira Njattuvela will have rains and 101 sunshine days, indicating that it is a perfectly balanced climate suitable for plants. There is a saying that even if a human finger is planted during this time, is will sprout. It is said that even the weeds become paddy if it rains in Pooyam. The star of seed is Atham. The Choti rain makes sun of the next year's food and the harvest will be week if it rains in the early days of the month of Chingam.

Seed Treatment

There are many rituals and beliefs that are followed in preparing seeds for sowing. Most of these traditions reflect the sanctity and biodiversity concerns of the local people. The major practices that bear a direct relation with organic farming in terms of equity are discussed.

Kaikkottu Chaliduka

This is tradition associated with the starting of agricultural operations in a new year. It is observed on the afternoon of medam first (mid-April) every year. Rice flour is used to draw a ladder with seven steps in the northern courtyard of the house where a new agricultural implement (usually hand hoe) is kept cleaned and decorated with flowers. Ganapathy worship is performed by eldest male member after which some seeds for sowing in the next season are placed in the banana leaves in the ladder. Rice flour is also used to decorate the implements used to inaugurate the ploughing operation, which is performed by digging the sectors in side the drawn ladder. Water is sprinkled over this and then digging is performed and some more seeds are added. Then the worshipped implements and seeds are handed over to the head labour. He takes this to the field to perform the *Chalidukka*. The practice is that in the courtyard the Owner sows and in the field the labour does this.

Chalidukka

This is a similar custom associated with the land preparation in rice observed on the day of Vishu (Medam 1). Farmers' family and farm labourers collectively do this worship of land and seeds. Some earth is piled at the right hand corner of the field and the top of the pile is leveled. Seven leaves of *kanjiram* (*Sttychnos nllxvomica*) are taken and each Jeafis made into a cone. These cones are placed in seven pits in the leveled top of the mount. Some seeds are put into each cone and are covered with sand. Some, *konna* (*Cassia fistula*) flowers are put on it and water is sprinkled. In certain areas, there is a practice of rubbing gold on the new plough before it is used. The belief is getting a golden harvest. The practice can be considered as an *in situ* pre-germination test for the seeds, which are sown over these small mounts taken in the right hand corner of the field.

Malian Vithiduka

This practice is offered to protect the crop from all natural calamities and is associated with sowing of seeds. *Mallan* is believed to be the security gatekeeper of Goddess Earth. A stone half submerged under the earth near the agricultural fields represents *Mallan*. *Thulasi* plants are planted near this and just before sowing the stone is worshipped and seeds are placed over it.

Birth Day of Paddy

Ayilliam Day of the month of *Kanni* (September–October) is observed as the birthday of rice and on the day, paddy seeds are not touched in the field or store. The day is observed as an Agricultural Holiday.

Pest Control

Bio-pesticides assumed larger significance in the context of recurring pest incidence and pest resurgence due to indiscriminate use of plant protection chemicals. Indigenous bio-pesticides that continue to be in use by the rice farmers of Kerala provide significant potential for organic plant protection. Local availability was the major criteria that determined the use of bio-pesticides, which favoured *Azadiracta indica, Glyricidia sepum* and *Hydnocarpus wighitiana, Lantana camera, Calotropus gigantia* and *Cleistanthus collinus* were also popular in prophylactic pest control measures and its potential could be used in contingent measures also if proper commercial formulations were made available. Other bio-pesticides that had the potential to evolve as alternatives to chemical pesticides include *Glycosmis pentaphylla, Mangifera indica, Ricinus communis, Cleistanthus collinus, Anacardium occidentale, Ailanthus tryphysa, Macaranga indica, Strychnos nuxvomica, Centella asiatica* and *Maranta arundinaceae*.

Chapter 25

Rice Diseases in Organic Rice-Fish system

About 40 diseases attack paddy in the field and of these the most prevalent and important are blast, sheath blight, brown spot, sheath rot and bacterial blight. The severity of these diseases from year to year is unpredictable since environmental factors, seed borne inoculums etc. influence disease incidence. Even though it is difficult to estimate the part that disease play in the economy of rice production, diseases such as blast, brown spot, bacterial blight to mention a few have from time to time caused great damage. Brown spot was the major fact contributing to Bengal famine, 1942.

Knowledge of the diseases and early diagnosis is the first and foremost step of disease management: Earlier disease management strategies mainly aim at prevention of out break of epidemics through the use of host plant resistance and chemical pesticides. The persistent injudicious use of chemicals has toxic effects on non-target organisms and can cause changes in the environment. Research during the previous two decades indicates another potential option for rice disease management, *i.e.* Biological control. This leads to the development of integrated disease management (IDM) practices in which different measures of disease management like host plant resistance, cultural, biological and chemical measures were used in combination. In the wake of serious pollution problems and bio-magnification of toxic chemicals in the various biological systems, organic farming which was practiced in India from time immemorial got significance. To manage diseases under organic farming situations, all practices of IDM except chemical measures can be utilized. Inclusion of fish into the organic system further diversifies the situation in which utmost care has to be taken in plant protection operations. Even though a natural reduction in the pest and diseases of rice were noticed in organic rice cultivation and in-organic rice fish integrated systems, an understanding of various diseases and different strategies for their management are essential. This paper deals with important rice diseases and measures which can be adopted for their management in organic rice fish system of cultivation.

Blast

Rice blast is probably the earliest recorded and most wide spread disease of rice. Disease symptoms appear on leaf as small brown specks of pinhead size which later turn larger. Typical leaf lesions are spindle shaped and often develop grayish centers with brown margin and coalesce on leaves of susceptible varieties leading to death of plants under favorable conditions. Highly resistant varieties have no lesions but only pinhead sized brown specks. Varieties with intermediate reaction have small round or short elliptical lesions with a small necrotic center and brown margin. Neck blast symptoms appear as dark necrotic lesions at the panicle neck or branch, leads to partially filled grains or no grains. Various studies in India has shown that 1 per cent panicle blast reduced yield by as little as 0.4 per cent or 1.4 per cent to as much as 17.4 per cent.

The disease is caused by *Pyricularia grisea*. The fungus is seed borne. Other main source of primary inoculum is the fungus carried over the straw or volunteer plants or on weeds functioning as collateral hosts. *Panicum repens, Leersia hexandra* and *Digiteria marginata* which grow in plenty near rice fields are the important collateral hosts of the organism. The air borne conidia from rice plants (volunteer/ cultivated) or from the collateral hosts are the important source of inoculums for secondary spread of the disease. There are considerable differences in the susceptibility of paddy varieties to the disease as well as of paddy growing under varying soil conditions. Plants are highly susceptible in dry soil and susceptibility varies inverse ratio to the amount of water contained in the soil. Environmental conditions like temperature and humidity are known to be the deciding factors in the epiphytotic outbreak of the disease. Temperature below 25°C and high humidity above 90 per cent favored the intensity of blast. Excessive application of nitrogenous fertilizers also predisposes the pants to infection.

Sheath Blight

Sheath blight was reported from India by about 1940. But the disease has reached a serious proportion with the introduction of high yielding and high nitrogen responsive varieties of rice. Symptoms first appear on the lower leaf sheath near the water line. They are greenish grey, elliptical or oval and enlarge and become grayish white with brown margin. Brown sclerotia, which are easily detached, form on these spots. Infection begins when the plants are in the maximum tillering stage.

The disease is caused by *Rhizoctoniasolani*. The disease appears to be soil-borne. The fungus survives in the soil either as selerotia or as dormant mycelium. The sclerotia can remain viable for a very long period and, may float on the surface of the water during puddling, levelling and other operations and infect the crop. Once the primary infection is established, the mycelium spreads rapidly over the surface of the sheath. The intensity of sheath blight was found to be more when the humidity was above 90 per cent and temperature was above 28°C.

Brown Leaf Spot

Brown leaf spot or *Helminthosporium* blight has been found in all rice growing countries. It is a fungus disease attacking germinating seeds, leaves and developing mains. Typical leaf spot is quite small and oval or circular. Small spots are dark brown, the larger spots are same color on the edges but have a pale usually grayish center. Most spots have a light yellow halon around their outer edge. Larger spots are seen on susceptible varieties. A black velvety appearance is seen when the fungus attacks the glumes and grains.

The disease is caused by *Helminthosporium oryzae*. The primary infection may be by seed borne inoculums. The secondary spread will be by air borne conidia. Collateral hosts also serve as a source of inoculums. Several grasses like *Cynodon dactylon, Digitaria sanguinalis, Echinochloa colona* and *Leersia hexandra* are known to harbor the organism. It is also reported that the pathogen can survive as a facultative saprophyte on organic matter in the field. Infection of seedlings by the seed borne inoculums takes place only when the soil temperature is between 18 and 22°C. Generally high humidity favors the infection. It is a general observation that the disease is more prevalent in fields which are not managed properly. Poor or lack of mannuring, disturbance of nutrient balance etc. are often associated with the disease.

Sheath Rot

The rot occurs on the upper most leaf sheaths enclosing the young panicles. Oblong or irregular spots develop in to a lesion with brown margin and grey centre or it may grayish brown throughout. The young panicle remains inside the sheath or rather checked and emerge only partially. A whitish powdery growth of the fungus may be found on the panicle inside the sheath. The grains become chaffy.

The disease is caused by *Acrocylindrium oryzae.* The disease is seed borne. Climatic conditions particularly cloudiness and incessant rains during flowering period helps in the spread of the disease.

Foot Rot

The disease known as foot rot, Bakane or Fusarium blight was 1st reported from India in 1931, from Andhra Pradesh. The disease occurs in the nursery and in the main field. In the nursery the affected seedlings are chlorotic and pale, thinner than the healthy ones and abnormally elongated. Such seedlings may wilt from the top downward and finally die. Seedlings which survive the attack in nursery, may die soon after transplanting. In the main field, the most conspicuous symptom is the appearance of stray, tall, lanky tillers, which are abnormally elongated. Such plants show the symptoms of fungus infection at the collar region and die out within 2–6 weeks. The leaves of the infected plant dry up from below. There is profuse branching of the roots, which otherwise appear healthy. Some times the fungus causes stunting of plants. Adventitious roots are also produced from the first two and three nodes above the ground level. The development of bakane symptom is influenced by the fungus strain and the environmental conditions such as temperature and moisture.

The fungus causing the disease is *Fusarium moniliformie.* The disease is seed borne and usually the pathogen is externally present. Generally infection takes place at the flowering stage. Severely infected seeds develop a reddish discoloration due to the development of the fungus.

Stem Rot

The disease occurs during the later stages of plant growth. Main symptoms of the disease are the excessive production of late tillers at the base of the stem, which becomes discolored and rotten in severe cases. The first symptom of the disease is the appearance of small, black, irregular lesions on the outer leaf sheath near the water line. As the disease advances the discoloration spreads up and down from the water line. When the affected plants are cut open and examined, the interior of the culms may be filled with grey fungal mycelium. Numerous sclerotia may be found either loose or embedded inside host tissue of the culm. Seedlings are also affected, resulting in drying up of leaves. Masses of sclerotial bodies can also be seen on the drying older basal leaves.

The causal organism of the disease is *Sclerotium oryzae.* The sclerotia of the fungus survive in soil and stubble during the off-season all.dserve as the primary source of inoculums. They are spread from field to field by irrigation or' floodwater, which' accounts for the secondary spread of the disease.

Narrow Brown Leaf Spot

Large number of brown linear spots is found on leaves. As the name suggests, the spots are narrower and light coloured than those of the brown leaf spot disease. Though rice plants are susceptible to the disease throughout the growth period, lesions develop in large numbers during the latter periods of their growth. The disease causes considerable reduction in the effective photosynthetic area of the leaves. The disease is caused by the fungus *Cercospora oryzae*. The fungus is internally seed borne. The disease spread in the field due to air borne conidia.

False Smut or Green Smut

The disease appears on the ears, where individual ovaries transformed into large, velvety green masses. Only a few spike lets in the ear are affected. The fungus infects young ovary, which transplanted into a masses of closely united, fine colourless hyphae. It produces spores and the immature yellow colour of which becomes brownish green at maturity.

The fungus, *Ustilaginoidea virens*, causes the disease. The disease is not seed borne. Primary infection is believed to be caused mainly by ascospores produced from sclerotia. The fungus has been recovered on many grasses and wild rice and the spores produced on the collateral hosts become air borne and serve as the main source of inoculums. Chlamydospores are air borne and are profusely present at the time of heading of rice plant.

Stack Burn Disease

The symptoms of infection are mainly noticeable on the seedlings, on the leaves and on the ripening grains. On the leaves of mature plants circular to elliptic brown spots with dark brown margin, encircling the spots like a ring are formed. The centre of these spots is light brown or white bearing very minute dark round dots, which are the fruiting bodies of the organism. Similar spots with irregular margins may be formed on leaf sheath also. On the grains, brown or sometimes pink to reddish brown spots are observed on the lemma and palea. They bear dark dots in the center.

The disease is caused by the fungus *Alternaria padwickii*. The pathogen is internally seed borne. Secondary spread may be by air borne conidia.

Leaf Scald

The disease makes its appearance usually at the tip of mature leaves. The spot, initially appear as water soaked blotches, develop into large ellipsoid or oblong brown olive areas, which are encircled by dark brown narrow bands. These areas are often surrounded by light brown halo. A characteristic zonation develop in the effected area by successive bands of dark brown margins and lighter inner areas. Under favourable conditions the lesions enlarge and coalesce resulting in blighting of a large area. The fungus *Rhynchosporium oryzae* causes the disease, The conidia formed on the infected lesions spread the disease in the field.

Udbatha Disease

The symptoms of the disease are evident only at the time of panicle emergence. The infected panicles present a bethy like appearance. The fungus *Ephelis oryzae* causes the disease. It is purely a seed borne disease and the pre disposing factors are cool climate of high altitude regions, where low temperature and high humidity is prevalent.

Bacterial Blight

In India Bacterial Leaf Blight was first reported in 1959 and little attention was paid to it until 1963, when a severe out break occurred in Bihar. Krezek and leaf-blight are the two types of symptoms observed in tropics. Krezek is the systemic form of BB and symptoms appear 2–3 weeks after transplanting. The infected leaves become grayish green and begin to roll along the midrib. Leaves wilt and plants die rapidly. Krezek is often mistaken for dead heart caused by stem borers. In leaf blight, lesions may appear on the leaves as water soaked stripes starting from the edges near the tip. The young lesion is pale green to gray green. It turns white or yellow after a day or two. It maybe seen on one side of the midrib or on both sides. They are asymmetrical. In susceptible varieties the infection may continue in to the leaf sheath.

The disease is caused by *Xanthomonas oryzae.* The infected seed is considered to be the most important source of inoculums. The bacterium survives in the infected seed till the next season. The organism invades the rice plant through natural openings of through wounds. The bacterial exudates formed over the surface of plants will dry up form small beads and drop into irrigation water. This contaminated water flowing from field to field will spread the disease. A number of weeds function as collateral lasts and *Leersia sayanuka, Cyperus rotundus, C. defformis* etc. are important among them.

Tungro

This disease, first observed in Philippines, was named as 'Tungro', which in the local language means degenerated growth. Stunting of the plant, the discoloration of leaves from various shades of yellow to orange and rusty blotches spreading downwards from the leaf tip are the symptoms. The young leaves may show a mottled appearance and older leaves appear rusty colored. If infection of highly susceptible varieties takes place at very early stages, the plants may die before flowering.

The tungro virus is transmitted by green leaf hopper *Nephotettix virescens.* Since the virus is non-persistent, the victor must repeatedly acquire the virus from diseased plants to be continuously infective. Hence destruction of diseased plants will help to eliminate the source of inoculums to a great extend.

Grassy Stunt

The virus disease was noted in 1963 for the first time in Philippines. Characterized by severe stunting, excessive tillering and erect growth habit with pale green leaves and rusty spots. These spots spread and give a bronze colour to the older leaves. Several small tillers are formed giving the plant a typical grassy appearance. No panicles are formed or inorrned, they will be sterile.

The brown plant hopper *Nilaparvatha lugetis* transmits it. After the acquisition of the virus, the insects retain infectivity throughout their life and hence the virus is of persistent nature.

Rice Disease Management

Host Plant Resistance

This is a cheap, practical and environment friendly method of disease control. Even if resistant varieties are not available, the use of susceptible varieties can be avoided. A list of rice varieties with resistance to diseases are given in Table 25.1.

Many modern cultivars share same genetic background, which has contributed to the instability of resistance against several rice diseases with high epidemic potential and sustainable approach in managing this problem is the use of cultivars with diversified functions to limit the development of disease. Inter planting modern high yielding hybrid rice varieties that have high disease resistance (blast) with susceptible traditional taller glutinous varieties is a rapidly expanding practice in China. Mixtures of cultivated varieties, growing simultaneously on the same land is another practices to prevent breakdown of resistance. Use of mixtures of genotypically identical lines that differs only in specific disease or pest resistant gene (multilines) is widely used to prevent break down of resistance against blast in Japan.

Table 25.1

Blast	*Sheath Blight*	*Sheath Rot*	*Bacterial Blight*
National release			
Tulasi	Swarnadhan	Mashuri	Ajaya
Aditya	Vikramarya	Janaki	Swarna
Suraksha	Radha		Sona Mashuri
IR-64	Manasarovar		Gobind
	Pankaj		
KAU release			
Swarnaprabha	Kairali	Karthika	Karthika
Suvarnamodan	Karthika	Aruna	
Karuthamodan	Aruna	Makom	
Kairali	Makom	Pavizham	
Karthika	Aathira	Remya	
Aruna	Aiswarya		
Makom	Pavizham		
Bhagya			
Bharathi	Lakshmi		
Jayathi	Dhany		
Aathira	Sagara		
Aiswarya			
Pavizham			
Remya			
Kanakom			
Lakshmi			
Dhanya			
Sagara			
Neeraja			

Cultural Practices

By adopting certain changes in the cultivation practices, pathogen population, host susceptibility and suitable environmental conditions for the disease development can be decreased and the disease can be managed.

Use of Disease Free Seeds

Many diseases of rice are transmitted through seeds, *e.g.* blast, brown spot, bacterial blight, sheath rot, foot rot, udbatha, stack bum etc. Hence always collect seeds from disease free plots. Proper drying of the seed will destroy many fungi and bacteria pathogens in the seeds. During germination, rotting of emerging roots and shoot of infected seeds was also reported.

Planting Time

Environment plays an important role in the development of rice diseases. So adjust planting time to avoid coincidence of susceptible stage of crop with favorable environment for pathogen, *e.g.* Late planting decreases blast incidence. Uniform planting helps to avoid the spread of disease from one plot to the next, *e.g.* Incidence of sheath rot is more in the late planted crop.

Nutrient Management

Application of organic manures reduces disease incidence in rice. Reduction in brown spot disease by the application of cattle manure was reported. Incorporation of green manures like vengai (*Pterocarpus marsupium*), mango (*Mangifera ilndica*) nux-vomica (*Strichnos nux vomica*) and Eupatorium (*Eupatorium odoratum*) @ 5t/ha significantly reduced sheath blight incidence and increased yield. Soil amendments have important role in disease management. Zacharia reported that neem cake and lime are useful soil amendments for reducing intensity of sheath blight disease. Soil application of neem cake is also recommended for management of rice tungro. Application of silicon as calcium silicate @ 2t/ha is recommended enhance plant to resistance to diseases and to increase yield. As rice hull, is rich in silicates, the applications of rice hull ash @20 kg/cent is recommended for the management of blast in the nursery.

Spacing in the Nursery/Main Field

Seeds should be sown in adequate nursery area. Otherwise seedlings will grow closer and diseases like blast, brown spot and bacterial blight may flair up. In the main field, spacing is to be given as per the duration of the variety. Tewari reported the effect of spacing on sheath rot disease. Higher disease severity was observed in 15 × 15 cm and 15 × 20 cm than in 15 × 25 cm and 15 × 30 cm spacing.

Water Management

Soil moisture play an important role in rice disease management. Bacterial blight pathogen spread through irrigation water from diseased plots to nearby plots. Hence drainage of water and prevention of irrigation water from diseased plots to nearby plots are important for management of the disease. Sheath blight and stem rot also flair up in the water stagnant condition and drainage of water is important for management of these diseases. Susceptibility of rice plant to the blast pathogen is high in dry soil. Hence increasing soil moisture through irrigation is one way to manage the disease.

Weed Management

Many graminaceous weed act as collateral hosts of rice pathogens (Table 25.2).

So removal of weeds from the field as well as from the bunds is a good practice of disease management.

Summer Ploughing and Destruction of Crop Residues

Rice pathogens dwelling in the soil can be destroyed by deep summer ploughing *e.g.* Sclerotia of *R. solani*. Burning of stubbles after harvest is helpful for the destruction of pathogen in the crop debris *e.g. H. oryzae.*

Table 25.2: Rice Pathogens

Weed	Pathogens
Leersia hexandra	*Pyricularia grisea*
Cynodon dactylon	*H. oryzae*
Cyperus rotundus	*Xanthomonas campestris*
C. difformis	*X. campestris*
Digiteria sanguinalis	*H. oryzae*
Echinochloa colona	*H. oryzae*
Panicum repens	*P. grisea*

Biological Control

Biological control could be defined as any condition under which or practice whereby survival and activity of a pathogen is reduced through the agency of any other living organism (except man himself) with the result that there is reduction in the incidence of the disease caused by the pathogen. The mechanisms by which biocontrol agents interface with the life processes of plant pathogens are competition, parasitism, predation and induced systemic resistance. There are many bacteria and fungi live in association with roots of rice crop and they can be used for the control of fungal/bacterial pathogens of rice. A list of biocontrol agents against rice pathogens are given in Table 25.3:

Table 25.3: Major Rice Diseases and their Biocontrol Agents

Disease	Casual Organism	Biocontrol Agent
Blast	*Pyricularia grisea*	*Pseudomonas fluorescens*
Brown spot	*Helminthosporium Oryzae*	*Pseudomonas* sp., *P. aeruginosa* *Bacillus* sp., *B. subtilis*
Bacrerial blight	*Xanthomonas oryzae* pv. *oryzae*	*Bacillus* sp.
Sheath blight	*Rhizoctonia solani* Kuhn	*P. fluorescnes*, *P. putida* *Bacillus* sp., *B. subtilis,* *B. laterosporus*, *B. pumilus* *Serratia marcescens* *Pseudomonas* sp., *P. aeruginosa*
Sheath rot	*Sarocladium oryzae*	*P. fluorescens*, *B. subtilis* *P. aeruginosa*, *Pseudomonas* sp.
Stem rot	*Sclerotium oryzae*	*P. fluorescens*, *P. aeruginosa* B. subtilis, B. pumilus
Tungro	Rice tungro virus Vector–*Nephotettix* spp.	*P. fluorescens* (for vector)

The major concepts involved in biological control are 1) Introduction of microbes in the phylloplane, rhizosphere or soil; (2) Stimulating indigenous antagonists; (3) Biorational approaches.

Pseudomonas fluorescens is a potential bio control agent and has broad spectrum of action against fungal and bacterial pathogens of rice. It can produce antibiotics, siderophores and can induce systemic resistance in plants. The talc based formulation can be applied through different methods like seed treatment (10 g/kg seed), soil application (2.5 kg/ha), seedling root dip (20 g/l) and foliar spray

(2 g/l). The symbiotic association of vesicular-arbuscular mycorrhiza (VAM) is known to improve the growth and disease resistance of host plants. Sivaprasad reported that intensity and incidence of sheath blight disease reduced markedly by with inoculation of *Glomus fasciculatum*.

Indigenous antagonists can be stimulated by the use of organic amendments, green manures etc. Baby and Manibushan Rao reported that organic amendments increased mycorrhizal spore density, per cent infection and intensity of infection and decreased sheath blight severity.

Biorational approaches combines host plant resistance and use of biocontrol agents. They should complement each other in their activity against pathogens.

Use of Botanicals

Many botanical products are reported to have antifungal/antibacterial properties and can be used effectively for the management of rice diseases in rice fish system. A list of botanicals with antifungal properties against rice pathogens is given in Table 25.4:

Table 25.4: List of Plant Extracts with Antifungal Properties Against Rice Pathogens

Plant Extracts	*Pathogen*
Ophiopogon japonicus	*P. grisea*
Prosopis julifiora	*P. grisea*
Zizphus jujuba	
Ocimum sanctum	
Abutilon indicum	
Cynodon dactylon	
Cyperus rotundus	
Catheranthuss roseus	*H. oryzae*
Polyalthia longifolia	
Azadirachta indica	
Vitex negundo	*S. oryzae*
Acacia leucocephala	
Polyalthia longifolia	
Azadirachta indica	

Some botanicals formulations are available for the management of rice diseases. Raji reported that of Achook (0.5 per cent), Spictaf (1 per cent), and Neem Gold (0.15 per cent) were effective in reducing sheath blight and improving yield of rice. Thapak reported that Neem Gold (2 per cent), Neem azal (0.3 per cent) and Raze (1 per cent) (*Lantana camara* based biopesticide) are effective to sheath rot.

Turmeric (*Curcuma longa* L.) has long been used as a germicidal ingredient. The powder of dried rhizome of turmeric can be safely used as a fungicide and bactericide as it is non-hazardous, cheap and ecologically friendly. Seed treatment of 10 kg seed with 100 g turmeric powder and 10 g sodium bicarbonate is effective to control seed borne diseases and foliar spray of the same in 10 litre water is effective to control blast, sheath blight etc.

The action of above mentioned disease management measures without chemicals is slow and has to apply in the initial period of disease development. Hence early detection of the disease and selection of suitable measured for their control are important. This necessitates regular field monitoring.

Chapter 26

Permanent Manurial Experiments in Rice

Introduction

Stabilisation of Crop production at the desired level requires proper assessment of the direct, residual and cumulative effects of various inputs over a period of years on crop yield and soil conditions. The combined effects of various inputs may be highly beneficial or some times harmful for soil-plant system. Although with the recognition of fertilizers as the most valuable additive to the soil for increasing crop yield, there have been appreherision on the ill effects of fertilizer on soil fertility and crop yields owing to their prolonged use. Moreover the impact of intensive land use in association with the continuous use of high doses of chemical fertilizers and organic manures requires careful and detailed investigation. The chance of exhaustion of any nutrient by continuous cropping without the addition of that nutrient and the toxic effects of any particular nutrient by continuous application are some of aspects which are worthy of investigation in Permanent Manurial Experiments (PME). Accordingly the PME was started in Rothamsted Experiment Station, Harpeden, England in 1843. These classical experiments gave results of value, not only in respect of crop production but also on the depletion and enrichment of plant nutrients in soil as result of cumulative and residual effect of annual dressing of manures and fertilizers and crop removal of nutrients. India had first series of three such PME laid out in Rothamsted model at Kanpur in Uttar Pradesh, Coimbatore in Tamil Nadu and Pusa in Bihar. These experiments were laid out in large sized unreplicated plots not in conformity with any statistical design. But they definitely did provide some information on the wide range of effect of the continuous use of manures and fertilizers on soil productivity. Discontinuance of many of these experiments at different stages without documenting the findings resulted in loss of many valuable research results in the field of soil fertility.

Late sixties was the period of introduction of high yielding varieties and beginning of green revolution. The heavy withdrawl of nutrients from soil by high yielding varieties and the need for heavy use fertilizers made it imperative to examine the sustainability of modern intensive cropping systems which is based on high external inputs of fertilizers, agrochemicals and high yielding cultivars. This leads to the initiation of long term studies at fixed sites for monitoring the changes in physical, chemical and biological properties and crop yields as influenced by continuous application of nutrients, singly and in combination, and in organic and inorganic forms.

Paddy lands in this area fall under three geographic categories namely rain fed uplands or "Modan", "Pallyals" or single crop wetlands and "Iruppu lands" or double crop wetlands. Two permanent manurial experiments one using Tall indica variety Ptb-2 in the first cropping season and Ptb-20 in the second crop season from 1961 onwards and another experiment using photo insensitive medium duration dwarf indica variety-Jaya from 1973 onwards are being initiated at Regional Agricultural Research Station, Kerala Agricultural University, Pattambi representing Brown Hydromorphic lateritic alluvium in double crop Wetlands in the 'A' Block of the station. The station is located at 10°N latitude and 76°E longitude at an elevation of 25 m from MSL. The annual rainfall in this 'area is about 2750 mm. 70 to75 per cent of this annual precipitation being received from the South West monsoon coinciding with the first crop season (Kharif), 15 to 20 per cent from the North East monsoon during the second crop season (rabi) and rest during the summer months. The temperature ranges from 15 to 40°C, a maximum of 40°C during April-May to a minimum of 15°C during December to January. The relative humidity varies from a minimum of 75 to 80 per cent during the summer months to 90 to 95 per cent during July–August.

The soil is lateritic sandy loam with moderate fertility, and the soil reaction is in the acidic range. These soils were included under a soil association namely Karakurissi-Koonthara-Vadanamkurissi association which is very deep dark grayish brown identified by the State soil survey wing of the state department of Agriculture.

Table 26.1: Initial Soil Properties at PME

Physical Properties		*Chemical Properties*	
Sand (per cent)	70–80	pH	5.0–5.5
Silt (per cent)	8–10	Organic Carbon (per cent)	1.5–2.0
Clay (per cent)	12–15	C.E.C Cmol (p+) kg/ha soil	12–15
Texture	Sandy loam	Total N (%)	0.2–0.3
Water holding capacity (per cent)	40–45	Total P (%)	0.0058–0.0060
Pore space (per cent)	45–48	Total K (%)	0.06
Apparent density	1.04	Available N (per cent)	0.038
Absolute specific gravity	2–2.5	Available P kg/ha	10–15
		Available K kg/ha	100–150
		Available Ca ppm	125–150
		Available Mg ppm	20–30
		Available Fe ppm	250–300
		Available Mn ppm	15–20
		Available Zn ppm	3–6
		Available Cu ppm	2–3
		Available Si ppm	100–200

The salient results PME upto 2002–03 are presented.

Permanent Manurial Experiment (Tall Indica)

Objectives

To study the effect of continuous application of cattle manure, green leaves and ammonium sulphate individually and in combination with and without P and K on Tall Indica rice.

Technical Programme

Date of commencement: 1961 Virippu

Design: RBD

Replication: 4

Test variety: PTB 2 (Ponnaryan during Kharif); PTB 20 (Chuvanna chitteni during Rabi)

Treatments

1. Cattle manure (CM) 8000 lbs/acre to supply 40 lbs N/acre
2. Green leaves (GL) 8000 lbs/acre to supply 40 lbs N/acre
3. CM + GL each 4000 lbs/acre
4. Ammonium sulphate to supply 40 lbs N/acre
5. CM 4000 lbs/acre+ NPK 20 : 20 : 20 lbs/acre
6. CM + GL each 2000 lbs/acre + NPK 20 : 20 : 20 lbs/acre
8. NPK alone 40 : 20 : 20 lbs/acre

Inorganic source of N, P and K were ammonium sulphate, super phosphate and MOP respectively. Organics were applied 10 days before transplanting and incorporated by digging. P and K were applied fully as basal, while half the dose of N was applied as basal and the remaining half, one month before flowering *Glyricidia* was used as green manure.

Observations: Grain and straw yield

Observations on ancillary characters: Observations on panicle characters.

Brief Summary of the Findings

The results of the past years, clearly indicated that the response of *tall indica* rice to produce higher yields are slowly shifting towards cattle manure alone (TI) from the combined application of cattle manure and chemical fertilizers. The results also emphasised the detrimental effects of applying either green leaves alone or nitrogenous fertilizers alone on the growth and yield of rice.

Results Obtained

In Kharif season the treatment receiving continuous application of cattle manure alone (T1, CM), recorded the highest grain and straw yield. Yield was lowest for the treatment T4 receiving ammonium sulphate alone.

But in Rabi season, highest grain yield was recorded by T5 (CM+NPK) and the straw yield was highest for the treatment T7 (CM+GL+NPK) but it was on part with T1 (CM), T3 (CM+GL) and T5 (CM+NPK). Continuous application of ammonium sulfate has detrimental effects on the performance of tall indica series as indicated by the lowest grain and straw yield recorded by this treatment.

Summary

The results obtained so far confirmed that the productivity of *tall indica* rice varieties on long term basis can be maintained only on combined application of organic manures and chemical fertilisers.

This is the 43 year of the trial. As a permanent manurial experiment, the trial is continued.

Table 26.2: Yield of Traditional *Tall indica* Rice Varieties as Influenced by Long Term Manurial Practices

Sl.No.	*Treatments*	*Grain Yield (kg/ha)*		*Straw Yield (kg/ha)*	
		Virippu 2002–03	*Mundakan 2002–03*	*Virippu 2002–03*	*Mundakan 2002–03*
1.	CM	1550	2029	4865	4008
2.	GL	1257	2016	3477	3247
3.	CM and GL	1495	2070	3728	3869
4.	Ammonium sulphate	997	1786	2418	3151
5.	CM + Ammonium sulphate + Super + MOP	1354	2259	3688	3977
6.	GL + Ammonium sulphate + Super + MOP	1219	1962	3441	3815
7.	CM + GL + Ammonium sulphate + Super + MOP	1050	2029	4205	4113
8.	Ammonium sulphate + Super + MOP	1371	1894	3136	3598
	CD	NS	78.6	67.9	265.5

CM: Cattle manure; GL: Green leaves; NS: Non significant.

Permanent Manurial Experiment (*Dwarf indica*)

Objectives

To study the effect of continuous application of cattle manure, green leaves and ammonium sulphate individually and in combination with and without P and K on *Dwarf indica* rice.

Technical Programme

Date of commencement: 1973 Virippu

Design: RBD

Replication: 4

Test variety: Jaya

Spacing: 15 × 15 cm

Plot size: 7.80 × 5.25 M (Gross)

7.50 × 5 4.95 (Net)

Treatments

1. Cattle manure (CM) 18,000 kg/ha to supply 90 kg N/ha.
2. Green leaves (GL) 18000 kg/ha to supply 90 kg N/ha.

3. CM + GL each 9000 kg/ha
4. Ammonium sulphate to supply 90 kg N/ha
5. CM 9000 kg/ha + 45 kg N, P_2O_5 and K_2O/ha
6. GL 9000 kg/ha + 45 kg N, P_2O_5 and K_2O/ha
7. CM + GL each 4500 kg/ha + 45 kg N, P_2O_5 and K_2O/ha
8. NPK alone @ 90 : 45 : 45 kg N, P_2O_5 and K_2O/ha

Inorganic source of N, P and K were ammonium sulphate, super phosphate and MOP respectively. Organics were applied 10 days before transplanting and incorporated by digging. P and K were applied fully as basal, while half the dose of N was applied as basal and the remaining half, one month before flowering Glyricidia was used as green manure.

Observations

Grain and straw yield

Biometric observations

Brief Summary of the Findings

In the earlier period of experimentation application of cattle manure in combination with NPK (T5) recorded highest grain yield. Thereafter the treatment T1 having cattle manure alone took its place with highest mean grain yield over 19 years. Continuous application of cattle manure alone (T1) recorded significantly higher grain yield during Kharif season, while it was on par with T3 the joint application of cattle manure either with green leaves or T5 with NPK fertilizers during Rabi season. From 1992 onwards continuous application of cattle manure alone recorded significantly higher yield both in Kharif and Rabi seasons. Later on it was observed that this treatment (T1) was on par with treatments receiving cattle manure and NPK fertilizers (T5), cattle manure and green leaves (T3) and cattle manures with green leaves and NPK fertilizers (T7). Continuous application of either ammonium sulphate of NPK fertilizers without any organic manure resulted a drastic reduction in the yield of grain and straw. Incidence of pest attack and disease was the lowest for the treatments having cattle manure either alone or partially. Moreover nutrient status were improved in treatments having cattle manure.

Results Obtained

In Kharif season, the grain yield was highest for the treatment receiving cattle manure and NPK fertilizers (T5) which was on par with the treatment T7 (CM+GL+NPK) and lowest yield obtained for the T4 ammonium sulphate alone treatment. Straw yield obtained was highest for T5 (CM+NPK) and lowest for T4 (AMS). During Kharif season, the grain was highest for the treatment receiving cattle manure, green leaves and NPK fertilizers (T7). Straw yield was highest for T3 (CM+GL) which was on par with the treatments T1 (CM) and T5 (CM+NPK). Lowest yield was recorded byT4 (AMS).

Summary

The results over the years indicate that the productivity level of *dwarf indica* genotype can be maintained at substantial level through application of organic manure at least to the tune of 50 per cent of the nitrogen requirement.

This is the 31st year of the trial. As a permanent manurial experiment, the trial is continued.

Table 26.3: Yield of *Dwarf indica* Rice Variety as Influenced by Long Term Manurial Practices

Sl.No.	Treatments	Grain Yield (kg/ha)		Straw Yield (kg/ha)	
		Kharif 2002–03	Rabi 2002–03	Kharif 2002–03	Rabi 2002–03
1.	CM	1865	2283	4322	2714
2.	GL	1466	2263	3220	2445
3.	CM and GL	1778	2202	4607	2929
4.	Ammonium sulphate	1526	1913	2374	2081
5.	CM + Ammonium sulphate + Super + MOP	1837	2479	4470	3260
6.	GL + Ammonium sulphate + Super + MOP	3214	2256	3470	2377
7.	CM + GL + Ammonium sulphate + Super + MOP	3777	2459	3467	2889
8.	Ammonium sulphate + Super + MOP	1451	1946	3070	2613
	CD	26.18	81.05	784.7	42.37
	CV%	0.84	2.49	15.11	1.09

CM: Cattle manure; GL: Green leaves; NS: Non significant.

Table 26.4: Permanent Manurial Experiment (*Tall indica*) 1978, 1999, 2003 Data

Treatments	1978			1999			2003		
	Grain Yield	Straw Yield	% OC	Grain Yield	Straw Yield	% OC	Grain Yield	Straw Yield	% OC
CM	3247	5925	1.16	3020	2776	2.09	2029	4008	1.42
GL	3165	6385	1.06	2695	2814	2.3	2016	3247	1.44
CM and GL	3233	5966	1.02	2950	2993	1.9	2070	3869	1.5
Ammonium sulphate	2840	5480	0.92	2327	2560	2.1	1786	3151	1.37
CM + Ammonium sulphate + Super + MOP	3463	6588	1.08	3209	3289	2.1	2259	3977	1.43
GL + Ammonium sulphate + Super + MOP	3003	6493	0.97	2814	3182	2.1	1962	3815	1.6
CM + GL + Ammonium sulphate + Super + MOP	3341	6710	1.15	3004	3166	2.1	2029	4113	1.7
Ammonium sulphate + Super + MOP	2962	6277	1.11	2760	3204	2.0	1894	3598	1.7
CD	372.6	790.8	NS	240.8	320.12	NS	78.6	265.5	0.29

Highlights of Above Two PME are Summarised Below

1. Green leaf of 4000 kg/ha along with 20 : 20 : 20 kg/ha N, P_2O_5 and K_2O as ammonium sulphate, superphosphate and muriate of potash had given better yield in the first 10 years of experimentation *i.e.* for *tall indica* variety Ptb-2 in first cropping season. But on continuous cropping drastic reduction in yield was observed in this treatment in the later years.
2. In the later years either 4000 kg/ha alone with 20 : 20 : 20 kg/ha N, P_2O_5 and K_2O or cattle manure, alone were found to be better than all the other treatments for *tall indica* variety, during first crop season.
3. Sustained grain yield was obtained for *tall indica* variety in the first crop season by a manure schedule of 4000 kg cattle manure along with 20 : 20 : 20 kg/ha N, P_2O_5 and K_2O.
4. Application of green manure alone or inorganic fertilizers alone resulted in higher gall fly incidence in the first crop season.
5. Inorganic fertilisers @ of 40 : 30 : 30 kg/ha was found to give good yield from 1961–71 for the *tall indica* variety Ptb 20 in the second crop season.
6. Cattle manure alone (8000 kg/ha) or cattle manure and inorganic fertiliser combination (4000 kg cattle manure + 20 : 30 : 30 kg/ha N, P_2O_5 and K_2O as inorganic fertilisers) were found give better yield in the later years from 1971 onwards. But a combination of cattle manure, green leaf and inorganic fertilisers (2000 kg cattle manure + 2000 kg green leaf + 20 : 20 : 20 N, P_2O_5 and K_2O kg ha as inorganic fertilisers) were found to maintain good yield from the beginning in the second crop season.
7. Cattle manure alone (18000 kg/ha) has given the highest grain yield for the photo insensitive variety Jaya during the first crop season. Half the dose of N as cattle manure (9000 kg/ha) along with inorganic fertilizers (45 : 45 : 45 N, P_2O_5 and K_2O kg/ha) or green leaves (9000 kg/ha) were effective for grain production.
8. Inorganic fertilizers either alone (9000 kg green leaves + 90 : 45 : 45 N, P_2O_5 and K_2O kg/h) or in combination with green leaves (9000 kg green leaves + 45 : 45 : 45 N, P_2O_5 and K_2O kg/h) were found to record low yield for Jaya in the first crop season.
9. Green leaves alone (18000 kg/ha) and ammonium sulphate alone were found to increase Gall fly incidence for Jaya in the first crop season.
10. Cattle manure either alone on along with inorganic fertilizers or green leaves were found effective for giving good yield for dwarf indica variety Jaya during the second crop season as in the first crop season.
11. Green leaves either alone or with inorganic fertilizers were found to give low yield for Jaya from the beginning.
12. Grain yield of *tall indica* variety is ten per cent more in the second crop season than in the first crop season whereas for dwarf indica variety it is 22 per cent less.
13. Cattle manure either alone or along with inorganic fertilisers beneficially influenced straw production of all the varieties tried in all the seasons.
14. In all treatments involving cattle manure, the organic carbon and available P, Ca, Fe, Mn and Zn status were found to be higher in the soil.

15. Long duration observed for the photo insensitive variety Jaya in the crop season might be due to low mean maximum temperature observed during this season.
16. Low yield moisture index was found to be associated with higher yield especially in the first crop season which can be attributed to nutrient losses consequent of heavy rain fall.
17. Response of grain yield to weather alterations were more pronounced in the case *dwarf indica* variety when compared to *tall indica* varieties.

Long Term Fertilizer Experiment (LTFE)

The ICAR launched an All India Coordinated Research Project on Long Term Fertilizer experiment (LTFE) in September 1970 at 17 centers in irrigated and intensively cropped areas representing different Agro-Climatic regions of India. The thrust in all these experiments is soil quality, environment safety and sustainability. LTFE at RARS, Pattambi was started during the 1997 in double crop Wetlands representing Brown Hydromorphic lateritic alluvium in the IV Block of the station. This is the 8th year of this experiment.

Objectives

1. To study the effect of continuous application of plant nutrients in organic and inorganic forms and in combination on sustainable crop production (in multiple cropping systems).
2. To study the effective of omission of secondary and micronutrients on sustainable crop yields and also on the assessment of the need for these under an intensive cropping programme.
3. Working out the amount of nutrient removal by the crops.
4. Monitoring the changes in soil chemical and biophysical properties as a result of continuous manuring and cropping and their effect on soil health.
5. To investigate the effect and intensive application of biocidal chemicals (herbicides and pesticides) on the build up of residues and on soil productivity.
6. To make an assessment of the incidence of soil borne diseases and changes in pests and pathogens under the proposed manuring and cropping programme.

Technical Programme

Name of the crop and variety: Rice–Aiswarya

Cropping sequence: Rice–Rice

Design: RBD

Plot size: 125 m^2

Replications: Four

Treatments: Twelve

The experiment was laid out for the first time during Rabi 1997 with 14 treatments.

The treatments in the technical programme have been modified and finalised by the Co-coordinator of LTFE during June 1998 to form 12 treatments instead of 14 as reported for Rabi 1997. In Rabi 2000 again some modifications were made by Project Coordinator. The treatments followed for the experiment from Rabi 2000 onwards are given below:

T_1: 50 per cent NPK (as per POP recommendation of KAU)

T_2: 100 per cent NPK

T_3: 150 per cent NPK

T_4: 100 per cent NPK + Lime @ 600 kg/ha

T_5: 100 per cent NPK (As suggested in the workshop, Copper sulphate not applied from Rabi 2000 onwards, will be done only if deficiency symptoms appear)

T_6: 100 per cent NP

T_7: 100 per cent N

T_8: 100 per cent NPK + FYM @ 5 t/ha to the kharif rice only

T_9: 50 per cent NPK + FYM @ 5t/ha to the kharif rice only

T_{10}: 100 per cent NPK + *in situ* growing of *Sesbania aculeata*, green manure crop for kharif rice only

T_{11}: 50 per cent NPK + *in situ* growing of *Sesbania aculeata*, green manure crop for kharif rice only

T_{12}: Absolute Control.

Initial Soil Physical Properties

1. Organic Carbon (per cent): 0.9
2. pH: 5, 4
3. CEC (Cmol kg^{-1}): 8.9
4. Bulk density (g/cc): 1.3
5. Particle density (g/cc): 2.5

Initial Soil Available Nutrients

1. Available N (kg ha^{-1}): 303.97
2. Available P (kg ha^{-1}): 15.57
3. Available K (kg ha^{-1}): 35.12
4. Exchangeable Ca (Cmol kg^{-1}): 1.60
5. Exchangeable Mg (Cmol kg^{-1}): 0.52
6. Available Fe (kg ha^{-1}): 123.7
7. Available Zn (kg ha^{-1}): 4.17

Lessons Learnt from LTFE KAU

1. Maximum grain and straw by T8 treatment 100 per cent NPK + 5 t/ha FYM for Kharif and Rabi.
2. It is on par with T11 100 per cent + NPK + *in situ* green manure crop–*Sesbania aculeate*. It is
   (*a*) Low cost, beneficial–IPNM
   (*b*) Easily adopted: Scarcity FYM
   (*c*) High cost of FYM
   (*d*) Extra labour cost in FYM transportation and application.

Table 26.5: Pooled Yield Data

Treatment	Pooled Yield Data from 2001–03			
	Kharif (Yield kg ha^{-1})		Rabi (Yield kg ha^{-1})	
	Grain	Straw	Grain	Straw
T_1	2139.67	3023.79	3175.66	3268.00
T_2	2621.16	3242.00	3210.16	3511.25
T_3	2343.16	3125.33	3158.25	3314.41
T_4	2179.66	2990.00	3239.83	3369.58
T_5	2602.83	3466.50	3360.50	3517.50
T_6	2292.83	3253.08	2970.75	3241.00
T_7	2413.25	2980.75	3210.58	3193.58
T_8	2886.67	3762.33	3842.66	4007.16
T_9	2572.58	2796.17	3300.33	3450.00
T_{10}	2831.08	3831.25	3652.41	3844.25
T_{11}	2323.50	3311.83	3242.83	3403.91
T_{12}	1511.50	1935.42	2244.50	2316.16
CD	234.70	302.10	200.20	217.20

Table 26.6: Balance Sheet of N, P and K

Balance Sheet P	Total Added P	Total Uptake	Apparent Change	Initial Available P	Final Available	Changes in Soil P
Treatment			TAP–TUP			FP-IP
T_1	270	69.06	200.94	15.58	17.682	2.10
T_2	540	124.28	415.72	15.58	21.252	5.67
T_3	810	76.20	733.80	15.58	21.042	5.46
T_4	540	71.01	468.99	15.58	20.58	5.00
T_5	540	70.71	469.29	15.58	20.706	5.13
T_6	540	68.66	471.34	15.58	21.252	5.67
T_7	0	70.01	–70.01	15.58	14.112	–1.47
T_8	540	92.69	447.31	15.58	22.176	6.60
T_9	270	75.31	194.69	15.58	19.824	4.24
T_{10}	540	78.34	461.66	15.58	21.966	6.39
T_{11}	270	70.42	199.58	15.58	20.622	5.04
T_{12}	0	48.76	–48.76	15.58	10.542	–5.04

Table 26.7: Balance Sheet

Balance Sheet K	Total Added K	Total Uptake	Apparent Change	Initial Available K	Final Available	Changes in Soil K
Treatment			TAK–TUK			FK–IK
T_1	270	736.02	–466.02	30.10	43.96	13.86
T_2	540	820.33	–280.33	30.10	51.24	21.14
T_3	810	870.32	–60.32	30.10	63.56	33.46
T_4	540	762.52	–222.52	30.10	55.44	25.34
T_5	540	822.04	–282.04	30.10	53.76	23.66
T_6	0	666.25	–666.25	30.10	41.44	11.34
T_7	0	665.89	–665.89	30.10	46.20	16.10
T_8	540	961.10	–421.10	30.10	54.60	24.50
T_9	270	793.37	–523.37	30.10	52.92	22.82
T_{10}	540	875.02	–335.02	30.10	58.24	28.14
T_{11}	270	748.36	–478.36	30.10	48.72	18.62
T_{12}	0	495.15	–495.15	30.10	33.04	2.94

Highlights of Results

1. Mean response to fertilizers were in the order of NPK + GM = NPK + FYM > NPK > NP > N. The rate of response to NPK declined after few years.
2. Continuous use of N alone produced decline in yield
3. Integrated application of 100 per cent NPK+GM *in situ* = 100 per cent NPK + FYM gave the highest sustained crop productivity.
4. Continuous application of graded levels of NPK significantly increased the content of N fractions. Integrated use of 100 per cent NPK+ GM and 100 per cent NPK+ FYM improved organic N fractions as compared to 100 per cent NPK alone.
5. Crop yield remained low when fertilizers are applied at 50 per cent level. The application of high quantity of nutrients *i.e.* 150 per cent NPK though gave higher yield than 100 per cent NPK but the same is not sustainable at present level of crop productivity.
6. The imbalanced use of N, NP or no fertilizer caused lower yield over years and indicates that they do not sustain high crop productivity as compared to the balanced use of N, P and K along with lime or GM or FYM. Integrated use of manures and fertilizers help to maintain soil fertility and sustainability of crop yields.
7. An increase in soil organic C content of the soil under 100 per cent NPK alone or 100 per cent NPK + GM or FYM is noticed over years.
8. This indicated that organic pools of C, N, P might be maintained in the rhizosphere zone and there by sustaining soil quality and productivity.
9. Particulate organic C (organic C in different aggregates) is a slow pool of C that increased with decreasing particle size aggregates and helped sequestering higher amount of C under 100 per cent NPK and NPK +GM/FYM.

10. Passive pool of HA-C is comparatively lower than FA-C in the treatments receiving NPK and NPK+GM/FYM possibly due to higher root biomass and regular application of newly humified manure into the soil. Here soil organic C over seasons and over years are positively build up by the above treatments. The fulvic acid N, P and S content were relatively higher than their content in humic acid indicating that low molecular weight humus restore more quantity of nutrients than higher molecular weight of humic acid. The restoration of SOC is recorded more in this rice-rice sequence in this experiment and in Palakkad rice tract indicates that cropping system and ecosystem played more important role than soil type in this unique pattern of SOC build up over years.
11. Carbon mineralization rate is maximum in treatments receiving both 100 per cent NPK+GM/FYM followed by 150 per cent, NPK, 100 per cent NPK and N indicating that regular application of organic manure sustain soil carbon mineralization process and thereby regulates desired nutrient transformation during crop growth.
12. A perusal of the data on nutrient applied and the nutrient uptake with optimal NPK dose indicated a positive balance of P and negative in respect of N and K. The uptake of nutrients followed similar trend as that observed in respect to yield, the highest being with 100 per cent NPK + FYM which is on par with 100 per cent NPK + FYM and 150 per cent NPK alone.
13. The uptake of K from non-exchangeable K pool was higher under 100 per cent NP and it reduced considerably with K application. It indicated that application of K is required still higher to reduce the mining of soil reserve K sources. Potassium fixation is found to be higher in higher K fertilized treatments and GM and FYM addition reduces K fixation. Non-exchangeable K release were improved with the level of K fertilization and GM/FYM addition.
14. Loss of N and K in flowing water and as well as percolating water observed in this experiment warns split application of the same for their efficient utilization. There is build up of soil P, loss of P can be reduced or the P application can be omitted or skipped of in high P build up plots for alternate years without significant reduction in yield.
15. The status of Fe, Mn, Cu and Zn were increased with the application of 100 per cent NPK + GM/FYM.
16. There was little effect of continuous application of fertilizers on as, Pb, Ni, Hg and Cd accumulation in the soil.
17. The population of bacteria, fungi, actinomycetes tend to increase with NPK fertilizers with GM/FYM *i.e.* there is improvement in biological health of soil due to regular use of GM, FYM and even crop residue. The microbial biomass C and N increased significantly with application of 100 per cent NPK + GM/FYM. Similarly the dehydrogenase and phosphatase activity also increased with these treatments.
18. The continuous application of 100 per cent NPK + GM/FYM decreased the bulk density and improved infiltration rate and hydraulic conductivity of the soil.
19. Continuous cropping with balanced and integrated use of organic and inorganic fertilizer did not change the weed flora, however application of N alone lead to emergence of more types of weeds.
20. Here is the tract the soil health suffered due to increased Fe toxicity and this lead to a major constraint in sustained lower productivity of rice-rice sequence.

It is important to know what is expected in organic farming. Organic agriculture voids or largely excludes the use of synthetic fertilizers, pesticides, growth regulators and livestock feed additives and even the use of genetically modified crops. Organic farming system solely depends on the use of crop residues, animal manures, green manures, off-farm organic wastes, crop rotation incorporating legumes and biological pest control to maintain soil productivity. The philosophy is to feed the soil rather than the crops to maintain soil health. We have to exercise caution against extremism of organic farming, rightly termed as ecological fundamentalism, that leads to the total exclusion of synthetic chemicals in farming.

Organic farming are to be dealt with Indian context of food security. We have to learn lessons from the organic farming disaster in China which suffered an appalling famine with 30 million starvation deaths in 1959–60 and still having nearly 200 million malnourished citizens by late 1970's. During the green revolution increase in crop yields resulted primarily from the introduction of fertilizer responsive HYVs and increase in area under irrigation and farm mechanisation. Organic farming in the form it is being promoted is primarily aiming at demolishing these pillars of green revolution. What organic farming should aim is to mobilize nutrients from all-available organic resources. It is the INM and IPM which hold the key for the future.

There are several myths propagated about organic farming and organically produced food. Few of them are:

1. Organic food tastes better and is of superior quality.
2. Organic food is more nutritious and safer.
3. Organic farming is ecofriendly.
4. Organic farming improves soil fertility and chemical fertilizers deteriorate it.
5. Organic farming sustains higher yields.
6. Enough organics are available to replace chemical fertilizers.

All policies and practices may supplement the use of chemical fertilizers but remember their use cannot be totally eliminated. Policy initiatives at reducing or eliminating the use of chemical fertilizers will in the long run will prove disastrous for Indian Agriculture. Organics cannot replace chemical fertilizers. They are not at all substitute for chemical fertilizers.

Chapter 27

Success Story of Organic Farming

In most of the developing countries, the subtropical and tropical climate makes it practically difficult to maintain organic matter status at higher levels under traditional agricultural practices, which necessitates the orientation towards organic farming. The important reasons which emphasis the need for organic farming in these countries are as follows:

1. Most of the arable soils contain organic carbon below the threshold level.
2. Majority of the farming community is resource poor and purchase of fertilizers and chemicals in adequate quantities is beyond their capacity.

In this system, a large fraction of farm by-products of plant and animal origin is utilized for non-farm use that is, fuel or other domestic purposes. Small and scattered land holdings of a large farming communities compel them to leave the crop residue in the farm itself rather than recycle it for composting. Lack of location specific technology to recycle organic wastes and lack of awareness to recycle organic waste in agriculture are the main reasons for its slow adoption even though it is native technique for the farmers which got lost during the period of green revolution.

Use of chemical in aquculture is considered by our planners and also by sizeable section of elite farmers as the most efficient way of augmenting food production. The need to produce more and more food to satisfy the increasing population has led man to adopt new method of agriculture which have now proved to be (1) harmful to the soil in the long run and (2) harmful to the people who consume its produce. This distorted, system of agriculture called *Modem Method* of *Farming*.

The harmful practices adopted in the modern farming are:

1. Excessive tilling
2. Excessive use of water
3. Use of inorganic fertilizer
4. Use of chemical pesticides
5. Neglect of natural allies of farming.
6. Use of varieties inconsistent with local factors.

Since 1950s, farming methods derived from the West are being followed. These techniques are adopted regardless of whether they suit our climatic, economic and social conditions or not. Use of Hybrid seeds is one such technique. These seeds may yield a bountiful crop but their nutritive values, taste, and satisfying capacity is doubtful.

The adoption of conventional farming obviously did succeed in achieving higher production rather than with traditional farming but it is capital intensive. The cost of inputs are so high that sometime it includes as much as 80 per cent of the realizable values of the produce. When farmers work on such a thin margin of profit, even one or two adverse seasons can put them into a tremendous debt burden. It is out considered opinion that if farmers switch over to *Organic Fanning Methods*, they will never suffer an economic loss over a period of time.

Organic Agriculture: An Attractive Alternative

Organic practices are more environmentally sustainable than most conventional practices. Emissions of toxic substances and dependence on non-renewable resources are limited. Organic practices are inherently better-adapted to prevailing agro-ecological conditions.

Export-oriented organic agriculture is attractive because of premium prices paid by North American and Western European consumers for certified organic products such as coffee, tea, vegetables, and cotton. The export income of these farms and estates is supplemented by a cash income from the sale of some products in the local market, primarily animal products introduced into the farming system to cover the need for manure. Furthermore, export-oriented organic farms produce food for home consumption, as well as by-products. Because of this increased product diversity, dependence on export income is lower than in comparable conventional export crop production.

Subsistence-oriented organic agriculture is attractive for marginalized farmers who primarily produce food crops for home consumption because food self-sufficiency and self-reliance for soil fertility and pest management are improved. Several cases are located in marginal areas which are in various stages of degradation. Organic practices combine well with efforts to regenerate soil, control erosion, or harvest water because of the many positive ecological effects of simple methods.

Compost for example, improves soil fertility, nutrient recycling, water retention, infiltration, and soil life. Contour farming with leguminous shrubs and trees replaces NPK fertilizers while at the same time improving soil structure and the microclimate and reducing weeds and erosion. In those situations organic agriculture allows sustainable productive use of marginal lands, combining improved food security and food self sufficiency with minimal capital investment.

The available material tends to support the conclusion that both in high potential areas and marginal lands, organic agriculture offers agronomically feasible includes a positive microeconomic effects for most of the producers involved, and possibility leads to positive effects on regional and national economics as well.

Not a Miracle Solution

However, Organic Agriculture Systems do not offer miracle solutions for all problems related to agricultural development. Social and economic problems like marginalization, prices, gender conflicts, and land tenure largely surpass the extent of this survey. Further marginalization may result from organic practices if the poorest sections of the population cannot afford the costs or labour needed for conversion. Better-off farmers might gain most when the approach towards organic farming is purely technocratic, just as they did earlier from Green Revolution technology. However, organic agriculture offers a better perspective for equitable development because it tends to increase farmers' self-reliance for food and inputs, and contributes to their autonomy and self confidence. Furthermore, it increases biological and labour diversity, thus spreading risks in the agronomic as well as the financial sense. Unequal development of richer and poorer farmers in organic agriculture systems is not inherent to the technology, but remains related to the given *resources endowment* (quality and acreage of land).

Adoption in the Farming Systems Context

The history of agricultural development over the last several decades contains many examples of failures in adoption or sustenance of technologies that were judged scientifically sound. The resource management technology may be failure due to water-management schemes in the inland, minimum tillage practices, poor take-up of improved clearing methods, soil conservation methods, and spread of alley cropping. The other masons for failure to adopt or persist with technology are social, cultural or economic factors as to technical deficiencies. This necessitates an approach to technology development and transfer than integrates the criteria used by the farmer in deciding the actions, as well as those assumed in the scientists models.

Soil management is only one small component of a farmer's spectrum of activity. The knowledge and perceptions of soil-based constraints should be combined with a holistic appraisal. Farmer's should have balanced approach between various alternatives based on whole range of available environmental, biological, and economic information.

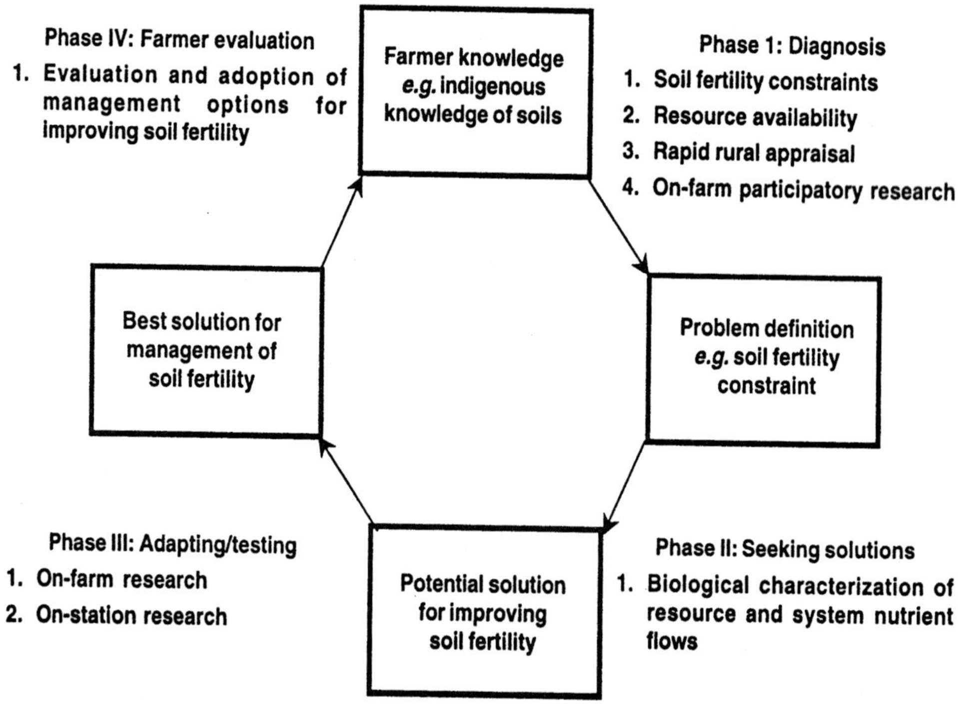

Figure 27.1: The Farmer-back-to Farmer Approach to Agricultural Problem Identification and Solution as Applied to Soil Fertility Research

Technological change is thus more likely to succeed in terms of both adoption and of impact on production when it is adapted to farmers' circumstances rather than merely optimised in terms of scientists' criteria. This is most likely to be achieved when scientists and farmers work together throughout the research and development process.

Figure 27.2 illustrates stages in a research processes that is initiated by joint diagnosis of soil fertility problems. This information based both on farmers' perceptions and on biophysical and socio-

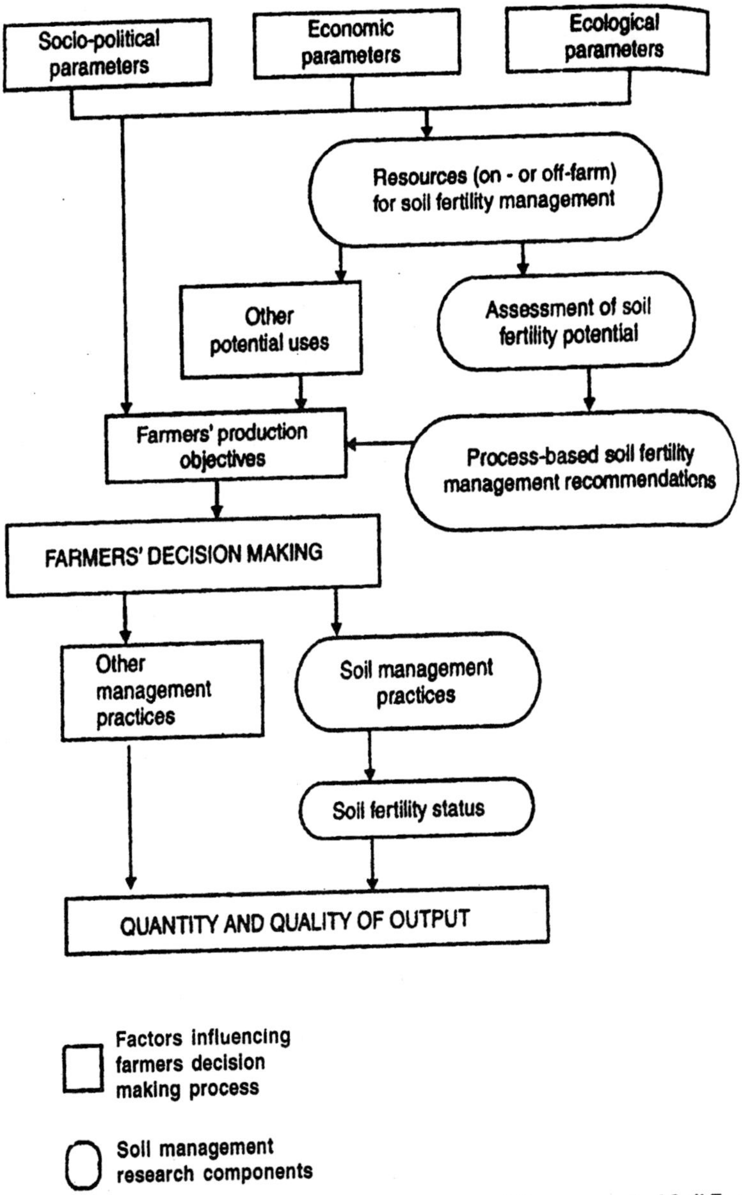

Figure 27.2: Integration of the Research Results for Biological Management of Soil Fertility with System Level Information on Farmers' Decision Making

economic characterization of the farming system. It should be used to guide scientific investigation of the opportunities for improvement in soil management practices.

The process-and-system approach places farmer decision making at the centre of the technology development process, highlighting the need for detailed characterization of the environmental context that influences those decisions.

The capacity for increasing and sustaining productivity through improved soil fertility management is subject to a variety of limits of physicochemical, biological economic, social, and cultural dimensions. The determination of these limits, the establishment of interactions-including hierarchical relationships, and the opportunities for modifying them is one of the major purposes of system characterization

Soil Management and Biological Processes

The range of management practices that a farmer can employ to regulate soil fertility is limited.

Table 27.1: Farmers Management Practices for Influencing Soil Fertility Manipulation of Biological Process

Management Practice	*Constraints to Use*	*Biological Processes Influenced*	*Soil Fertility Effects*
Biological Inputs	Cost		
Rhizobium inoculation	Availability of inoculum	N-fixation	Increase N-acquisition
Mycorrhiza inoculation	Environmental adaptation	Nutrient uptake by mycorrhiza	Increased efficiency of uptake of P and other nutrients
Soil fauna inoculation	Competition with fertilizer		Increased efficiency of H_2O uptake
	Competition with indigenous biota		Increased heavy metal tolerance
		Fauna burrowing	Soil structure/porosity
		Decomposition	Stimulation of nutrient release
Organic Matter Inputs			
Crop residues	Recycles nutrient only	Decomposition	Increased short-term nutrient availability
Root residues	Inaccessibility to management	SOM synthesis	
Weed residues	Crop impact from competition		Increased nutrient storage/exchange
Tree litters/prunings	Land set aside		
Green manure	Land set aside		Soil physical structure improved
Farmyard manure	Livestock and fodder availability		Soil water regimes improved
Household waste	–		Acidity/toxicity diminished
Purchased organic input	Cost	Soil fauna/microflora growth	Macropore formation improved (macrofauna)
Pre-Composting	–		Soil aggregation improved (microflora)
	All: Labor availability Opportunity cost of other uses		

Contd...

Table 27.1–Contd...

Management Practice	Constraints to Use	Biological Processes Influenced	Soil Fertility Effects
Inorganic fertilizer inputs	Cost Availability	Mycorrhiza inhibited	Direct transfer of nutrient to plant increased
		N fixation inhibited	Nutrient losses increased
		Mineralization/immobilization balance changed	Acidification
Tillage Conservation	Soilborne disease	(Intensive tillage) Decomposition stimulated by OM incorporation	(intensive tillage) Short-term nutrient availability increased
Hand	Labor	SOM decay stimulated by aeration and particle size reduction	Root growth intilled layer promoted
Mechanized	Cost	Faunal and microbial populations diminished	Nutrient losses increased
			Long-term nutrient storage diminished
Pesticides	Cost Environmental and health impact	Non-target organism populations diminished or eradicated	Destabilization of nutrient cycles Loss of soil structure

The above practices are not unique to biological management but common to farming practice the world over. The use of biological inputs such as N-fixing bacteria, mycorrhiza, or soil fauna as a means of enhancing the endemic biological activities are the means of biological soil management. Direct management is also achieved by the use of organic matter inputs–in effect a means of selectively feeding the heterotrophic biological populations of soil–a practice of very ancient origin but at times eschewed in modern agriculture. Equally direct but usually unintentional effects are also achieved by the use of pesticides, which may kill particular groups of soil organisms that are involved in processes of significance to soil fertility. Management techniques such as tillage and fertilization also influence the activity of the biota indirectly by altering the physical and chemical environment of the soil.

The success of biological management practices depends on following two preconditions which must be satisfied:

1. The availability of a management practice that is practically and economically acceptable to the farmer.
2. The demonstration by the scientist that the practice leads to enhanced soil fertility.

Basic Recommendation for Initiating Adoption of Organic Farming

Agronomic

Fertilization

1. Promote soil organic matter management, balanced fertilization; and soil protection as basic elements of soil management.
2. Facilitate an efficient, complementary use of mineral fertilizers, preferably of local origin such as phosphate, in the conversion period in areas where this is necessary to initiate the nutrient cycling.

3. Avoid the use of easily soluble chemical fertilizers, especially near sources of drinking water, in densely populated areas, and in areas of high ecological value.

Regeneration

1. Develop and implement organic matter management as the technique for regeneration in areas where soils are degraded.
2. Develop and implement organic matter management in combination with water harvesting and water storage as the primary techniques in areas where water is the limiting factor for agricultural production.
3. Enforce techniques such as alley cropping and contour farming in hilly and mountainous areas.

Pesticides

1. Phase out the use of synthetic pesticides, beginning with those most harmful to the environment and human health.
2. Replace synthetic pesticides by alternative pest management and techniques based on traditional and modern knowledge with a strong preventive character such as cultural practices, biological control, and selective botanical pesticides.
3. Immediately restrict the use of herbicides in areas subject to erosion.
4. Phase out the use of herbicides and replace them alternative strategies and techniques such as rotation, mulching, intercropping, and mechanical weeding.

Post-Harvest

1. Emphasize human and environmentally friendly local storage and conservation techniques for agricultural products.

Socio-economic

Farm System

Promote mixed farms and diversification to improve resources efficiency, food self-sufficiency, and income security.

Human Resource Management

Develop education and training programs for extension officers and farmers for ecosystem management so that farmers can make choices base thorough understanding of their local agro-ecological production capacity and the market system.

Conversion

Assist farmers (especially women) intensively during the period of conversion towards sustainable agriculture systems (up to 5 years) and a compensate them financially when necessary.

Processing

Promote on-farm or communal processing.

Marketing

Develop appropriate domestic and international marketing structures to ensure outlets for all products of organic farms.

Factor Affecting Adoption of Organic Farming

The appropriateness and potential for adoption of organic farming practices depend upon a range of interrelated environmental and socio-economic factors. These are as follows:

The Degree of Intensification of Farming System

The potential for adoption of organic agriculture varies upon on the degree of intensification of the farming system. This, in turn, is related to population size growth and consequent land pressures.

Those farmers who practice unimproved traditional farming, where soil productivity tended to be sustained solely by fallowing. These farmers may not expending any serious effects from their current farming practice and are unlikely to be interested in any new techniques. However, they typically have limited access to agrochemicals and there may be scope for researching and promoting simple alternative organic techniques, especially for pest and disease control.

The second group of the farmers widely use agrochemicals. The farmers have reached limits to productivity through traditional means, and have had to adopt new technologies, the most common being use of chemical fertilizers. Traditional practices are still used, but they were supplemented with chemical inputs. Many of these farmers used a mixture of organic and inorganic techniques, but that they have generally not used chemicals to the extent of directly experiencing negative effects, and the medium to long term effects of using agrochemicals have not been explained to them. As such, the farmers tend not to appreciate any comparative benefits of using organic techniques, and are unwilling to adopt these unless artificial fertilizers and pesticides are unavailable or cannot be afforded. Many of these farmers have previously used chemical fertilizers, but had stopped when subsidies were removed and the fertilizers had become too expensive. Thus there is an economic incentive to develop cheaper alternatives. Considerable potential exists for optimizing the use of available natural resources, and for a detailed exploration of the compatibility of organic and inorganic inputs. This group of farmers is also likely to experience many of the general constraints to the adoption of organic techniques.

The third group of farmers practice intensive agriculture. They rely heavily on agrochemicals, producing mainly for an urban market. However, they are often also the farmers who use the most sophisticated and intensive techniques of organic soil fertility management. They have adopted these strategies due to the need to intensify their production systems, and have been exposed to extension activities and introduced to new improved technologies. This group of farmers shows a relatively limited interest in using exclusively organic techniques, although observation of the negative effects of chemical inputs and the economic reality of increasingly costly inputs has served as incentives for the adoption of organic techniques. Providing that there is not a relative drop in fertilizer prices, it is likely that farmers would continue to use chemicals, and organic strategies for improving soil fertility management have considerable potential with this group. In particular, manure management can be substantially improved and there exists great possibility for urban waste utilisation.

Organic pest and disease control methods are considered ineffective by this group, and therefore not viable alternatives to pesticides. This area requires further research to prove its effectiveness before any serious uptake by farmers can be expected. However, in the short and medium-term, if environmental aspects, food safety and health prove to be considerations at least for consumers, there is a need for the promotion of safer pesticides and pesticide use, and for the use of more benign techniques such as biological control and the use of resistant varieties.

Significant potential to raise productivity through the optimized use of locally available natural resources exists in farming systems at all three levels of intensity.

Agroclimatic Factors

Specific organic farming techniques may be expected to be more suitable and common in some climatic zones than in others, and the constraints to sustainable husbandry are also likely to differ according to agro-climate. For example, the extent of manure use is dependent on livestock numbers and management, and these vary with agroecological zone.

The wet and humid tropics are characterised by nutrient poor soils and by high oxidation and mineralisation rates, and this results in difficulties in ensuring the presence of sufficient nutrient and organic matter inputs in cropping systems. Transport of organic materials is difficult due to heavy rainfall and inaccessible paths. Although vegetation growth is prolific, the rate of decay is fast and the management system such that the quantities of crop residues are too low for farmers to rely on these for sustainable organic matter inputs.

In the semi-arid areas where, livestock numbers are high and the management regimes of relatively low intensity. With biological processes being slower than the humid tropics, biomass production is lower and many of the sustainable technologies are less efficient. Farmers experience alack of organic materials for soil fertility. This is aggravated by the fact that manure is used for fuel in some areas. The collection and distribution of organic material is especially time-consuming in the drier areas, where the farming systems tended to be extensive and cropped plots widely dispersed. Investment in land rehabilitation may not seem like a rational option when climate-related risks are high and returns to investment unpredictable. However, if the rehabilitation techniques available, organic manures are sometimes preferred by farmers in dry areas since chemical fertilizers require water to prevent crops scorching, and crops perform better without fertilizers under drought conditions. Organic manures also increases the water holding capacity of the soil.

Labour Inputs

A frequently cited constraint to organic soil fertility management is that of labour. This is true both in general and for individual organic techniques. For example, farmers identified lack of labour as a constraint to the adoption of composting. Organic soil fertility maintenance may also involve manure collection and spreading, composting, tree pruning and green manure incorporation. These techniques are likely to require more labour then either no maintenance traditional farming or modern high input mechanized agriculture. In areas with low natural resource potential, the labour requirement tended to be the highest and returns to the investment took longer to be realized. Cash poor farmers who could not afford to hire labour, and farmers who depended on off-farm work to supplement their farm income, may not have been able to adopt any agricultural practices which increased the labour burden; particularly if short-term increases in farm productivity is not guaranteed. Hired labour are primarily used for weeding, and as the cost of labour increased, so herbicides become an economic alternative. It is not just to availability of labour *per se*, but also unwillingness to spend the necessary time. Views such as 'the practice organic arming is tedious to undertake' and that 'collecting and preparing ingredients needed for organic farming is tedious and time-consuming'.

Environmental Concerns/Negative Experiences with Agrochemicals

In the more intensive farming systems, where artificial fertilizers have been promoted and used since the 1960's, there has been some experience of the negative impact of agrochemicals. These, farmers have an appreciation of the comparative benefits of organic soil inputs and many wish to reduce their fertilizer used. There is a widespread perception that chemical fertilizers are damaging both soil structure and soil moisture holding capacity and also creating an economic dependency on costly inputs. There is a clear understanding of the relationship between organic fertilizers and soil physical properties. A commonly known fact is that chemical fertilizer provide some concentrated nutrients for the plant but that manure or compost is required to feed and give structure to the soil, and that chemical fertilizers are effective for only one season whereas manure can last for two to four seasons. Manure is therefore seen as an investment.

Other farmer groups, who have not used fertilizers to a great extent, also have some reservations about artificial fertilizers for environmental reasons *e.g.* once they start using chemical fertilizer they have to continue to do so. With both limited cash and fertilizers supply in such areas, this creates feelings of fear and risk.

There is limited appreciation of alternatives to fertilizers and lower availability and more constraints to manure use.

Land Size and Ownership

The regional land allocation/inheritance systems influence both tenure right and the size and distribution of farms. This can directly or indirectly influence the current and potential development of organic techniques. Farmer willingness to adopt sustainable agricultural technologies which require investment in time and/or money and which will provide benefits is the long term, is directly related to land tenure.

Farmers have a choice of using manure or chemical fertilizer, rented land tends to receive less manure and more fertilizer than owned land. This is because the farmers do not wish to invest valuable manure for someone else's benefit, since the land could be repossessed by the landlord at any time. Furthermore, there is an incentive to use high chemical inputs on rented land to ensure a rapid return to cover the cost of land rental.

Cropping Systems and Sub-systems Within the Farm

Important differences also exist within the farm unit or among distinct holdings of one particular farmer. Many farmers operate a number of enterprises which are often spatially separated. For example, there is often more scope for sustainable soil management in home gardens, or in plots close to the homestead, than on the larger/more distant plots where the main cash and field crops are grown. This appeared to be the case with the current use of organic techniques. Farmers practices a bush fallow system without fertilizers on the main farm plots, whilst women use organic inputs to improve the soil fertility on vegetable plots. There is selectivity on the type of crop being fertilized by organic manures.

Proximity to Urban Markets

Farmers in remote areas with poor infrastructure have limited access to chemical inputs. In this situation, there is little option to achieve a more productive agriculture than to optimize the local resources available through organic farming techniques. Near urban areas, farmers tended to be commercially orientated and could generally afford to purchase inputs. The demand from urban markets increased the use of pesticides to attain a high standard of crop appearance, and the use of fertilizers for continuous intensive vegetable production. Off-farm employment in the city may be seen as an attractive alternative to farming, particularly when the farming system requires high labour inputs, long hours, or direct handling of agrochemicals.

However, farmers living relatively close to urban centres do have access to agro-industrial or urban municipal organic wastes, offering greater potential for organic soil fertility management. Intensification of animal production through zero grazing, due to high demand and lack of space, may increase the opportunities to produce and manage manure as a soil input.

Economic Factors

There is a link between the degree of commercialization and the use of agrochemicals, with farmers who supply well-developed markets having both the economic incentive and the ability to access inputs. Farmers using credit facilities is significantly higher near the urban areas than in rural areas, and there is a significant positive correlation between credit adoption and fertilizer used. Having limited funds to purchase fertilizers is seems to be an important reason for adopting alternative soil management techniques.

Organic fertilizers themselves are not always readily available at a low cost. The recycling of plant residues is rarely adequate on small intensive farms, and unless management practices include the concentrated use of green manures, N-fixing crops and zero-grazing by products, then external sources of fertilizer are necessary. This results in a market for organic fertilizers and composts, and they may become economic inputs with the same implications as chemical fertilizers for poor farmers. However, organic inputs on the market are more likely to be produced locally and often traded within the local exchange/bartering network.

There is a market for animal manure, it is sometimes transported over considerable distances. The same applied to urban wastes and poultry manure from intensive poultry farms in several countries. In the majority of cases, these markets are informal and with the little information available on the costs and value of the available material. These costs and values would be relative to that of chemical fertilizers but would need to take into account the potential long-term benefits from their use.

Education/Lack of Knowledge

A lack of knowledge was by far the most widely cited reason for not adopting organic farming techniques. This included a lack of awareness of the existence of organic methods, a lack of training and detailed instructions on application methods and rates, and lack of examples of success.

It is often assumed that farmers have a comprehensive indigenous knowledge about soil fertility management and pest control. This is frequently not the case. Farmers are not optimizing the natural resources at their disposal, nor are they clear as to why they were carrying out certain practices. Farmers stress that the use of organic inputs is not a positive decision based on value for soil fertility but a 'passive' or 'pretty much accidental use. These inputs and practices included poultry manure and wood ash thrown onto gardens, crop residues left in the fields and unplanned and informal composting of piles of sweepings. However, the farmers are unintentionally following the habits and customs of those around them and before them, whether they understood it or not, and they are likely to underestimate their innate ability as farmers.

Farmers commonly lack the appropriate skills to convert their farming terms to become more productive and yet sustainable. Furthermore, the predominance of agrochemical use over the last 20 years or so has directly encouraged the loss of traditional knowledge. Examples are specific techniques, such as pit composting, that have been used in the past but have not been adopted by the younger generation. The same is true for traditional pest, and disease control methods. Also, non-traditional crops are widely grown which are associated with pests for which there is no traditional knowledge. Many farmers feel that there are no effective alternatives to pesticide use; that the crop varieties they were cultivating are 'foreign' and therefore not strong enough to grow in the local environment without the use of pesticides.

The long-term benefits of organic fertilizers are often not clearly appreciated by farmers. The majority of farmers are aware that manure is of benefit to crops but not of the longer term effects of manure on soil fertility. As a result there is little understanding of the effect of cessation of manure application and the implications of this for the maintenance of soil fertility. Generally, manure management was poorly developed in many parts of the developing countries. The understanding and promotion of organic technologies for soil fertility maintenance was generally very low amongst extension staff. There was also little evidence of NGO work on organic agriculture among the random sample of farmers interviewed.

Availability and Cost of Organic Materials

Unavailability of organic materials as being the prime constraint to utilising organic soil inputs. The particular materials mentioned are animal manure and crop resides. This highlights a lack of knowledge on the wider range of techniques possible.

Farmers in many areas are faced with a choice of uses for available materials. In areas, where there is a shortage of fuel wood, cow dung is used for fuel, in some areas, crop stalks are collected as fuelwood, for construction and mat making. Crop residues are often too valuable as an animal feed to be used as fertilizer or for composting.

The cost of off-farm organic materials may prevent the poorest farmers from using them. The farmers perception of the benefits of manure increase, so too do the price, particularly as the cost of inorganic fertilizer has also risen. The use and cost of the various types of manure is dependent on the distance of the farmer from the source.

Cultures

In many areas, there is an unwillingness to handle waste, particularly human waste. Considerable social stigma is attached to manure handling as it is often mixed with human excreta and rubbish.

Gender

Consideration of the influence of gender the distribution of land, activities income and decision-making need to be taken into account in relation to the potential for organic farming strategies. Women are often responsible for cultivating the small vegetable gardens as a contribution to household food production or for the market. These gardens currently receive organic inputs and have considerable potential for the further development of organic techniques.

Aversion to the Use of Organic Methods

When converting from chemical to organic inputs, there is usually a fall in productivity over a transitory period, until fertility, structure and microbial activity of the soil have been restored and enhanced through organic matter inputs. If properly managed, this fall is usually thought to be no more than 20 per cent of crop yield, although this difference may be greater if very high yields are previously being obtained from over-use of agrochemical inputs. In agroforestry, techniques are being introduced, the effects from trees planted to provide products and improve soil fertility will usually not be realized in the first few years. It may not be feasible for resource poor farmers to adopt alternatives to fertilizers unless short-term benefits can be realized. Farmers express a fear of changing to organic farming methods for fear of failure or risks of reduced yields, and a lack of 'immediate' results as a reason for not using organic inputs.

The problems of increased weeds, pests and disease is also perceived by the farmers. Farmers believe that animal manure encourages weeds'. It was a common that manure from animals which had been fed infected residues could carry diseases such as tomato and potato blight, or 'compost heaps attract very difficult pests and animals' or 'some claim the methods, like application of mulch, harbors pests which destroy crops'. These misconceptions again highlight the lack of sound knowledge on the subject.

Perceptions of the value of organic inputs is very important. Some farmers are reluctant to use poultry manure in vegetable production systems as they perceived it to have slow nutrient-release properties, to encourage pests and diseases within the soil and to increase decay in vegetables during post harvest storage. In contrast, some places poultry manure is highly valued as fertilizer, and it is often used on vegetables.

Adverse comparisons of organic inputs with non-organic ones are also expresses in phrases which indicates that the former are considered to be *old-fashioned*.

Active Promotion and Preference for Agrochemicals

There is concern among farmers of the high costs of agrochemical pesticides, and also on their safety. The emphasis is clearly on agrochemical pesticide use. This, combines with the facts that pesticides are readily available, considered effective, and give the product a marketable appearance, encouraged the farmers to view natural pest control methods as having limited potential.

The uptake and increased use of chemicals has been directly influenced by active promotion. The influence of government agriculture departments and, more particularly, the extension services, on choice of farming techniques is a reason for use of fertilizers. Comments such as 'Government extension service advocates modern farming methods' are common, and several replies referred to the fact that the extension services are trained to use, and therefore promote, chemically-based methods. Government encourages the use of chemicals in their radio agricultural programmes. The promotion, through subsidy, of fertilizers and pesticides which has contributes to farmers becoming used to non-organic methods over a long period of time.

Adoption in India

For centuries, Indian farmers practiced a balanced system of agriculture, which was based on a thorough understanding of environments, seasons, crop mixing, rotation and gene preservation. Yet, over time practices became ritualized, without concern for the concept of balance. Hence, the systems stagnated and failed to respond to the changing needs of the times. After independence in 1947, traditional agricultural systems in India failed to meet the needs of a burgeoning population and suffered setbacks through a series of natural disasters. The government took on the responsibility of improving agriculture, by making huge investments irrigation development and research was on high yielding varieties. As a result, chemical intensive and water-intensive agricultural practices were actively encouraged. Heavy subsides for chemical inputs induced farmers to practice chemical farming.

The fall-out of such practices was soon amply evident. It is in this context the organic farming came to the fore. Many farmers took it up because of the ill-effects of modem farming. Though still in its infancy, organic farming has a strong ideological foundation and offers many fresh ideas. It is a response to the growing eco-awareness in agriculture.

By the late' 80s, three types of people were into organic farming. The first type consisted of urban-educated technocrats. They embraced this new ideology, for the sheer pleasure of feeling the soil. But, in several cases their enthusiasm was short-lived. The second type consisted of farmer with a modern education. Their farming practices reflect their earnestness to understand the strengths of traditional farming as well as their ability to assimilate scientific knowledge. A third group consists of ordinary farmers who by sheer observation, trial and error have tried to understand and practice organic farming. Though this third group is insignificant in number right now, the future of organic farming in India lies with them.

Two striking common factors underlie most of the success stories of organic farming in the country. Almost all the successful farmers have access to sufficient water, and enough organic inputs in and around their farms. Many of them therefore grow cash crops like sugarcane, areca, cocoa, coconut, pepper and various spices. These farmers have effectively allayed apprehensions that a switch-over to organic agriculture would affect yields and income. But small and marginal dry-land farmers are hesitant to change over precisely because they lack the resource advantages of the successful farmers (water and organic inputs close to the farms).

Causes of Low Adoption

1. Chemicals are easy to use and less costly
2. The benefit of organic practices are not seen immediately
3. Large quantities of organic inputs are required
4. Difficult to get organic fertilizers
5. Unorganized market for organically grown produce
6. Preferential behaviour of consumers towards organic food not yet established
7. Economic loss due to transition (from traditional agriculture to organic agriculture or conventional to organic)
8. No experimental evidence on the cost benefit ratio of organic farming
9. Government effort to propagate organic farming is very little
10. Scientific research programmes on organic farming are also scarce

The main element that is brought up time and time again is the fact that there lack of knowledge about the above. The above points show the areas in which education is needed. The government also needs to be informed, as this is one of the first steps in telling people about the benefits of organic agriculture, because many think whatever the government says is right and best.

Success with Organic Farming

The Organic farming has all round favourable effects, and these effects are documented which ultimately results in long term sustainability of the production system. To strengthen this fact some examples of the areas or farms where organic farming is being followed since long time, are as follows:

Improvement in the Environment

Organic farms require both to avoid all forms of pollution that may result from agriculture and to maintain the genetic diversity of the agricultural system and its surroundings, including the protection of plant and wildlife habitats. These principles are given equal status with those of production and economic viability, so that organic farming systems consider environmental protection as one of the key outcomes of farming practice. The complex interactions between farm management and the components of the farming system interact with natural eco-systems at the field, farm and regional scale.

Soil Health

Organic Matter Build-up

Organic matter build up in organically managed soils is widely reported. Such increases reflect the inclusion of leys and green manures in crop rotations and the incorporation of crop residues and manures into the soil. Higher levels of carbon have been measured in the *light fraction* of soils under organic management, which is thought to represent a more biologically active pool. Changes in organic matter drive/underpin many of the other changes in soil biological and physical properties.

Biological Activities

Higher levels of biological activity and increased microbial populations have been measured in cultivated soils managed organically. However, the differences between phases of the rotation in any system can be equally great; microbial activity was always higher under a ley than a wheat crop.

Bacterial biomass was similar in long-term grasslands, whether managed organically or conventionally measured an increased phospholipid fatty acid (PLFA) content in organically managed soils which was correlated to biomass C contents, but could not show any difference in the taxonomic composition or physiological status of the microbial biomass. However, increased abundance and diversity of the microbial community was observed in soils managed organically.

In long-term grassland the size of the active fungal population was increased under organic compared to conventional management. In cultivated soils increased fungal diversity is also seen, including more zygomycetous fungi and more species which are antagonistic to plant pathogens. The species richness of *Fusarium* was increased in soils under organic management, with the proportion of antagonistic *Fusarium* species higher than under conventional management. Increased numbers of nematodes have also been measured in long-term grassland soils. In cultivated soils in organic farming systems the use of composts and manures increases the numbers of non- parasitic nematodes.

An increase in earthworm abundance and activity was measured in cultivated organically managed soils. Increased in the biomass, density, and population diversity of earthworms we observed in organically managed soils, However, in long-term organic grassland. Rotational phases can have a large effect, especially in crops where intensive tillage is required such as potatoes.

Physical Conditions

Aggregate stability and porosity was increased under organic management. Such changes in soil structure largely result from changes in the management of crop residues and organic matter inputs and may lead to increased aeration and water holding capacity of soils under organic management. Lower rates of run-off and soil erosion have also been measured in organic systems. Increases in the depth of the a horizon of the soil have been recorded for organic systems, which may result from decreased bulk densities and/or decreased soil losses through erosion. However, several research studies have found no differences in the soil physical properties between organic and conventional farming systems and no differences in soil particle detachment between organic and conventional systems. Measurable changes in soil physical properties can take decades to establish.

Chemical Properties

Available phosphorus (P) and potassium (K) have been shown to increase and decline in soils under organic management. Increased soil pH has been an measured in some organic systems. In New Zealand lower levels of total S and total P was measure in soils under organic management, as conventionally managed soils often received large inputs of these nutrients. Most changes in soil chemical properties could be predicted from a consideration of the inputs and outputs of nutrients to the system.

Nutrient budgets are a useful tool to summarize the nutrient status of complete systems. Depending on the farm management and the balance of imports and exports of nutrients, nitrogen, phosphorus, and potassium budgets range from deficit to surplus in organic farming systems. Balanced budgets are often more easily achieved in mixed farming systems; where imports of manure were large in a horticultural system, N, P, and K budgets were very positive.

Water Quality

Where nitrate leaching losses are measured directly, they are generally significantly lower than comparable conventional systems measured leaching losses between 7 and 52 kg N ha^{-1} at all points in the rotation on organic farms and suggest that overall farm losses would average c. 20 kg N ha^{-1} $year^{-1}$. However, where nitrate leaching is expressed relative to crop production levels rather than on an area basis, differences between conventional and organic systems are reduced, lower nitrate concentrations was measured in porous cups under organic crops.

The management of residual nitrogen from legumes is a key to the minimization of nitrate leaching. Large leaching losses of N have been measured following the cultivation of grass-clover leys. About 33 per cent of the N accumulated above ground in cut-mulched green manures was lost by leaching, when they were incorporated in the autumn. The timing of incorporation and the nitrogen demand of subsequent crops must be carefully considered. Storage and composting of farmyard manure on nor - paved surfaces can also result in significant nitrate leaching. Covering manure or pre-composting on paved areas can reduce these losses. Leaching losses from manures and green manures can also represent significant losses of potassium, which while they may not be environmentally significant represent agronomically important removals of potassium from the farming system.

Measurements of soil mineral N concentrations tend to indicate no increase and often a decreased leaching risk in organic farming systems. Similar nitrate concentrations was measured in the soil in autumn in organic conventional farms and suggested that there was no inherent differenced the teaching risk from conventional and organic farming systems.

Nitrogen budgets are generally positive for organic systems. There is surplus nitrogen, which may be lost by nitrate leaching. Lower nitrogen surplus was measured on organic than conventional mixed dairy farms. However the nitrogen surplus of organic pig production was greater than that of conventional systems. In livestock systems, the nitrogen surplus increase of with increasing stocking density.

The water quality risks associated with the pesticides used inorganic farming systems remain largely unstudied. However, the permitted active ingredients pyrethroids and rotenone, are not persistent and are also highly non-mobile in soil. It is only, therefore, by spillage or mishandling that these might enter water resources. The absence of conventional synthetic pesticides in organic farming systems suggests that the risks of water contamination are: significantly lower than in conventional systems. Prohibition of the routine use of antibiotics organophosphates, in-feed supplements such as copper and zinc in livestock production reduce the impact of organic farming systems in comparison with conventional livestock systems. However, certain products such as synthetic pyrethroids permitted for ecto-parasite control in organic livestock, do represent a threat to water quality if spilled. The risks of contamination of water with pathogenic organisms are reduced in organic farming systems due to the reduced intensity of livestock grazing and housing.

Air Quality

There has been little work carried out on gaseous losses from organic farming systems. Calculated CO_2 emissions for crop and dairy systems suggest that emissions are generally lower in organic farming systems. CO_2 emissions and energy use are usually closely linked. Nitrogen surpluses in organic farming systems, discussed in tend to be lower than those for conventional systems, which indicate a lower potential for gaseous nitrogen emissions. Estimated NO_2 emissions, which also contribute to the greenhouse effect, show no clear differences between organic and conventional farming systems. Reduced emissions of ammonia from livestock housing in organic systems is likely due to the increased use of straw for bedding. However, handling and spreading manure can result in significant losses of ammonia. Lower stocking rates are likely to result in lower gaseous nitrogen emissions per hectare, even though emissions per animal are likely to be similar in organic and conventional systems.

Research on methane emissions is very scarce in organic farming systems. Methane emission per hectare is likely to be lower in organic farming systems, due to lower stocking rates.

Energy Use

For arable crops, lower energy consumption in organic farming systems in commonly calculated. Lower energy input to direct crop protection operations in organic farming systems was partially offset by greater investment of energy in cultivations and showed that although three times as much machine energy was require to produce an organic crop, when the external energy inputs of fertilizer and pesticide production were taken into account organic farming systems required only half the energy input of a conventional system. For crops requiring repeated mechanical weeding, *e.g.*, potatoes, the energy consumption per unit output may be the same or greater than that of conventional farming systems. For livestock farms, 70 per cent lower energy consumption, mainly due to reduced imports of feed. Variation between organic farming systems in terms of energy consumption is high, which is controlled by the soil type, proportion of fodder production on the farm, and proportion of weed control achieved by secondary cultivations.

Biodiversity

Flora

The use of cultural controls and the targeting of weed control to specific periods in organic systems results in a greater abundance and diversity of weed species in organic systems than conventional arable systems. Weeds were able to complete their development in organically managed fields of winter barley, while in conventional systems, weeds were confined to margins and adjacent areas.

Un-cropped areas [sown grass strips (beetle banks), grass margins, uncropped wildlife and flower strips, hedges, ditch and bank habitats] are intrinsic in organic regimes where their management is central to the regime. These areas tend to be the reservoirs for both faunal and floral diversity, and therefore, the nature and extent of these habitats are the key in determining the overall biodiversity of agricultural areas showed that organic farms had a greater diversity of landscapes and in general had "more forms, colors, smells, sounds and spatial experiences." Because of the diversity of land-use types and the adaptation of land-use to the diversity of physical features, organic farms allowed a greater diversity of landscape elements.

Hedges tend to be higher and wider, field boundaries contain more tree and field sizes are smaller on organic farms. In N. America the scarcity of hedgerows and herbaceous field margins has reduced much of the potential of for wildlife integration within agricultural landscapes, which remains in much of Europe. However, field margins in organic farming systems did not contain significantly more diverse plant species than conventional margins. Floral species richness, and abundance of weed species in field margins is heavily influenced by management. In the short term at least, management techniques which increase biodiversity in the field margin do not result in increased weed ingress to the adjacent field edge.

Management of grassland, particularly the intensity and balance between cutting and grazing and the level of external inputs, rather than the system to which it belongs controls potential to provide more stable, less disturbed environments for the establishment of biodiversity. Established organic pastures showed more early flowering species. However, there were few major differences in flora between organic and conventional mixed lowland grassland, Experimental grass leys managed for biodiversity conservation by sowing with a species-rich seed mix and without nutrient inputs were less productive and tended to harbor fewer pernicious weeds than conventional leys, which gradually became dominated by rye-grass.

Landscape

Issues relating to scale and landscape structure (*e.g.*, patch fragmentation and heterogeneity) are increasingly important in ecology. They are particularly pertinent to arable farmland, which is uniquely characterized by a mosaic of distinct habitats and by severe and abrupt perturbations *e.g.*, harvesting, cultivation. In comparing organic and conventional farming systems we are looking at large scale landscape processes. These are often also the hardest to study and indeed most of the studies reviewed here which inform us about system differences focus at a much smaller level on differences between individual fields. Overall the combination of management and structure practices adopted under an organic fanning regime clearly have the potential to enhance the biodiversity of arable farmland. However, in terms of the regional biodiversity, effect of organic farms is small since they currently occupy les than 1 per cent of agricultural land.

Sustainability of Food Production System

The global population is expected to reach 8.5 billion by the year 2025. A frequently expressed concern is whether, if organic farming were adopted on a large scale, global food requirements could still be met. This issue is highly complex. It is, of course, a different question than whether malnutrition will be prevented, as it is clear than even in countries with ample food supply quantities and of appropriate quality, malnutrition may still exist. Malnutrition has many roots: inadequate food supply, limited purchasing power, poor health conditions, and incomplete knowledge about nutrition. Food supply is also somewhat divorced from the technical ability of farming systems to provide crop and animal products but is largely controlled by the systems and infrastructure for distribution, markets, finance, and political structures.

The limited studies that have been carried out to date to study the implications of increasing use of organic farming systems on global food production and security suffer from serious problems with underlying assumptions, availability of data, and the limited range of factors analyzed. For example, in Europe, we know that crops yields are commonly 20–40 per cent lower in organic systems compared with conventional systems. Rotational constraints prevent crops being grown so frequently on arable farms, so that the overall yield reduction for cereal crops may be 40–50 per cent. In northern Europe domestic food demand could be met with organic methods, but food would be more expensive. Similar conclusions were reached in the studies in the U.S. The proportion of grain legumes grown would increase substantially, while some studies suggest the vegetables crops would stay stable or increase slightly, and other crops such as oilseeds and sugar beet would stay stable or increase slightly, and other crops such as oilseeds and sugar beet would decline by as much as 75 per cent. But most of these assumptions reflect current demand patterns for organic products, which place more emphasis on horticultural crops and less emphasis on crops for processing such as oilseeds and sugar beet. The extent to which these crops are produced organically will depend on the market for them, and this is still very difficult to predict.

Future livestock production levels are equally difficult to determine. It is reasonable to suppose that the production of ruminant livestock might decline by 10–30 per cent and that pigs and poultry production levels might be substantially lower as a consequence of reverting to land-based, extensive systems. In addition, an implicit assumption is often made that we actually need to maintain current patterns of crop use. Organic farming, with its emphasis on farm-produced feed for livestock, is likely to result in significant reductions in the total quantities of cereals and other crops used to feed livestock, emphasizing instead forages produced as part of maintaining soil fertility or in areas not suited to crop production. By relying more on home-produced proteins, organic farming is also likely to reduce the need for protein crops, like soya, produced in developing countries, thus releasing resources to meet domestic food needs. Consumer demand patterns may also change to favor low meat diets as awareness of the environmental, welfare and health costs of intensive livestock production systems increases.

In addition, the assumption is often made that the yield reduction associated with organic farming in western Europe will also apply in other parts of the world. The evidence is to the contrary. Where resource-poor farmers are unable to afford purchased inputs, the information-intensive, agro-ecosystem management approaches which characterize organic farming may be critical to achieve production increases. However, such approaches require changes from teaching and dissemination approaches to agricultural extension to the development of methods which develop the farmers capacity to think for themselves and develop their own solutions.

Many authors have argued that it would be possible to feed double the current global population only if we continue to intensify production through increased use of fertilizers, energy, bio-technology and water. Yet it is increasingly clear that the pressures on some of these resources, particularly water, and the negative environmental impacts of intensive input use on soil water and air quality, are such that further intensification is not a sustainable option. Instead a strategy that increases total food supply over time, while doing this with lower levels of resource use and environmental pollution per unit of food produced, is a fundamental requirement-increasingly organic farming is demonstrating its capacity to do this.

Balance Sheet to Decide Adoption

1. Most studies find that organic agriculture requires significantly greater labour input than conventional farms. This is especially true in areas of low ecological potential. However, when labour is not a constraint organic agriculture can benefit underemployed labour in rural communities. Furthermore, the diversification of crops typically found on organic farms, with their various planting and harvesting schedules, may distribute labour demand more evenly which could help stabilize employment. Land tenure is also critical to the adoption of organic agriculture. It is highly unlikely that tenant farmers would invest the necessary labour and sustain the difficult conversion period without some guarantee of access to the land in later years when the benefits of organic production are attainable.
2. Soil-building rotations need to be designed both from the economic and the technical points of view-uses must be identified for all the crop and livestock products produced. As in all agricultural systems, diversity in production increases income-generating opportunities and can, as in the case of fruits, supply essential health protecting minerals and vitamins to the family diet. It also spreads the risks of failure over a wide range of cops. It is possible that, even on those farms where organic crop yields are lower than those produced under systems which use high levels of inputs, the overall economic yields of the farm will be competitive since organic systems benefit from market premiums and sometimes lowered input costs.
3. In the situations where farmers cannot produce competitively for the market or are resource poor, the case income they obtain from farming is too small to allow them to buy external inputs. High transport and credit costs are amongst the factors that make these inputs expensive and risky to use. Where ecological conditions are not particularly favourable for farming or the soil has been degraded, using synthetic fertilizers is relatively inefficient. In these conditions, synthetic fertilizers do not generate enough profit to make them a viable option in compensating nutrient losses. Increasing numbers of farmers are being confronted with these problems as the relative cost of synthetic fertilizer increases and soil degradation accelerates. If these farmers continue to produce for the market and export/loose nutrients in quantities higher than they can afford seen their options for compensation of nutrient losses, further nutrient depletion, soil degradation and yield decline will be unavoidable. At the moment this process can be seen in many regions in Africa and elsewhere in the tropics. Organic farming based on ecologically sound practices, therefore, seems to be the only sustainable basis for livelihood within the reach of these farmers. This does not entirely exclude production for the market but must be kept to levels that farmers can afford.
4. The insights generated by organic farmers in their search for site- specific production strategies can be of great benefit to non-organic farmers interested in expanding their management options. However, organic farmers still face huge uncertainties. A lack of information is an obstacle to organic conversion (*e.g.* surveys find that 63 per cent of sub-Saharan African farmers and 73 per cent of North American (US and Canada) organic farmers cite a lack of knowledge as the greatest barrier to adoption). Extension personnel rarely receive adequate training in organic methods and studies have shown that they sometimes discourage farmers from converting. Furthermore, institutional support in developing countries is scarce. Professional institutions with capacity to assist farmers throughout the production, post-production and marketing processes are non-existent in many developing countries.

While there are helpful research results that immediately could be extended to farmers, much more are needed. In 1990, FAO sponsored a conference at which organic research needs were identified (*e.g.* economics of stockless farms, animal husbandry, nitrogen cycling) however these challenges have largely gone unmet.

5. The explicit goal of organic agriculture is to contribute to the enhancement of sustainability. Nevertheless, negative impacts may occur and organic agriculture is not an exclusive method for sustainable farming. The soil and water protection and conservation techniques of sustainable agriculture used to combat erosion, compaction, salinization and other forms of degradation are evident in organic farming. The use of crop rotations, organic manures and mulches improves soil structure and encourages the development of a vigorous population of soil microorganisms. Mixed and relay cropping provide a more continuous soil cover and thus a shorter period when the soil is fully exposed to the erosive power of the rain, wind and sun. Terracing to conserve moisture and soil are used in appropriate situations and particular attention is paid in irrigated areas to on-farm water management. Properly managed organic farming reduces or eliminates water pollution and helps conserve water and soil on the farm (although improper use of manure can seriously pollute water). A few developed countries compel or subsidize farmers to use organic techniques as a solution to water pollution problems (*e.g.* Germany, France).

6. Organic farmers rely on natural pest controls (*e.g.* biological control, plants with pest control properties) rather than synthetic pesticides which, when misused, are known to kill beneficial organisms (*e.g.* natural parasites of pests, bees, earthworms), cause pest resistance, and often pollute water and land. Reduction in the use of toxic synthetic pesticides, which the World Health Organization (WHO) estimates to poison three million people each year, should lead to improved health of farm families.

7. Organic fanners aim to make the maximum use of the recyclable fertility in on-farm crop residues (straws, Stover's and other non-edible parts) either directly as compost and mulch or through livestock as farmyard manure. Eliminating the use of synthetic nitrogenous fertilizer greatly lowers the risks of nitrogen contamination of water. Crop rotation is a widely used method of fertility maintenance and pest and disease control; which is used in large- and small-scale farming in both developed and developing countries, especially under intensification. Fodder legumes are well-known fertility for the dual purpose of feeding livestock and adding nitrogen to the farm fertility cycle. Grain legumes may also produce a reasonable amounts nitrogen to the overall farm system through biological fixation; other nitrogen-fixing plants such as *Azolla* may also be used.

8. Biological nitrogen fixation is a powerful technique but it often requires some addition of minerals to the soil, especially phosphorus. Most certification programmes restrict the use of mineral fertilizers which may be necessary to supplement the organic manure produced on the farm. Natural and organic fertilizers from outside the farm are used (*e.g.* rock phosphate, potash, guano, seaweed, slaughterhouse by-products, ground limestone, seaweed, wood-ash). While most certification programmes prohibit the use of sewage sludge and night soil they are still used in some places. However, sludge may contain many contaminants including heavy metals which can have a deleterious and cumulative effect on the soil, while night-soil contains human pathogens and must be carefully composed before use.

9. Crop rotations encourage a diversity of food crops, fodder and under- utilized plants; this, in addition to improving overall farm production and fertility may assist the on-farm

conservation of plant genetic resources. Integrating livestock into the system adds income through organic meat, eggs and dairy products, as well as draught animal power. The crops and on-farm forestry integrated into the system provide shade and windbreaks while providing food, income, fuel and wood. Integrated agri-aquaculture may also be found within diverse organic agricultural systems. Economic objectives are not the only motivation of organic farmers; their intent is often to optimize land, animal, and plant interactions, preserve natural nutrient and energy flows, and enhance biodiversity, all of which contribute to the overall objective of sustainable agriculture to preserve natural resources and ecosystems for future generations.

Prospects

The increasing development of organic production systems and markets illustrates that this represents a valid alternative approach, often in direct opposition to intensive biotechnology agriculture. Certainly organic farming systems fit well with a view of the countryside where agricultural production is tempered with conservation and where less-intensive more environmentally friendly agriculture is practiced through the whole countryside rather than the North American model, where agriculture is separated into highly intensive bread-basket regions, leaving large nature reserves / wildernesses on the less productive areas.

Organic farming systems can deliver agronomic and environmental benefits both through the structural changes and the tactical daily management of the farming system. Organic farming aspires to a complex mix of agronomic, environmental, social and ethic objectives, which set the target for the development of a farming system in its particular location. In well managed organic farming systems are relevant both to developed nations (environmental protection, biodiversity enhancement, reduced energy use and CO_2 emissions) and to developing countries (sustainable resource use, increased crop yields without over-reliance on costly external inputs, environmental and biodiversity protection). With greater political will and investment in research, more of this potential could be realized.

To be the successful organic farmer an applied ecologist with a broad rang of skills and knowledge is required. To provide and enhance those skills and knowledge, the development of research in organic agriculture has to focus on a much closer association with many aspects of ecological research develop the tools necessary for understanding complex interactions at different systems levels. Such developments should help not only to further the development of organic agriculture but also inform and improve the sustainability of other forms of agriculture.

Success Stories of Organic Farming

Organic Food (Gujarat, India)

Organically grown food is now a popular concept everywhere particularly in the markets of the developed countries of the world. Bhaskar Hiraji Save, a farmer in a tiny village called Dehri in south Gujarat, has been well awarded of the benefits of organic farming four decades ago. While India was widely propagating the use of modem methods of agriculture, B.H. Save was making good progress on perfecting organic farming techniques.

B.H. Save, has not used fertilizers or pesticides in his farm 'Kalpavruksa', and still manages to earn a record profit of nearly 400 per cent. The coconut and chiku plants in his farm have yielded more fruit than in farms using pesticides and fertilizers. In fact the rice grown in his farm fetched higher prices in the market than rice grown using chemical fertilizers owing to its superior quality.

B.H. Save started farming back in 1950 when he was a primary school teacher. He bought half an acre of land in Dehri. He took up fun-fledged farming only in 1956. He began by using chemical fertilizers and pesticides like the others. *I not only used fertilizers and pesticides, but also had an agency through which I supplied the items to other farmers,* recalls B.H. Save. It didn't however, take him much time to realize that the input requirements as far as pesticides and fertilizers were concerned, were increasing steadily. In fact his cost of production was increasing at the rate of nearly 10 per cent per year. He says, *I was quick to realize that if I continued to put in this extra amount, in 10 years my cost would go up by 100 per cent, leaving with no profits.*

The setback prompted B.H. Save to look at other means of improving the yield. He then completely revamped his approach and conducted an experiment on his farm. He stopped using any kind of chemical fertilizers and pesticides. The immediate consequence of this sudden move was the production fell and the yield witnessed negative growth for a period of nearly five years. Due to this some people dubbed his new approach as a *mad experiment.*

However, after five years the yield stabilized and the production slowly started to rise. People had by now realized that B.H. Save had proved his point. They said that he had come up with a new method of farming. According to him, he had not done anything new and was only doing what his forefathers had done for centuries before him.

B.H. Save says that in agriculture, the emphasis should not be on increasing production. Instead one should focus on reducing the cost of production. The four things, which are considered by him as unnecessary for his method of farming are fertilizers, pesticides, weeding and unnecessary tilling. An important aspect of agriculture in B.H, Save's opinion is that farmers should learn to work with nature rather than attempt to control it through conventional methods involving mechanization and inorganic substances.

In addition to B.H. Save's style of agriculture being low on production costs, it also reduces the water requirement of the area under cultivation by as much as 70 per cent. The nutrient requirements of the crops are met by using dead leaves and organic waste. The process also utilizes biological pest control. While earthworms and red ants carry out the tilling in his farm, the grasshoppers and frogs carry out the job of insect control.

Economic Viability

In his "Kalpavruksha" farm he switched over to "Organic Farming" which did not show spectacular results in the very first year. The soil was recovering from the negative effects of chemical farming to which it was subjected earlier. Once again the soil regained its life, production increased and the cost of input nose-dived. The coconut trees which were planted 35 years ago and which were yielding 200 to 250 coconuts per year are now yielding 350 to 400 coconuts per year. "Kalpavruksha" farm was started with "Organic Farming" after full development he has moved over to "Natural Farming".

Recently, he has started "Sanghavi Farm" where he is willing to experiment every possible method of organic, irrespective of the risk involved, and based solely and fully on the principles of Organic Farming. That was a unique experiment for the first time in India 4 years ago. Today, one just has to visit this farm and get convinced about the efficiency of organic farming.

In any type of farming there is a principle of keeping certain distance between two plants. Normally, coconut trees are planted on 25′ × 25′ basis. So that 4 trees occupy 625 sq. feet of land, but they require sunlight falling on 625 sq. feet space.

He therefore, planted 2 trees together leaving 50 feed land on one side 25 feet on the other. This way the trees were getting sunlight on 625 sq. feet of space. Benefits of such plantation however, were many:

1. Per acre we could accommodate 84 trees as against 70 in the normal method.
2. Because of the two trees being together, we could save expenses for maintenance, planting watering, etc.
3. On the vacant land of 50 feet between two rows of trees, we could plant vegetables and pulses (which improves nutrient content of the soil) and also planted Banana trees on the vacant land of 25 feet between two columns of coconut tree.
4. This cosmopolitan plantation of short-living, medium-living and long-living trees and plants helps all the trees and plants. Both the quality and the volume of production improves. Normally coconut trees yield first fruits after 5 to 7 years. But at "Sanghavi Farm", we got fruits in a period of 4 years. Isn't that a big hit!

The quality and volume of bananas grown at "Sanghavi Farm" is very good. Such products also have a much longer shelf life.

His experiences in this regard are:

	Weight when Grown Natural Farming Way	*Weight when Grown Modern Way*
First bunch of banana on a tree	18 kg.	25 kg
Second bunch	30 kg	30 kg
Third bunch	25 kg	20 kg
Fourth bunch	15 kg	Tree dies
Total	**88 kg**	**75 kg**
He sold this @ Rs. 2.50 per kg	**Rs. 1.75 per kg**	
Total Income	Rs. 220.00	Rs. 131.25
Less: Expenses	Rs. 66.00	Rs. 105.00
	Water, labour, transportation)	(Fertilizer, Pesticides, Labour Water, Weeding, Transportation
Net Profite per Tree	**Rs. 154.00**	**Rs. 26.25**

Secondly the satisfaction gained by consuming such products. Enjoying such fruits satisfies the hunger sooner, thus leading to overall savings of individuals and also leading to their better health.

Having touched upon the beneficial features of Organic Farming, he now move over to the principles of the system.

1. The soil needs no tilling. However, he would like to quality this statement to the extent that in any type of fanning, initial tilling is necessary to clear the site. But subsequent tilling is harmful. Initial tilling is to facilitate the spread to roots. Any tilling done after sowing is, however, harmful as it cuts off the roots. Further, even as regards the initial, mandatory tilling, it must be done using light implements as the use of very heavy tractors tends to compact the soil, making it less arable.

2. The most effective tilling is carried out for the farmers by earthworms through their burrowing action. earthworms multiply very rapidly in millions per hectare in 2 or 3 months time. Earthworms eat organic debris along with soil continuously and excrete digested material at frequent intervals on the soil surface. For excretion they have to come to soil surface about 8 to 10 time a day. In this process they mix various soil components and improve thereby the soil structure. The earthworm castings are very rich plant nutrients. The population of earthworms will be high in the soil receiving plenty of organic manures.
3. Chemical fertilizers are not at all used in organic farming. They have no beneficial influence on the activities of soil organisms. Only organic manures, bulky and concentrated, are used.
4. By adopting the 'platform and trench' method, both organic manures and water are kept at regulated distance from the plants/tree. They are never "fed" directly to the plants as manure is not the food to plans/trees but it is the food of the living organisms present in the soil like earthworms, bacteria, etc. These organisms make the soil rich and arable. The humus made by them also enables the soil to absorb and retain moisture.
5. When the plants/trees are being nourished organically, they develop strong roots which protect them against strong winds. Secondly, because no chemical fertilizers are used, their water requirement is also less to the extent to only 30 per cent of the normal requirement.
6. In organic farming, weed growth in plant basins is suppressed through organic mulching. In other areas weeds are not controlled as the provide surface protection against high temperature and the water run off.
7. No chemical pesticides are required in organic farming. It is not that pests do not attack plants if been grown by the way of organic farming, but the use of organic pesticides such as tobacco decoition, diluted cow urine etc. is effective against pest infestation. Wherever possible biological control using parasites and predators could be practiced.
8. As the principle of organic farming states "care of plants and not cure" it is possible to manage various activities with less workforce. Displacement of under employed work force from traditional agriculture to more productive sectors of economy becomes possible when organic farming is more widely practiced.

Having enunciated the principles of organic farming, he hasten to add that the above are illustrative and not exhaustive. In fact, he strongly believe that regular workshops be conducted by and for the actual farmers, right at the farms, to enable the idea of organic farming to be propagated.

Personally speaking he is getting minimum 400 per cent profit from his farms (without even treating farming as a business) even if he take in to account cost of all the inputs like seeds, water, organic manure (which is purchased only in the beginning of the farms) labour, electricity, transportation etc.

It is very important to understand that this increasing population is reducing the area of soil available for agriculture. Hence, unless we move to methods which renew the soil fertility even while getting maximum production from it, we can not hope to solve the problem of food for the people in the years to come. As we have always been saying, Oil may last till 2050, soil will not. Leading to people's better health, economic gain for our nation, for the society, for the individuals and for the farmers themselves and to help understand and propagate the ideas of Natural farming.

The Organic Way

1. Natural farming is a path, not a technique.
2. Weeds are a blessing, chemical fertilizers an interference in the life of the soil.

3. Co-operation is the basic principle of nature.
4. The practical philosophy for maintaining environmental balance is live and let live.
5. The life of all plants and creatures is dependent on the condition of the soil. Farming should be done with minimum interference in the ways of nature.
6. Nature's tillers are earthworms living earth.
7. Inorganic fertilizers are a threat to humus.
8. Humus play a vital role in growing tasty and whole some food.
9. Only healthy, living fertilizers can nurture healthy plants.
10. Only healthy plants can provide health to humankind.
11. Living soil is the foundation of plant growth.
12. Co-operate with the organisms of the soil.
13. Trees do not require much. Just keep the soil alive.
14. Trees need moisture, not a great deal of watering.
15. Natural farming is only possible through compassion.

Narayan Reddy's Farm

This 5-hectare farm is a half-hour drive from the city of Bangalore, in southern India. Reddy grows the same kinds of crops that his neighbours do: coconuts, fruits, vegetables, mulberry, finger millet, and rice. Also, like many other farmers in the area, he keeps dairy cows and rears silkworms.

Specialty

His goal is to make his farm self-sufficient. He and his family grow or make nearly everything, they need. They try out new ways of doing thing, to see if they can do the tasks faster or better or at a lower cost. Sometimes they find that the old method still work best. And they are always willing to share what they have learned.

Reddy's father began cultivating the land more than 20 years ago. He used chemical fertilizers and pesticides because that was the modem way. At first this worked well. But after a few years, he found he was losing money. The chemical were not always available. They were expensive. Crop yield were decreasing. What's more, the soil was deteriorating.

So Reddy decided to use different farming method. He knew about the old, traditional ways of cultivation. He felt that if he combined the best of the traditional methods with modem technology, he could improve his farm. And that is just what he did.

Reddy Tries many Ideas

Reddy learns all he can about modem scientific ways of farming. He chooses and toes out those ideas that he thinks will work for him. But he also makes sure that the methods will not pollute the soil and water or make him too dependent on outside sources for his supplies.

For example, one of the first things Reddy did was to stop buying chemical fertilizers and pesticides. At first this meant that he had to bring more organic materials onto the farm. For instance, he gathered animal manure from other farms. He collected leaves on roadsides and in the forest. He started to make lots of compost. He grew green manure crops, such as sesbania (*Sesbania sesban*) and sunnhemp (*Crotalaria juncea*). And he used the leftover leaves and stalks of harvested crops for compost or mulch instead of burning them.

He also reduced pesticide use. Reddy now makes his own insecticide from ground neem seeds, cow urine, and other materials available on the farm. His crops look perfectly healthy, and the fruit trees are laden with fruit.

During the time that Reddy was changing from chemical farming to organic farming, he needed more labour on the farm. And his yields went down a little, but because his costs also went down, his profit was about the same. With time, as the soil improved, and he started growing a number of different crops each season, his overall profit increased.

Reddy has Many Crops and Enterprises

The main crops Reddy grows are coconuts, fruits–especially sapota (*Achras sapota*), mango, and watermelon–finger millet (*Eleusine coracana*), and mulberry (*Morus* spp.). He also harvests two crops of rice a year. In between the coconut trees he grows vegetables to sell in the city.

All the food for the family is grown on the farm. They have their own cereals rice–and millet–and coconuts. They get plenty of milk from their cows. Even the coffee they drink at breakfast is their own, grown on a row of shiny coffee bushes just outside their house.

Over time Reddy has added other enterprises. One of these is rearing silkworms. In nature, silkworms live and feed on mulberry trees. They can also be reared indoors by farmer if they are fed the right kind of mulberry leaves regularly. Silkworms are valued for the cocoons they spin. These cocoons are reeled into silk yarn, which is used to make costly silk cloth.

On one side of Reddy's farmhouse is a large shed with shelves from floor to ceiling. Here the silkworms are kept in flat round bamboo trays and fed mulberry leaves. In about 4 weeks they spin their cocoons. These cocoons sell for a high price.

Rearing the silkworms does not cost much. Because Reddy has planted a large area of this farm with mulberry, he always has mulberry leaves to feed large silkworms. He has also planted several bamboo clumps. The bamboo is the split and woven to make the trays where the silkworms live.

Reddy has planned his farm enterprises carefully. Each one links with and supports the others. The many different farm activities ensure that Reddy and his family have a secure income all year round.

From Loss to Profit: An Organic Farming Success

With imagination and hard work, he combines modern technology and age-old farming methods. He and his family grow or make nearly everything they need on their 5-hectare farm. They are always trying ways to lower costs and work more efficiently. And every enterprise on the farm is carefully planned to serve several purposes.

One of Reddy's most important enterprises is livestock. He owns 14 excellent dairy cows. They are specially bred to yield a lot of milk. They cost more to buy at first. But he makes a good profit from selling the high-quantity milk.

To keep up that quality, Reddy looks after the cows well. They are not allowed to graze free in the fields where they might damage the crops. Instead, they are kept in clean stalls in a cowshed and fed properly. They are also vaccinated against diseases.

Reddy Built a Biogas Plant

The cows supply valuable manure. As every farmer knows, cow dung makes excellent crop fertilizer. But in this part of India, because firewood is scarce and costly, it is common to dry cow dung and use it as cooking fuel. This is a waste of good free crop fertilizer. Reddy learned that there was a way of using livestock manure to provide both fuel and fertilizer. Before it is spread in the field, it can be used to make biogas, a low-cost cooking fuel.

So, Reddy built a biogas plant near his house. This was costly to do. But it was only a one-time expense. The biogas plant has paid for itself many times over. It now costs very little to maintain, and it turns a waste material, cow, dung, into two essential products–fuel and fertilizer.

First, all the cow dung from the dairy is put into the underground biogas pit. As it decays there, it gives off gas. This gas is piped into the kitchen and supplies all the cooking fuel for the household. So there is no need to burn dried cow dung or cut down trees for fuel.

The best part is that the cow dung can still be used as fertilizer. The leftover material from the biogas plant, called slurry, is still full of good plant nutrients. Reddy adds this to his compost bins. From there, it is taken to fertilize the crops.

So Reddy's cows supply both milk for sale and manure for the crops. And, because Reddy has built a biogas plant, the dung from his cows also provides biogas fuel for cooking.

Reddy links all his farm activities. He uses the biogas slurry to fertilizer napier grass. This in turn is used as fodder for the dairy cows. And the whole cycle begins again. So you see how everything is interlinked? This is a good example of how Reddy has used new ideas to save money and farm better.

Reddy also makes compost to use as fertilizer. He uses earthworms to speed up and improve the composting process.

To make the compost, Reddy has built stone bins from granite slabs, which are easily available in his area. The bins are about 5 meters long, 1 meter wide, and 1 meter wide, and 1 meter high. In these, he layers the farm and kitchen wastes as they collect. He then covers them with a layer of soil and adds them earthworms. The worms feed on the waste and break it down quickly.

This compost is one of the main fertilizers now used on the farm. It nourishes the crops. It builds up the soil. It is safe to use year after year. Because it is made from wastes, it costs nothing! And it returns to the earth whatever was taken out of it.

These are some ways the Reddy family uses to keep their farming costs low. They also make sure they do not have to depend on anyone else for their materials.

Of course some of the equipment Reddy uses is costly to install. But only at first. He only chooses equipment that will save him money and work in the long run, such as the biogas plant.

Reddy did not have much formal schooling. He faces the same problems as any other farmer in the areas. But he has taught himself new ways of looking for solutions. He finds out where he can get information. He learns how to do things himself. And, perhaps best of all, this pioneer spirit has encouraged other farmers to feel more confident and try new ways of farming.

Breaking Out of the Pesticides Cotton Wedlock

Cotton is cultivated in over 60 countries on more than 80 million hectares which is equivalent to 5 per cent of all cultivated land, and its production uses 25 per cent of all pesticides consumed annually by the world's agriculture, with an estimated value of US$ 7.4 billion.

Of all the 300 million kilograms of pesticides used in Third World countries every year, half is used in cotton. In many African countries, over 70 per cent of pesticide imports are used on cotton; in

India 45 per cent of pesticides are used in cotton production. Considering the importance of the fibre in their economy, for some countries cotton accounts for 70 per cent of export earning, it is no wonder that poor countries spend so much of their foreign currency on the import of pesticides for use in cotton. Being a non-food crop cotton has attracted the most deadly pesticide formulations, classified under toxicity Classes Ia and II of WHO.

Pesticides such as monochrotophos, dichlorvos, quinalphos, dimethrin, endosulfan among many others are still in use, including herbicides such as parquet and diuron. In the last ten years the use of synthetic pyrethroids such as cypermethrin though to be safer, created problems with resistance building up in the whitefly (*Bemisia tabaci*). In Senegal, the government used to supply farmers with a formulation containing a mixture of endosulphan, DDT and methyl parathion, the empty containers are still to be seen dumped around small farmers' households even in the remotest locations. This season Senegalese farmers have complained that the pesticide Cypercal has led to the burning and shedding of cotton leaves. In Zambia, the multinational, Lonrho, supplies farmer son credit with a range of potent organophosphate pesticides which end up used in maize crop.

A range of pesticide is used in the control of a whole variety of insect pests attacking the plant during its vegetative stage (thrips, aphids, jassids and leaf-eating caterpillars such as *Sylepta derogata* in Africa, *Spodoptera litura* in Asia). But it is in the reproductive state that the problem is greater since damage to capsules directly affects the yields. Species involved are larvae and adult caterpillars for instance, of the Noctuidae family several species of *Earias* sp. (Spotted and spiny bollworm), *Pectinophora gossipiela* (pink bolloworm), and most important of all *Helicoverpa armigera* the famous American bolloworm. Due to the high toxicity of the formulations and the regularity of application, cotton areas the world over have seriously reduced the natural predator population. In many cases. Imbalances are so acute that there is a need to re-introduce entomoparasites and entomopathogens in order to hasten the biological recovery. The range of pests are clear symptoms of biological and plant nutrient/physiological imbalances originating from improper rotation and self fertilization practices, such as the one-sided use of soluble nitrogen, especially urea, which led to Increased attacks of sucking pests such as aphids and whitefly.

Other indirect problems arise when, as is the custom in most smallholder systems cattle put to graze on the leaves of cotton after the harvest or when cattle fed on de-oiled seed cake, a high protein concentration made from cotton seeds. This practice leads to contamination problems. In 1994, an entire consignment of Australian beef was rejected and returned by the American authorities after residues of pesticides used in cotton were found the in levels higher than permitted by the Codex. In Senegal, farmers have noted the intoxication and consequent death of livestock after feeding on the leaves of cotton. If high levels of residues were found in samples of meat from Australia (where pesticides regulations are strict) one can only imagine what is happening with residue levels in meat and milk from countries like India, Pakistan, and many African countries. Cotton cake is also exported for animal feed to Europe's dairy 'factors'. No data is available on residue levels in such products.

The list of environmental, health, socio-economic and structural problems associated with cotton production is exhaustive. The drying of the Aral Sea for cotton irrigation, hunger in Sudan while cotton occupies the best soils, total depletion of soils and contamination of the environment and workers in Colombia due to constant aerial sprays, are just a few examples. Similar problems arise out of cotton processing (blenching, mercerizing, dyeing, etc.).

These problems are also seriously affecting the economic viability of the crop. Costs of production have risen steadily, a rise not being matched by an increase in yield or income. Pest outbreaks have led

China to buy cotton on the international market for the first time in the past four years, which sent prices soaring upwards, only to fall again, destabilizing many textile industries which had already sold their yarn ahead of time with fixed prices and had suddenly to bay expensive cotton to complete their orders.

All of this is making cotton farming a more risky business with decreasing margins for the farmer and a long-term problem for the industry which tries to meet the new demand for natural fibres at affordable prices.

The Organic Solution–Maikaal Project, M.P., India

Conversion to organic methods can reverse this process, has been proved in the Maikaal Project. Rather going in detail the technological strategies are being given here.

Technical elements of a cotton-basedorganic farming system developed for the Maikaall Project, M.P. state in India

Cropping System

Rotations and intercropping systems have been designed to improve soil fertility, guarantee the economic viability of the process, but also to increase bio-diversity and homeostasis by reducing pest pressure.

Environmental Reconstruction

Multi-purpose shrub and tree planting around field boundaries for fodder, microclimate and biodiversity improvement, reduction of contamination risk.

Soil Fertility Management

Mycorhyza favourable rotation: Azotobacter or *Azospirillum* and PSB (Phosphorus Solubilising Bacteria) as seed inoculants. Locally made bio-dynamic humus preparation. FYM compost enriched with rock phosphate applied on the basis of soil analysis, top-dressings with a special compost made of oil-cakes and bio-fertilizers treated with CCP (Compound of bio-dynamic compost preparations). Nome-made bio-fertilizer foliar sprays. Bio-dynamic silica spray for fibre quality improvement.

Pest Monitoring and Mass Trapping

Training for pest identification and pest counts to establish economic threshold levels. Pheromone traps for monitoring and mass trapping of bollworm complex.

Pest Controls and Management

Environment management for the increase of natural predators, release of entomoparasites and entomopathogens when necessary, commercial and locally identified botanical pesticides and formulations. Natural predators develop in bio-farms such as *Chrysoperla carnea,* ladybrid-beetle, spiders, birds and become indicators for the restoration of the ecological balance. Soil fertility, pest management and natural controls result in minimal presence of aphids, and whitefly, reduced attack of jassids, presence of bollworms, at an economically tolerable level.

Weeding

Cross cultivation using oxen and hand weeding.

Harvest

Hand-picked, segregation of damaged capsules.

Yields

Organic seed cotton average: 500 kg/acre, equal to 1200 kg/ha. Best organic farmer's yield registered at the Maikaal project in India using the above recommendations 3000 kg/ha (irrigated).

Eco-farming of Medicinal Plants in the Arid Tropics of India

The worlds largest arid tropical region is spread from Northern Africa to the Western part of Asia. Its eastern end which covers about 12 per cent (32 million ha) of India is popularly known as the *Thar* desert. Low (100 to 300 mm/yr) and erratic rainfall, high wind and solar regime and sandy soils make the Moisture Deficit Index–85 to 90 per cent in this region, which generates high probability of drought and desertification. Under such conditions the production of conventional agricultural/foodgrain crops is highly uncertain resulting in poor economic conditions for the majority of marginal farmers.

Within this ecosystem there exists a large degree of heterogeneity in terms of soil, climate and other economically relevant natural systems, which provides a rich biological diversity and, within species, a range of genetic diversity suitable for different conditions. About 657 species of plants belonging to 351 genera and 87 families, are found in this region, with 8 per cent of these endemic to the region. About 194 species are used for economic or medical purpose.

Medicinal Plant Production

Increase in human and animal populations in the region have put pressure on the natural vegetation of the ecosystem, causing deforestation, overgrazing of rangelands etc. Furthermore, some developmental activities, *e.g.* mining of sandstone, granite, marble etc., cultivation on marginal lands, increasing use of tractor in agriculture and the irrigation network of the Indira Gandhi Canal, are rapidly destroying natural habitats. Some introduced plants like *Prosopis juliflora* are becoming biological invaders.

Thus destruction of natural habitats and over-exploitation without replanting are threatening the population and diversity of medicinal plants (MP), in particular, for example, *Commiphora wightii* and *Asparagus recimosus*. It is thus imperative to protect and increase the production of medicinal plants.

Ecological Production of Medicinal Plants in the Thar Desert of India

The plants which are used for medicinal purposes should not have any residues of fertilizer or pesticides–otherwise they become more of poison than a medicine, Moreover, the MPs have several active ingredients (biochemical) and growing them with fertilizer, pesticides and irrigation may alter or deteriorated their quality. Thus, as far as possible, they have to be grown in near-to-near condition so that both the quality of MP and the eco-balance can be maintained, For that there are two possible ways: (*i*) *in situ:* to make conditions suitable in their natural habitat to increase population; (*ii*) *ex situ:* to cultivate as an organic crop.

In situ Production

Nature has immense potential for restoration of an ecosystem provided the conditions are favourable. The arid ecosystem has been sustained since time immemorial and it is only the over exploitation of natural resources which makes the system unsustainable. In general, man-made reforestation replanting is difficult, as one would need to study the habitat and micro-environmental requirements of each and every species for their cultivation. Given these problems *in situ* conservation projects have been started, for example, a project near Jodhpur. The area has been protected by stone wall fencing, watershed soil and water conservation measures are in place and the participation of local people has been gained. The naturally regenerated plant population of *C. wightii* has increased by 813.3 per cent in 7 years. *C. wightii* is an endangered medicinal plant in the region, however, despite much research into macro/micro propagation, experiments in cultivation have not been comparable with the *in situ* conservation project.

Agriculture Production

Although it is better to increase the production of MP by conserving them *in situ* there are plants which are required or demanded in large quantity such is Senna and Psyllium. For meeting this demand, cultivation is necessary. Under the arid tropical conditions cultivation of medicinal plants has economic advantages for farmers, helps the conservation of natural resources and can ultimately contribute to the sustainability of this fragile ecosystem.

Harnessed the Potential by N.D. Prajapati

Arid regions have huge possibilities and potential for the cultivation of medicinal plants and Mr. Narayan Das Prajapati an innovative farmer of the Thar desert has very well understood and concluded that

1. The MPs of this region are drought hardy by nature, thus their production is less affected by drought compared to food grain crops. As the root, stem and leaf are the useful parts in most MPs, rather than the fruits, the yields of these plants' are less affected by drought.
2. The soils of the region are comparatively untouched by fertilizers, pesticides and irrigation (FPI). The use of FPI is highly risky due to climatic conditions, moreover the marginal farmers are unable to use costly inputs. On the other hand the farming systems of these regions are dominated by animals, and the *excreta* of animals can improve the properties of sandy soils, These, which are seen as limitations for food grain production, are opportunities for MPs cultivation. Furthermore, in organic farming quality (*i.e.* no residue of pesticides/fertilizers) is more important than the quantity.
3. A major part of the region is currently wasteland (12 million ha) due to various land forms *e.g.* deep sand/sand dunes, gravely, rocky areas, saline soils/water etc. where cultivation of food crops is not possible. MPs can, however, be grown in these areas, *e.g. Cassia angustifolia* (senna) ill deep sand/sand dunes; *Commiphora wightii* on rocky/gravely land; *Plantago ovata* and *Vinca rosea* in saline soils. Using wastelands for cultivation will not only increase the production of these plants but will also help halt desertification due to increased vegetative cover.
4. At present the farming community remains idle for 7–8 months, after the rainy season (November–June). Year round cultivation of MPs would provide more work and additional income, thus providing support to the economies of drought prone regions.
5. Because of the climatic uncertainty agroforestry systems have developed t in the arid tropics, MPs can easily be accommodated in these systems.

There are many MPs which have potential for cultivation in this region. After several years of experimentation Mr Prajapati has found that Senna is the most suitable plant for sandy arid lands because it grown well in bright sunshine and dry climates, its deep tap root system helps it to thrive in deep sand and no animals graze it.

Now only because of his efforts senna is being cultivated with completely organic method in about 3000 ha area of the *Thar* desert. The *Plantago ovata* (*Psyllium*) is already commercially growing in saline soil/water areas contributing about 60 per cent of the world's production.

SEKEM in Egypt: Bio-dynamic Production and a Way of Life

The SEKEM farm in Egypt is situated in the desert near the south tip of the Nile delta, not far from Cairo. It was established in 1977. The total farm area is 55 ha, of which about 30 ha are under cultivator. Additional farms, which supply products on a contract basis to the main farm cover about 500 ha. And this area is steadily expanding agriculture. The products are certified in accordance with the Demeter standards for biodynamic agriculture. The main crops grown can be divided into three categories: fruit trees, herbs and spices, and vegetable consumption in Western Europe. Spices and herbs are designed for both domestic consumption and exports. The farm also includes animal production.

From the very beginning SEKEM aimed at supplying local people with high quality organic food. SEKEM has contributed to getting organic production onto its feet in Egypt. It has established several organic shops in Cairo and delivers its products to about 7,000 pharmacies and 2,000 shops. In a tea-drinking country like Egypt it is significant that certified organic herb teas from SEKEM are amongst the most popular in the country. Because of the success of the SEKEM products on the West European. market, exports have now risen to 40 per cent of the total turnover instead of the 10–20 per cent planed initially.

There is little doubt that the SEKEM initiative is a show case of holism within the organic movement. Well-rooted in anthroposophic and bio-dynamic philosophy, all SEKEM's projects and enterprises have developed very rapidly. Today, SEKEM's has more than 120 people in full-time employment and cooperation with over a hundred bio-dynamic farms marketing their products through SEKEM provides another 250 employment places for Egyptian men and women. SEKEM has also played a pioneer role in developing organic cotton production.

What began as a farm on sandy desert soil on the outskirts of Cairo, has now become a center of cultural and social development. The SEKEM kindergarten and school provides education for 350 children some of who are handicapped. A health clinic provides medical care and a mobile ambulance station reaches those who are unable to get to the clinic. Art and culture within the SEKEM community are expressions of SEKEM's approach to life and the design of its building and the conditions and environment reflect these principle.

Areas, Need Attention to Popularize Organic Farming

1. Adopt the interests of the farmers and their families as the most important criteria when developing sustainable agriculture.
2. Put more emphasis on securing food production in the long-term rather than the short-term. Emphasis should be on a source-directed policy, which prevents problems rather then trying to solve them in a non- sustainable way.
3. Secure long-term land tenure rights. This would demonstrate the government' s interest and facilitate the farmer's interesting soil protection.
4. Develop a differentiated policy about the introduction of Organic Agriculture Systems in ecosystem variability and ecosystem users.
5. Introduce Organic Agriculture Systems in ecologically sensitive, biologically and genetically diverse, and nature-protected areas.
6. Promote the development of organic agriculture projects by facilitating organic food imports from developing countries.

7. The technique of mineral and organic matter balance being in equilibrium or positive for certified organic products should be enforced by the organic agriculture movement if it wands to substantiate its claim that organic agriculture is sustainable.
8. Promote the consumption of certified organic food and non-food products.
9. Research projects for formulating organic farming practices may have to be framed and implemented as national projects.
10. Incentives for production of good quality organic manure, biopesticides, biofertilizers and green manuring crops may be strengthened.
11. Development of pesticides of plant origin (such as neem) and use of agents especially under Integrated Pest Management Systems need to be promoted.
12. The indiscriminate use of inorganic fertilizer and pesticides needs to be discouraged.
13. Steps are needed to reduce hazardous chemical residues in seeds, fodder food products and milk.
14. District-wise inventory of recyclable wastes (both on-farm and off-farm) needs to be prepared in order to know the potential of organic agriculture as also to develop location specific technologies.
15. Depending on the nature of wastes available, the following composting techniques need to be standardized and publicized among the farmers.
    (*a*) Rapid composting of crop residues with suitable microbial strains such as *Trichoderma* spp. or other microbes having cellulolytic and lignolytic actives.
    (*b*) Vermicomposting using earthworms.

The nutrient content and microbial population of organic manures should be enhanced using rock phosphate and pyrites: biological agents such as N. fixes, P-solubilizer for preparation of good quality compost material.

Cropping System

Efforts should be made to maintain yield stability and suppress the weeds and pests through adoption of appropriate cropping systems and minimizing weeds and pests incidence through adoption of appropriate the term crop rotation systems such as:

1. Introduction of legumes as break crops in intensive cereal-cereal system resulting in yield stability and restoration of soil fertility.
2. Periodical substitution of wheat with a fodder crop in rice-wheat system is a remedy for phalaris minor.
3. Inter cropping of garlic with sugarcane minimizes incidence of shoot borer.

Alternative Land Use

Location specific alternative land uses should be identified and popularized for adoption in marginal lands as well as the lands under fragile eco-systems. Depending on the situations, these land uses may under fragile ecosystems. Depending on the situations, these land uses may be agri-horticulture, agro-forestry, silvi-pastures etc.

Governmental Efforts

1. The changing scenario of increasing cost of fertilizers gradual removal of subsidy on fertilizers and increasing concern of likely pollution of ground waters by fertilizers through leaching from soils calls for proper use of fertilizers.
2. The encouraging results obtained by conjunctive use of fertilizers, organic manures, and bio-fertilizers are providing leads to future strategies of rational use of fertilizers and manures for continually enhancing productivity without being detrimental to the environment.
3. To promote the use of organic manures, the existing potential of about 650 million tonnes of rural compost and 16 million tonnes of urban compost may be exploited intensity. As per the existing estimates, only about 135 to 240 million tonnes of rural compost and 4.8 to 6.5 million tonnes of urban wastes are being used as manures.
4. To demonstrate and popularize the scientific method of compost making and for balanced fertilization though soil testing, a scheme of Balanced and Integrated Use of Fertilizers is being implemented by the Government of India. Under this scheme, Rs. 18 crores were spent during the Eight Plan. The outlay has been enhanced to Rs. 46 crores during the Ninth Plan for this scheme which is expected to be further increased.
5. Another new area, which requires special thrust, is the promotion of biofertilizers. Starting from the Seventh Plan, a total of 58 bio-fertilizers production units have been established with Government support, having a production capacity of 8300 tonnes per annum. Besides this, 29 biofertilizer units have been set up by private entrepreneurs and State Departments of Agriculture and others having an estimated production of 8,000 tonnes. Thus the total biofertilizer production capacity is about 16,000 tonnes/year.
6. Continuous Governmental emphasize on the usefulness of promoting integrated use of organic and chemical forms of nutrients for better results.
7. Need of continuous efforts to popularize organic farming with the main aim of ensuring sustained productivity, environmental protection and making available food and food products raised without chemicals of any king. The limitation of the system is the difficulty in supplying the required amounts of nutrients to the crops in large areas only through organic sources and hence the integrated nutrient management system is a better option. However, organically grown food is expected to fetch higher price and this can offset any loss due to lower yields.

International Efforts

Policy

1. The environmental and economic benefits of organic agriculture have captured the attention of several countries; however, only a small number have enacted policies to assist the organic sector. Most assistance has developed in the private sector, especially by NGOs. This private sector infrastructure is not only recognized by countries, but also encouraged.
2. Farmers and consumers in almost all countries rely on a system of private self-organized producer organizations and independent certifiers which have, over the years, provided an economically-efficient mechanism of certification. The degree to which private organization are significant is indicated by the fact that IFOAM has some 650 individual and institutional members in over 100 countries, 75 per cent of which are in developing countries. However,

the network of private certifiers needs to expand; as mentioned earlier, many developing countries still lack certification organization.

3. Organic programmes, maintain evolving "input lists" of acceptable inputs for organic production, processing aids, and ingredients. Countries, in consultation with civil society organizations, can propose their own input lists to Codex and negotiations will probably take place over what ultimately is acceptable in the international marketplace. Many developing countries have not drawn up material lists and generally lack the resources and training to effectively participate in international standard-setting processes.
4. Engaging in international trade requires quantities of goods well beyond 1he production capacity of individual small organic farmers. NGOs have supported farmers in their efforts to establish appropriate organisations for collecting, processing and marketing organic products. Countries could support NGOs in building organizations which play a key role in disseminating best practices and ensuring that farmers use adequate production and post-production techniques.

Technical Support

1. Northern developed countries have invested by far the most in organic agriculture research but even so, the contribution is minimal compared to overall research agriculture (*e.g.* less than 0.01 per cent of the US Department of Agriculture research budget is directed to organic agriculture). The lack of extensive formal organic research with the highly site-specific nature of organic agriculture, suggests that it would be most advantageous for farmers themselves to participate in locally-based, applied field research. Experience with FAO-initiated Integrated Pest Management (IPM) Farmer Field Schools and community forestry projects has shown that farmers, whether owners or tenants, large or small, can practice good scientific methods if they are given orientation and technical support. Countries could also undertake market research. The USA, for example, analyzed the market for organic goods in more than 20 countries, with a focus on opportunities for US exports.
2. Most national and private organic standards, disallow genetically engineered organisms (GEOs), including transgenic crops, in organic production. By providing consumers with a non-GEO produced alternative organic agriculture has secured an important market, at the same time relieving much of the pressure to impose out-right bans on GEO-produced food. While current biotechnology applications have been rejected for organic production, it is possible that future scientific development and furthers discussions may produce technologies acceptable to consumers (*e.g.* the value of tissue culture in producing disease-free planting materials).

Direct Assistance

1. Interest in environmental protection and the preservation of small family farms has led developed European countries to subsidize organic production to varying degrees.. The subsidies can provide significant encouragement (*e.g.* UK provides up to £450 per ha during the conversion period for designated land areas; European Community subsides have helped 15 per cent of Irish organic farmers develop dairy operations). Interested communities would have to be advised about organizational aspects and on producing and commercializing the necessary inputs as well as on marketing any surplus produce (*e.g.* seed and planting material of adapted varieties, suitable livestock genitors).

2. Consumer confidence in the integrity of organic claims is essential if goods are to be sold at a premium. Because organic products cannot be distinguished from conventional products, consumers depend entirely on certifiers to truthfully distinguish organic from non-organic goods.

Countries could better enforce organic standards by punishing those who engage in fraudulent activities as well as undertaking systematic tracking and measuring of fraud and its impact on the market.

International Organisations

Scientific Congress on Organic Agricultural Research (SCOAR)

Scientific Congress on Organic Agricultural Research (SCOAR) is a collaboration of producers, scientists, and others. Its mission is to plan and promote research and information-exchange for understanding and improving organic agricultural systems.

The Goals of SCOAR

1. Cultivate a peer-level process of mutual learning and democratic collaboration among a diverse community of organic farmers and ranchers, scientists, extensionists, consultants, consumers and others for pursuit of the project mission.
2. Create a shared understanding of organic agriculture as an ecological approach to managing agricultural systems.
3. Identify and characterize research and information-exchange priorities pertaining to the mission through various activities and documentation of those activities.
4. Advance the state-of-the-art of organic systems research, including on-farm, participatory, whole systems, and multi-farm studies.
5. Increase access to the knowledge, skills, and practices of successful organic producers by developing and implementing models for networking and information sharing among organic producers and scientists.
6. Encourage the use and implementation of SCOAR's results.
7. Potential SCOAR Themes: Core Research Needs and Some Big Issues.
8. Scientific interpretation and elaboration of traditional *Best Management Principles:* (i) Feed the soil to feed the plant, (ii) Building health in the soil and animals builds healthy farms.
9. Importance of conducting research in a bonafide organic context.
10. Emergence vs. extraction of beneficial organisms/behaviours: Systematic biological control paradigms.
11. Applied microbial ecological management: regional soil food web labs, micro-livestock husbandry, and *recombinant ecologies.*
12. Biological capital formation: lowering the costs of soil organic matter.
13. Getting past weeds vs. soil: reducing/mitigating impacts of tillage in systems without herbicides.
14. Explicit strategies for both conversion and post-conversion.
15. Applications of agricultural *rocket science* in an ecological context.

16. Surviving success: sustaining profitability as price premiums to farmers fall requires optimisation of small-medium size production systems for long-term cost-efficacy.

International Federation of Organic Agricultural Movements (IFOAM)

Organic agriculture alone can ensure that we are assured a future that is both ecologically sound and economically sustainable. Basic Standards of Organic agriculture and Food Processing have been set by International Federation of Organic Agricultural Movements (IFOAM). Its main function is to coordinate the network of organic movement around the world.

The Organic Farming Research Foundation (OFRF)

The Organic Farming Research Foundation (OFRF) is a producer-directed organization dedicated to the improvement and adoption of organic farming methods. As analysts of U.S. policy for food and agricultural issues, we appreciate the opportunity to comment on FDA's policies and performance regarding *bioengineered foods*. OFRF not categorically opposed to every use of *bioengineering* in agriculture, but OFRF believes that the federal government has fundamentally failed to provide an adequate regulatory regime for transgenic technologies and products. In particular, the division of responsibilities among FDA, EPA and USDA ensures that there is no comprehensive, public interest oversight governing the release of transgenic organisms into the environment and the food supply. The environmental, agricultural, food safety and food security issues cannot be considered in isolation from each other, and sound public policy cannot be formulated or Implemented in such a fragmented system. Furthermore, under the existing regulatory framework there is no definite point of accountability for the consequences of this regulatory failure. There is nobody ultimately in charge except the biotechnology industry itself. The Administration should immediately begin the process of forging a new, comprehensive regulatory process, with extensive participation from consumers, farmers and the public interest community.

Organic Farming in the Indian Sub-Continent

Background

Organic farming involves an approach to food production. It seeks to create sustainable agricultural system that relies first and foremost on ecological interactions and biological processes rather than chemical inputs. The aim of organic farming is to produce optimum yields with minimum harm to the surrounding environment.

Organic farming is a production system which avoids or largely excludes the use of synthetically compounded fertilizers, pesticides growth regulators and livestock feed additives. To the maximum extent feasible, organic farming systems rely on crop rotations, crop residues, animal manures, legumes, green manures, off farm organic wastes and aspects of *biological pest control to maintain soil productivity and tilth to supply nutrients and to control insects, weeds and other pests.*

Farming before the western world colonized the developing world had, in many ways, been organic as there were no artificial inputs.

History

The ISC was involved in sustainable fanning for several thousand years. Farming was not just practised for food production or income. It was a way of life. Agricultural activities and techniques were tailored to suit the soil conditions and climate. People were content with what they had, the land was natured with care, pesticides were unheard of. It was self-sustaining. Farming before the arrival of the western world into the east was at one with nature.

Half a century ago, Sir Alfred Howard, the father of modern sustainable farming wrote in this classic, An Agricultural Testament that; *In the agriculture of Asia we find ourselves confronted with a system of peasant farming which, in essentials, soon became stabilized. What is happening today is in the small fields of India and China took place many centuries ago. The agricultural practices of the Orient have passed the supreme test, they are almost as permanent as those of the primeval forest, the prairie or the Ocean.*

Status of Organic Farming in India

Green Revolution in India was mainly realized with the introduction of high yielding varieties of various crops and by following intensive cultivation practices with the use of fertilizers, pesticides and other inputs. The fertilizer consumption in the country accordingly has increased from 66,000 tonnes in 1951–52 to about 12.16 million tonnes in 1992–93. In other terms, fertilizer consumption in India was only 0.5 kg/ha in 1951–52, and increased to 67 kg during 1992–93. Same situation was with pesticides. All these agro chemicals are produced from non-renewable fossil fuel. In future, diminishing availability of these inputs may not only cost heavily on our foreign exchange but may also limit agricultural production.

With the increase in the country's own population, compulsion would be not only to stabilize agricultural production but to increase it further in sustainable manner. The scientists have realized that the green revolution with high input use has reached a plateau and is now sustained with diminishing return and falling dividend. The intensive use of inputs has not only polluted the soil, water and the environment causing their slow degradation but also affected the human beings. It is because of these the health conscious countries like, Germany, Japan, USA, Netherlands etc. and the affluent domestic population are ready to pay a premium for food raised without the use of chemicals. Thus, a natural balance need to be maintained for survival and well being of the human beings and plant and animal kingdom. The obvious choice for that would be adaptation of organic farming practices without compromising agricultural production.

Reports of deleterious effects arising with the use of agrochemicals was being observed by the Government of India for quite sometime. This situation, coupled with the degradation of cultivable land and increasing agricultural pollution has created an unhealthy situation in the country. In order to balance this situation, it was felt to have a viable programme which may not only check soil degradation, underground water pollution, high residues of hazardous chemicals in food, but may sustain the food production and improve the soil health without effecting crop production. Organic farming, which aims at cultivating the land and raising crops in such away as to keep the soil alive and in good health may be an alternative to the present system of farming solely depending on chemicals.

To study the viability of organic farming in India, a Technical Team constituted by the Ministry of Agriculture visited several research centres and farmers practicing organic farming during 1993 and made the following observations:

1. The country at present is not in a position to completely eliminate the use of chemicals especially fertilizers. However, it would not be difficult and unrealistic to phase out the use of these chemicals systematically.
2. The doses of fertilizer need to be gradually reduced and be balanced by increasing the use of optimum quantity of organic manures and biofertilizers.

3. Commercial sale of organic manure is uncontrolled and needs proper quality control standards so that the poor cultivator may not be cheated on one hand and on the other the eco-system may not be disturbed.
4. The use of farm yard manure even in their package of practices.
5. Development of knowledge in farmers about the hazards of pesticide use. They should minimise its use.
6. Neem and other materials of plant origin for pesticide should be used and should be readily available.
7. Integrated Pest Management (IPM) practices should be followed. Though, the plant extracts have been used by various farmers for controlling insects and diseases with sufficiently high efficiency and various combinations may be tried.

The Extension Staff of many States should be aware of the beneficial Use of organic manure and deleterious effect of agro-chemicals.

Major meetings on organic farming have been held towards the end of the 80s. In 1994, a sum of 20 lakhs each is to be given to 10 medium sized municipalities to be identified by state governments to set up composting units.

Nodal Agency of IFOAM-India

Institute for Integrated Rural Development (IIRD)

IIRD is involved in various national and international networking programs. The institute is an active member of The World Conservation Union (IUCN), International Federation of Organic Agricultural Movements (IFOAM), Asian South Pacific Bureau of Adult Education (ASPBAE), Associated Country Women of the World (ACWW), Environmental Liaision Centre International (ELCI) and World Rural Forum (WRF), Approtech Asia and ARISE. The Institute has about 65 full time staff and 103 part time staff. Besides, IIRD has about 300 volunteers in the project area.

IIRD and Organic Agriculture

Institute has promoted organic agriculture through awareness programmes, training of farmers, community action programmes on organic agriculture, networking with like minded organizations for the cause of organic movement and lobbying for organic agriculture.

Organic Agriculture Training

Since IIRD has involved women as primary actors in development processes, IIRD's organic agriculture training programmes are primarily addressed to women farmers and leaders of rural women's groups.

The *Vikas Sevikas* (development animators) are women who organize *Mahila Mandals* (women's groups) in the villages covered by IIRD. At resent 72 *Vikas Sevikas* attend every week, two days of issue based participatory learning sessions at IIRD. In these training sessions, various issues relating to agriculture like soil fertility management, cultural practices, crop management and animal husbandry are solved by using organic agricultural practices.

IIRD with the involvement of women's groups in Paithan *Taluka*, has organized 48 *Pariyavaran Samitees* (environment clubs) consisting of women in different villages. Most women in these *Pariyavaran Samitees* are small and marginal farmers. As an effort to restore their local environment in order to sustain them, these *Pariyavaran Samitees* resolved to adopt organic agricultuie in their farming practices. In May 1996, 15 *Pariyavaran Samitees* decided to send 15 women organic farmers for a one year full time course on organic agriculture to IIRD'S newly established School for Organic Agriculture.

IIRD's School for Organic Agriculture is in Babhulgaon village in Paithan *Taluka*. The school offers a number of short need-based training on organic agriculture like composting, soil fertility management, livestock management etc. to farmers and a one year full time course for women organic fanners. The training sessions in this full time course include.

Community Action Programmes on Organic Agriculture

1. Home Gardens
2. Establishment of *Pariyavaran Samitees*
3. Eco Development Programmes
4. Marketing of Organic Produce

IIRD has played a significant role in the development of National Standard for Organic Agriculture along with other IFOAM-India members. Presently IIRD is involved in developing a domestic marketing structure suitable for the area, based upon *Teikei* principles.

Networking

IIRD is an active member of International Federation of Organic Agriculture Movement (IFOAM) and is the seat of the Asia regional office of IFOAM. IIRD has conducted a number of networking workshops for the members of IFOAM-India and is constantly involved in lobbying the government agencies for formulating favorable policies toward organic agriculture.

Current Organic Farming Practice in the Indian Sub Continent

Bangladesh

There are 75 NGOs working towards an organic farming basis, they are however, facing difficulty in getting policy changes at government level for the promotion of organic agriculture.

Nepal

Organic agriculture began with post-political changes in 1990 and workshop was held in 1991. The concept of sustainable agriculture is appreciated but it is not high on the agenda of most NGOs. There are only 4 NGOs actively promoting organic agriculture NESCOS and CWDS.

Pakistan

The Pakistan Organic Farmers Association follows the approach of Sir Albert Howard, stressing self-sufficiency form on-site resources. There are the total of 30 farms all over the country producing cotton, sugarcane and vegetables. Their claims of success are however being ignored by the scientific community.

Sri Lanka

Since 1960 funds from the World Bank, IMF and FAO have been importing agrochemicals to assist farmers. Attitudes towards agricultural practices are now changing. As subsidies on fertilizers have been withdrawn farmers can no longer afford to rely on these and there is now an increasing interest in compost making.

Organic teas by the Stassen Group has promoted other tea plantations to convert to organic methods. There are 3 direct marketing co-operatives established for organic produce.

Conclusions

There is considerable demand and scope for development of 'organic technologies', either individually or as a package, without necessarily aiming at full adoption of organic system. The potential to increase productivity by either traditional soil fertility management techniques, is limited. Modern organic techniques have, however, shown more potential than traditional techniques and long-term benefits. With the organic system working on the principle that the whole is greater than the sum of its parts; the optimum output achievable through organic agriculture may not be reached in situations where organic techniques are combined with non-organic ones.

With the present technologies as well as availability of organic manures, plantation crops can be solely brought under organic farming, while in agriculture crops, organic agriculture can be practiced on a limited scale and in pockets, Practicing exclusively organic farming will have the limitation of being a low input-low output system.

A Chinese experience of success in achieving high growth in agriculture with high amounts of organic sources of plant nutrients like pig manures and crop residues is often cited as a useful model. Farmers know about the benefits of organic manures but in the present context, a lot of scientific support is needed to make it a success. A well established composting technique to ensure mineralization of nutrients from all types of high and low cellulose containing bio-degradable materials including composting of urban waste has to be popularized. The production and supply of biofertilizers having efficient soil and crop specific strains of bacteria need special attention and popularization.

Each organic techniques holds a range of potentials and constraints that differ between agroecological zones. Individual organic techniques require to be supported and promoted by both basic and adaptive research in order to obtain information for more specific usage recommendations. Organic pest and disease control strategies are a major priority for basic research, while organic soil fertility techniques require both strategic and adaptive work.

Bibliography

Chandra S., and Ali, M. 1986. Recent achievements in pulses production. Tech Bull. No. 1. Directorate of Pulses Research, Kanpur, pp. 44.

Dargan, K.S. *et al.* 1975. In alkali soils–green manuring for more paddy. Indian Fmg., 25(3): 13–15.

FAI. 1990. Fertilizer Statistics for 1989–90. New Delhi. p. 1–199.

FAO. 1988. Bio- and organic fertiliser: Prospects and progress in Asia. FAO/RAPA publication 1988/10, Bangkok, pp. 78.

Gaur, A.C. *et al.* 1984. Organic manures. ICAR, New Delhi, pp. 159.

Goel, B.B.P.S. *et al.* 1973. Availability and disposal of dung in India. Indian J. Animal Sci., 43: 671–676.

Grewal, J.S. and Trehan, S.P. 1990. Micronutrients for potato. Tech. Bull. 23. CPRI, Shimla, pp. 24.

Gupta, R. and Sandhar, N.S. 1990. Biogas wet slurry can be an effective fertiliser. Indian Fmg., 40: 13–14.

Hazra, C.R. 1992. Fertilizer use in forages, pastures and grasslands. In: Non-traditional Sector for Fertilizer Use (Ed. HLS Tandon). FDCO, New Delhi (in press).

Hegde, B.R. *et al.* 1988. In Natn. Symp. Recent Advances in Dryland Agriculture. CPIDA, Hyderabad.

IARI. 1978. Algal technology for rice. Res. Bull. No. 9, New Delhi, pp. 9.

IARI. 1980. Soil fertility maps or India. Res., Bull. No. 22, New Delhi, pp. 16+maps.

ICAR. 1987. Biological N-fixation in groundnut. Technologies for Better Crops No. 37. ICAR, New Delhi.

Juwarkar, A.S. *et al.* 1991. Biological and industrial wastes as a source of plant nutrients. In Fertilisers, Organic Manures, Recyclable Wastes and Biofertilizer (ed. HLS Tandon). FDCO, New Delhi.

Katyal, J.C. 1985. Research achievements of All-India Coordinated Scheme on Micronutrients in Soils and Plants. Fert. News, 30(4): 67–81.

Leelavalhi, C.R. *et al.* 1986. Revised yardsticks of additional production of rice due to improved measures. IASIU, New Delhi, pp. 108.

Mahapatra, B.S. and Shamra, G.L. 1988. Effect of *azospirillum*, cyanobacteria and *azolla* biofertilizer on productivity or lowland rice. Indian J. Agron, 33: 368–371.

Mahapatra, I.C. *et al.* 1981. Fertilizer use in rice-rice cropping systems. Fert. News, 26(9): 3–15.

Meelu, O.P. and Morris, R.A. Green manure management in rice-based cropping systems. In: Proc. Symp. Sustainable Agriculture–the role of green manure crops in rice farming systems. IRRI, Los Banos, Philippines, p. 208-222.

Mishra, M.M. and Tauro, P.M. 1983. Use of organic vis-à-vis mineral fertiliser. In: Recycling Organic Matter in Asia for Fertiliser. APO, Tokyo, p. 29–39.

Nambiar, K.K.M. and Abrol, I.P. 1989. Long-term fertilizer experiments in India: An overview. Fert. News, 34(4): 11–20, 26.

Nambiar, K.K.M. and Ghosh, A.B. 1984. Highlights or research or a long-term fertiliser experiment in India (1971-82). IARI, New Delhi, pp. 100.

Palaniappan, S.P. 1991. Green manures and their management. In: Fertilizers, Organic Manure, Recyclable Wastes and Biofertilizers (ed. HLS Tandon). FDCO, New Delhi.

Patil, A.J. and Kulkarni, S.D. 1988. Effect or organic recycling of Subabul (*L. leucocephala*) on yield, nutrient uptake and moisture utilisation by Rabi sorghum. In: Natn. Symp. Recent Advances in Dryland Agriculture. CRIDA, Hyderabad, Abstracts, p. 30–31.

Pillai, K.G.K. *et al.* 1980. Biofertilizers in rice culture-problems and prospects for large scale adoption. Fert. News. 25(12): 40–45.

Pillai, K.G.K. *et al.* 1985. Crop responses to fertilizers and fertilizer use efficiency in different soil and agro-climatic regions. Bull. No. 2. AICARP, Bangalore, pp. 50.

Prihar, S.S. 1986. Fertilizer and water use efficiency through the aid of mulching. Proc. FAI-NR Seminar Fertilizer Use Efficiency Through Agronomic Management, p. 31–44.

Rai, P. *et al.* 1979. Effect of various sources and levels of N on the botanical composition, dry matter production and forage quality of Sehima-Heteropogon grassland. Forage Res. 5: 37–48.

Randhawa, N.S. *et al.* 1985. Current status of yardstick of crop response to fertilizer. Proc. FAI Group Discussion on Measures to Increase Crop Response to Fertiliser, p. 25–46.

Randhawa, N.S. and Tandon, H.L.S. 1982. Advances in soil fertility and fertilizer use research in India. Fert. News, 27(2): 11–26.

Roy, R.N. and Braun, H. 1987. Development of integrated plant nutrition systems. Proc. FAI Seminar Fertilizer Industry Challenges and Strategies, New Delhi, S-IV 12/1–12.

Rupela, O.P. and Saxena, M.C. 1987. Nodulation and nitrogen fixation in chickpea. In: The Chickpea (Eds.) M.C. Saxena and K.B. Singh. CAB International, UK, p. 191–206.

Sarkar, A.N. 1990. Using biogas slurry with chemical fertilisers is beneficial. Indian Fmg, 40(4): 23–27.

Singh, N.T. 1983. Green manuring in cropping systems in recycling organic matter in Asia for Fertiliser. APO. Tokyo, 11–28.

Singh, P.K. 1982. *Azolla* as an organic nitrogen fertilizer for medium and lowland rice. In: Review of Soil Research in India. Part I. ISSS, New Delhi, p. 236–242.

Singh, P.K. 1991. Biofertilizers for flooded rice ecosystem. In: Fertilisers, Organic Manures, Recyclable Wastes and Biofertilizers. (ed.) H.L.S. Tandon. FDCO, New Delhi.

Swaminathan, M.S. 1982. Biotechnology research and third world agriculture. Science, 218: 967–972.

Tandon, H.L.S. 1980. Soil fertility and fertiliser use research on wheat in India: A review. Fert. News, 25(10): 45–78.

Tandon, H.L.S. 1983. Fertilizer use efficiency systems: components and their status. Fert. News, 28(12): 46–56.

Tandon, H.L.S. 1987. Phosphorus Research and Agricultural Production in India. FDCO, New Delhi, p. 160+viii.

Tandon, H.L.S. 1989. Long-term fertiliser experiments in India: Lessons and practical expectations. Fert. News, 34(4): 21–26.

Tandon, H.L.S. 1991a. Sulphur Research and Agricultural Production in India, 3rd edition. TSI, Washington D.C., U.S.A., pp. 140+viii.

Tandon, H.L.S. 1991b. Secondary and Micronutrients in Agriculture. FDCO, New Delhi, pp. 122+viii.

Tandon, H.L.S. and Pratap Narayaran. 1990. Fertilizers in Indian Agriculture: Past, present and future (1950–2000). FDCO, New Delhi, pp. 160+viii.

Tandon, H.L.S. and Sekhon, G.S. 1988. Potassium Research and Agricultural Production in India. FDCO, New Delhi, pp. 144+viii.

Tauro, P. and Khurana, A.L. 1986. Problems and prospects of growth, extension and promotion of biofertilizers. Proc. FAI Seminar Growth and Modernisation of Fertilizer Industry. PS1/1/1–4.

TNDA. 1989. Crop Production Guide 1989–90. Tamil Nadu Directorate or Agriculture, Madras, pp. 299.

Verma, L.N. and Bhattacharya, P. 1991. Production, promotion and distribution of biofertilizers. In: Fertilizers, Organic Manures, Recyclable Wastes, and Biofertilizers (ed.) H.L.S. Tandon. FDCO, New Delhi.

Wani, S.P. and Lee, K.K. 1991. Biofertilizers for upland crop production. In: Fertilizers, Organic Manures, Recyclable Wastes and Biofertilizers. (ed.) H.L.S. Tandon. FDCO, New Delhi.

Index